NATURAL SCIENCES IN AMERICA

NATURAL SCIENCES IN AMERICA

LIFE-HISTORIES OF NORTHERN ANIMALS

AN ACCOUNT OF THE

MAMMALS OF MANITOBA

BY

ERNEST THOMPSON SETON

VOLUME II.—FLESH-EATERS

ARNO PRESS
A New York Times Company
New York, N. Y. • 1974

Reprint Edition 1974 by Arno Press Inc.

Reprinted from a copy in the Princeton
University Library

NATURAL SCIENCES IN AMERICA
ISBN for complete set: 0-405-05700-8
See last pages of this volume for titles.

Manufactured in the United States of America

Publisher's Note: The illustrations in this book have been reduced by 15%.

Library of Congress Cataloging in Publication Data

Seton, Ernest Thompson, 1860-1946.
Life histories of northern animals.

(Natural sciences in America)
Reprint of the ed. published by Scribner, New York.
Bibliography: p.
CONTENTS: v. 1. Grass-eaters.--v. 2. Flesh eaters.
1. Mammals--Manitoba. 2. Mammals--Behavior.
I. Title.
QL721.5.M3S47 1974 599'.09'7127 73-17845
ISBN 0-405-05767-9

LIFE-HISTORIES OF NORTHERN ANIMALS

PLATE XLVII.—THE LEADER OF THE PACK.

Life study by Ernest Thompson Seton.
The property of the State of New York, reproduced by courtesy of Governor C. E. Hughes.
Exhibited in Salon of 1895.

LIFE-HISTORIES OF NORTHERN ANIMALS

AN ACCOUNT OF THE
MAMMALS OF MANITOBA

BY

ERNEST THOMPSON SETON

Naturalist to the Government of Manitoba

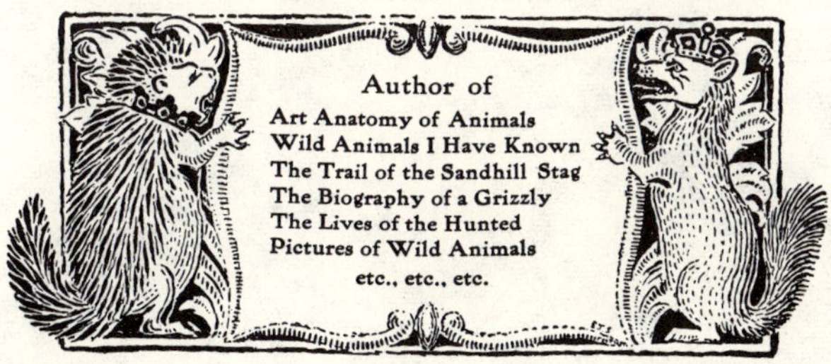

VOLUME II.—FLESH-EATERS

WITH 68 MAPS AND 560 DRAWINGS BY THE AUTHOR

Published by CHARLES SCRIBNER'S SONS, New York City :: 1909

Published October, 1909

CONTENTS

PLATES.

FULL-PAGE AND OTHER ILLUSTRATIONS.

Full-Page and Other Illustrations

MAPS.

Maps

LIFE-HISTORIES OF THE FLESH-EATERS

ORDER CARNIVORA

XXX.

Lynx, Canada Lynx, Bobcat, Gray Wild-cat, Lucivee or Loup-cervier.

Lynx canadensis Kerr.

(L. *Lynx*, the ancient name of its European kinsman; *canadensis*, of Canada.)

Lynx canadensis Kerr, 1792, An. King., I, pp. 32a, 157.
Type Locality.—Eastern Canada.

French Canadian, *le Loup-cervier, le Pichu, le Lynx ou le Chat.*
Ojib., Cree, & Saut., *Pee-shoo'.*
Chipewyan, *Chee'-say.*
Yankton Sioux, *Ee-hee'-mo.*
Ogallala Sioux, *Ig-mu-ho'-ta.*

By an unfortunate error the Canada Lynx is sometimes called 'Wolverine' in Quebec and in the Adirondacks.

The Cat Family or *Felidæ* comprises digitigrade carnivores of medium or large size; they have 5 toes in front, 4 behind; tail, various; head, short and round; claws, sharp, curved, and retractile; teeth, 28 or 30.

The genus *Lynx* (Kerr, 1792) comprises large Cats, with very short tails (*i. e.*, less than one-half the length of the body), very long legs, large feet, usually with tufted ears, and with the following dentition:

$$\text{Inc. } \frac{3-3}{3-3}\text{; can. } \frac{1-1}{1-1}\text{; prem. } \frac{2-2}{2-2}\text{; mol. } \frac{1-1}{1-1}=28$$

In youth there are 2 additional premolars above.

To these generic characters the Canada Lynx adds the following: Length, about 36 inches (915 mm.); tail, 4 inches (102 mm.); hind-foot, $9\frac{1}{2}$ inches (242 mm.). SIZE

WEIGHT

An extremely lean male from Halifax, Nova Scotia, weighed 16 pounds;[1] another from Petersburg Mountains, east of Troy, was 22 pounds.[2] The full-grown but lean female whose feet appear in Fig. 186, weighed 13 pounds. A young female, taken on Great Slave River, I weighed at 15 pounds. A small but adult female which I examined at Calgary, Alberta, weighed 19 pounds 11 ounces. S. N. Rhoads accepts[3] and records the following weights for Canada Lynx in Pennsyl-

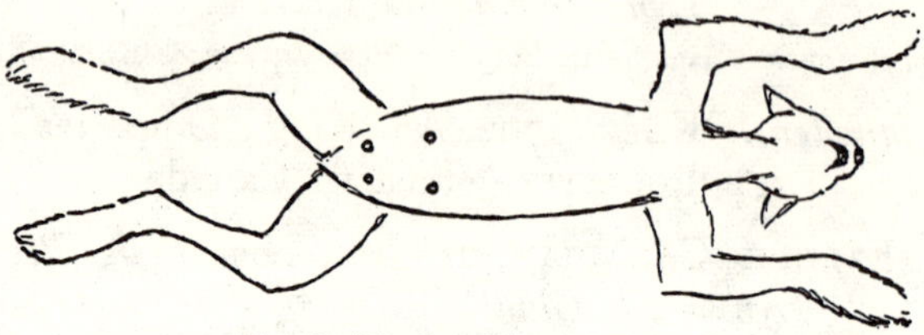

FIG. 183—Mastology of Canada Lynx ♀.
Taken Athabaska River, May 19, 1907.

vania: "about 40 pounds, as high as 44 pounds," but these are exceptionally heavy.

COLOUR

In summer: Grayish-brown, much darker on the head and back (where the long hairs are black with occasional white tips), and shaded into dull whitish below; the ears behind are black, with a central spot of whitish; a spot at the corner of the mouth, the bars on the ruff, and the *whole end of the tail black;* a few dusky spots show on the inside of each limb.

In winter: The colour is much paler and grayer; at all times the tuft of hairs on the ears is long and black.

When seen alive it looks and behaves exactly like a huge gray cat. Its tufted ears and short bobtail will distinguish it from its near relatives. It might be mistaken for the American

[1] Aud. & Bach., Quad. N. A., 1849, Vol. I, p. 138.
[2] *Ibid.*
[3] Mam. Penn., 1903, pp. 137–8.

PLATE XLVIII.—THE LYNX AT BAY.

Wild-cat or Bay-lynx (*Lynx ruffus*), but its tail is very different and furnishes a sure guide; while the Lynx has the tip of the tail wholly black above and below and the rest of it grayish-white, the Bay-lynx has the tip black above and white below, and also has other broken bars on the upper part

FIG. 185—Lynx Tails. (Half life size.)
1. Canada Lynx.
2. Bay-lynx.
(Both from Muskoka, Ont.)

FIG. 184—Head of Canada Lynx ♀ in summer. (Half life size.)

Nothwithstanding the fact that the Bay-lynx is said by Herrick[4] to be the prevailing species in Minnesota, I have failed to determine its occurrence in Manitoba.

Three races of Canada Lynx are recognized:

canadensis Kerr, the typical form.
mollipilosus Stone, a browner race.
subsolanus Bangs, a darker race.

[4] Mam. Minn., 1892, p. 73.

Life-history.

RANGE This species ranges over the whole of Manitoba wherever there is cover. In autumn it is often found three or four miles out on the prairies. Premier Roblin has supplied me with the record of a Lynx killed on his farm near Carman, among some willows, 3 miles from timber and 10 from woodland of any extent. In the end of October, 1883, I met with a Lynx on the open prairie 20 miles west of Shellmouth. In the fall of 1905, E. W. Darbey says 2 Lynxes were killed within the limits of the city of Winnipeg. But its usual haunts are the woods, the thicker the better.

HOME-RANGE The Lynx is generally believed to be a wide ranger. While the young are unable to travel it would be impossible for the mother to go more than four or five miles from home, but, in the autumn and winter, there is reason for believing they will go fully ten times as far. I remember meeting with a Lynx near Toronto in December, 1875, although it was commonly believed that they were no longer found within 30 or 40 miles of that city.

ABUNDANCE I met with but three or four Lynxes during as many years in Manitoba, so that in the poplar region about Carberry and westward they cannot be called abundant. In the sandhill tract between Carberry and the river, about 20 miles by 15, I doubt if there are ordinarily a dozen Lynxes resident. In the thickly wooded regions northward, they are said to be much more plentiful, and in the Peace River country, during the great Rabbit year of 1904, the Lynxes so abounded that nearly every hunter and trapper in the country got from 20 to 50 that season.

SOCIABILITY So far as known, the only approaches to sociability in this animal are the bands of four or five that are seen together in autumn and winter, and it is the opinion of most hunters that

MAP 39—RANGE OF THE CANADA LYNX AND ITS THREE RACES.
Lynx canadensis Kerr.

Founded chiefly on records by J. Richardson, E. W. Nelson, O. Bangs, C. Hart Merriam, J. Fannin, W. H. Osgood, R. Bell, A. P. Low, R. MacFarlane, W. Stone, S. N. Rhoads, E. A. Preble, and E. T. Seton.

these are the family of the year, still with the mother and occasionally accompanied also by the father. George Linklater assures me that he has often seen in the snow signs of Lynxes gathered together to chase each other and play, at a time when sex instincts were out of the question. But what the nature of the game was I have failed to learn. E. W. Nelson says:[5] "The fur-traders and Indians of the Upper Yukon claim that the Lynxes sometimes unite in parties of 5 or 6 and make Rabbit drives on the small islands in the Yukon. They claim to have heard the Lynxes utter a sharp whistling noise, and to have found their tracks in the snow where the line had swept the island, until each secured its prey, near the farther end."

INTERCOMMUNICATION

The stripes on the face, the black ear-tufts, the whiskers, and the little nervous twitching black-tipped tail, are no doubt important direction marks to help the Lynx's own kind in recognizing it, but its voice is its chief means of communicating with its distant fellows. I never heard a Lynx purr, but all other sounds that a house-cat has, the Lynx has, and uses them in much the same way. I know nothing of the 'whistling' mentioned by Nelson in a previous paragraph. It has, however, another vocal effort which is even better developed than in the cat, and that is a yowling song. This begins with a long low '*me-ow,*' followed by others in quick succession, with rising pitch and volume, till after three or four minutes continuous performance the final '*me-ows*' are terrific screeches. I have heard this in August, October, and December, and do not know what it means, or which sex utters it. But the trappers tell me that the somewhat similar and frightful caterwauling of the males is mostly heard early in March and has a direct relation to the mating.

MATING

The species is generally believed to pair, but I have no evidence beyond the opinion of hunters.

The mating season is any time during the first half of

[5] Nat. Hist. Alaska, 1887, p. 235.

March. My Ottawa guide, Ned Crête, of Deux Rivières, tells me that, in 1904, he saw 7 Lynxes together on March 15 or 20. It was the regular running season, and the year before a similar group was seen at the same place.

There were 2 small and 5 big ones. They were caterwauling like cats; it was this noise that called his attention to them. Two were fighting exactly like tom-cats; the one down on his back had the better of it, being able to scratch with four feet instead of only two.

The hunters generally believe that the Lynx is monogamous, and Miles Spencer gives[6] it as the opinion of the Indians that the Lynx assists the female in rearing the young. Linklater takes the same view and maintains that though the male does not actually accompany the young, when they follow the mother, he is always found at no great distance, both in summer and in winter. This same trapper believes that Lynxes travel in families the year round, except in the spring.

The period of gestation, according to MacFarlane,[7] is about 3 months. This would bring the young into the world about the middle of June in Hudson Bay Territories, but in Pennsylvania, Rhoads says, they arrive in May.[8] The mother prepares a comfortable nest for them in some hole or hollow log. Whether the father assists in this, I cannot learn. The young are, according to all accounts, from 1 to 3 or 4 in number, but Linklater tells me that he has found up to 6 in the female. This discrepancy I have come across many times, the average number of young in the brood being less than the number of embryos in the female. It may mean that some are still-born, a parallel case being the addled eggs in nests; or, if too many for their food supply, the weak ones die. YOUNG

When born they are much like the kittens of the house-cat. In 1886 I made the herewith drawing of one (Plate XLIX) to illustrate a paper by Dr. C. Hart Merriam. The specimen

[6] Mam. H. Bay, Low's Rep. Geol. Surv. Can., 1888, App. III, p. 76 J.
[7] Mam. N. W. Terr., Proc. U. S. Nat. Mus. XXVIII, 1905, p. 692.
[8] Mam. Penn., 1903, p. 140.

was secured by Montague Chamberlain, of St. John, N. B. He says:[9]

"It was dropped on the 20th of March, 1883, when the mother had been in captivity about a month. She gave birth to 5 kittens, but this was the only one rescued from her unmotherly jaws. When the first was born she at once prepared to clean it, and seemed fond of it. After a short time, however, it gave vent to a weak squeal, which caused her to eye it curiously for a moment, when another squeal was delivered. This settled the kitten's doom—it was devoured at once. The mother did not exhibit any tenderness towards the other 4, and the keeper made two unsuccessful efforts before he was able to get one away from her. This kitten lived two days, and then died from injuries received in its removal from the cage. Its 'mew' was something like that of a domestic kitten, but stronger and harsher; it was almost fierce and very penetrating. The general strength of the animal was greater than that of a domestic kitten. Two hours after birth it stood firmly on its feet and turned around in its box, but it did not show any inclination to fight when teased. The eyes were open at birth."

Miles Spencer states[10] that they are born with closed eyes.

Dr. Merriam adds in his description of the specimen figured:

"It is but a trifle larger than the young of the domestic cat at birth, and may have been born a little prematurely, though the fact that its eyes were open argues against this supposition. I am unable to give many measurements of value, since I did not see the specimen till after it came from the taxidermist.

* * * * * * *

"The ground colour of the body is light fawn, paler below, and inclining to buff on the sides. It is much obscured above by the stripes and rows of concatenating brown blotches, and below by small dark stripes. * * *

[9] Bull. Nat. Hist. Soc. New Brunswick, 1886, No. V, pp. 10–13.

[10] *Loc. cit.*, note 6.

PLATE XLIX.—YOUNG OF CANADA LYNX (LIFE SIZE).
From drawing on stone by E. T. Seton, made for C. Hart Merriam in 1886.

"It is hardly necessary to call attention to the fact that this specimen is one of unusual interest, since its very decided markings, of which scarcely a trace remains in the adult animal, cast some light upon the genetic affinities of the genus to which it pertains. A critical study of these markings leads to the interesting conclusion that the genus *Lynx* was derived from the group of Cats of which the Ocelot (*Felis pardalis*) is the nearest living representative."

After the young Lynxes have been suckled for two or three months the family—mother, kits, and probably father—set out on their travels. At this time the young are weaned and have for a month or more been eating solid food, but now they begin to learn hunting for themselves. The instinctive habit of the race, stimulated by hunger and the mother's example, is doubtless the prime motive power.

FAMILY LIFE

Although usually a shy creature, avoiding a meeting with man, the Lynx mother is very ready to fight for her family.

On one occasion, while out on a camera hunt in Colorado, I heard a buck stamping in a little dale and, slipping off my horse, camera in hand, sneaked after the Deer. I found nothing but his tracks, and was peering across an open place, when I caught sight of a large animal close to me on the right. On passing into the clear space it turned to look at me. It was a Lynx, but it seemed very small, and its expression was one of innocent curiosity, entirely without menace. It paused at 30 feet. I hastened to adjust the camera, and as I did so a deep rumbling growl and a movement in a thicket close at hand made me jump. I turned around, to see within 15 feet a Lynx three times as big as the first, and eyeing me savagely from behind some willows. My first thought was to wish for a gun, for I realized that the Lynx in the open was only a kitten; now I had to meet the mother. My second thought was that the old one would do me no harm if I faced her, and did not molest the kitten. So I tried to get her photograph, but she disappeared, and when I looked around the little one also was

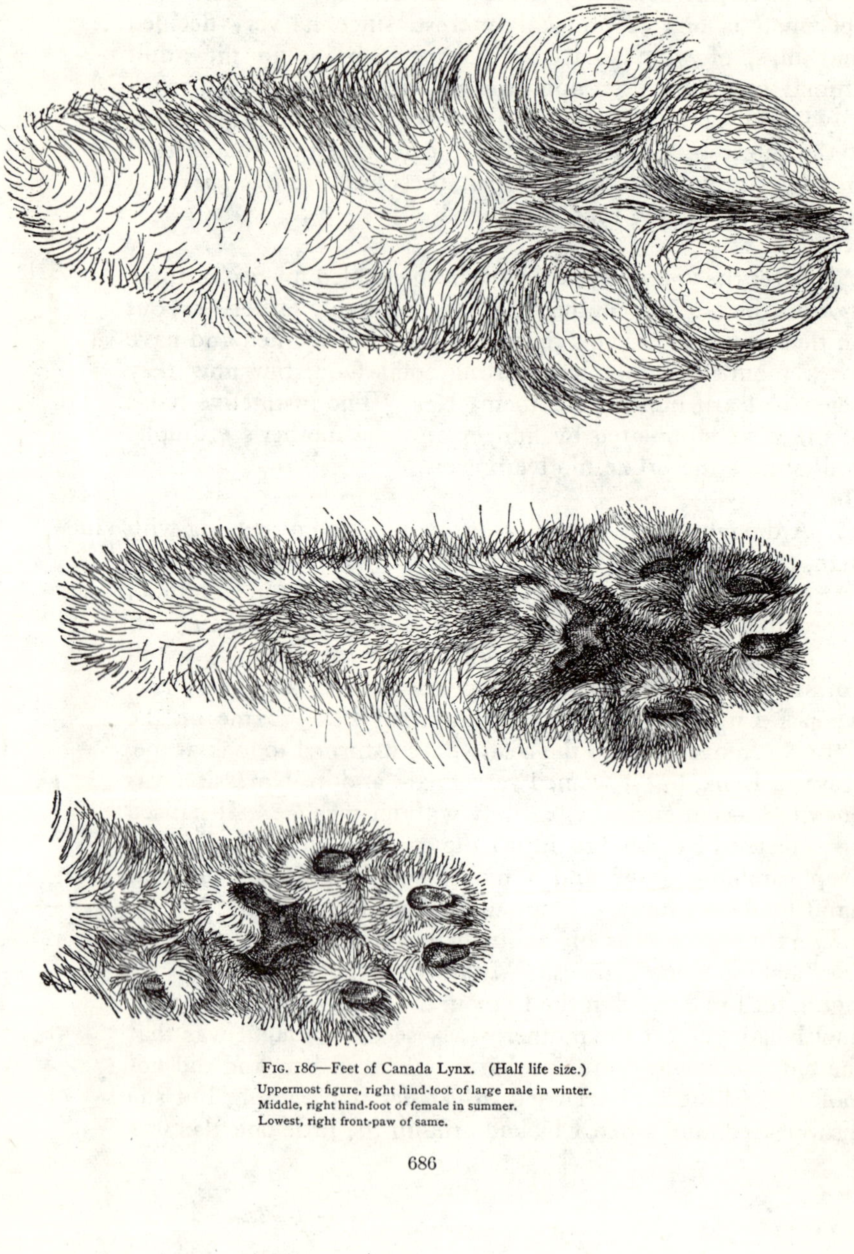

FIG. 186—Feet of Canada Lynx. (Half life size.)

Uppermost figure, right hind-foot of large male in winter.
Middle, right hind-foot of female in summer.
Lowest, right front-paw of same.

gone. On this occasion I saw nothing of a second old one, or indeed of any other young ones, for that matter; but they may have been there, as the undergrowth was very thick. The date was September 8.

The family continues together all autumn. As proof of this, Linklater tells me that in October, 1904, he saw 4 Lynxes together hunting at Desbarats; probably they were mother and kits. In 1894, at Green Lake, Ont., he saw 5 together about Christmas; all seemed fully grown. Charles G. D. Roberts informs me that in New Brunswick a band of 5 or 6 Lynxes

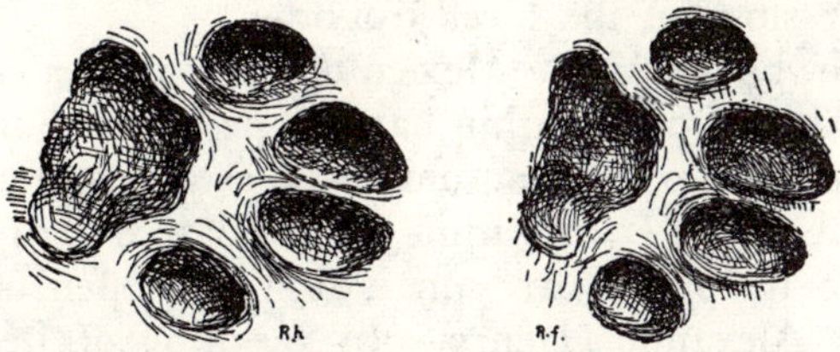

FIG. 187—Right hind and right front tracks of a domestic cat (natural size) for comparison with those of Lynx.

are sometimes seen in company. All these cases are, I believe, incidents of family life, and the Rabbit drives Nelson tells us of have a similar explanation.

The group may continue together until March comes, bringing with it that great disintegrator of the family band—the mating craze—which prompts the brother and sister to shun each other, and seek each one a helpmate for himself.

In hunting the Lynx a single small dog is enough to make PURSUIT
it take to a tree, but it is very apt to regain courage, come down and kill the dog, unless the hunter be close at hand to succour and support his noisy colleague.

Although a desperate fighter when cornered, this animal is easily killed. If it is taken in a snare, the trapper usually picks up a stout stick and dispatches the victim with a blow on the snout or back.

In following it in winter I have often been impressed by

the admirable adaptation of its feet as snow-shoes. Although its weight be 30 or 40 pounds, its feet are so large and so spread with stiff hairs that it walks lightly on soft drifts where a dog would flounder in utter helplessness. As it rambles through the woods it usually walks every log it comes to. Sometimes, in the midst of a slow walk, it will spring forward 12 or 15 feet, apparently without any object other than a wish to see how far it can jump.

RUNNING

Although a creature of superb activity among matted branches and labyrinthine logs or underbrush, the Lynx is surprisingly slow on the level ground.

The cowboys of New Mexico on their ponies could catch a Bay-lynx in the open within half a mile, even though it had a quarter-mile start. Not only will any common cur dog overtake a Canada Lynx within a few hundred yards, but even a man who is speedy can run it down in open country, as attested by Alexander Henry. In speaking of Le Boeuf, a famous Indian hunter and runner on the Red River, he says:[11]

"He came in to-day with a Loup-cervier that he had caught in the plains in a fair chase and killed with his small axe; he certainly is an extraordinary runner. He is a tall man, spare and lean, of a mild disposition, but wicked when provoked to anger."

During my journey to the Far North in 1907, I often heard of such exploits on the part of hunters, and at length, on Great Slave Lake, was eye-witness of this very achievement.

SWIMMING

On the other hand, the Lynx seems very much at home in the water. The garrulous and ever-entertaining Henry says, in his Red River Journal, April 22, 1804:[12] "Caught 15 sturgeons and a Loup-cervier; how the latter came into the sturgeon net I cannot say. We saw his track on the beach until he came opposite the net, which completely crossed the river; he appeared to have then taken to the water, for what

[11] Red River, November 2, 1802, Journal A. Henry, 1897, p. 206.

[12] P. 242.

reason I cannot tell. However, he was found drowned, entangled in the net about 10 feet from shore."

Richardson, in his Overland Journey, 1848, relates[13] that on June 26, at Buffalo Lake, "a Canada Lynx was seen swimming across a strait, where the distance from shore to shore exceeded a mile. We gave chase and killed it easily. This animal is often seen in the water," and elsewhere[14] he remarks, "it swims well and will cross the arm of a lake two miles wide." I have several times known Lynxes to take to the water without being in any sense driven, and was surprised to find this member of the cat tribe as good a swimmer as a dog and far better than a Fox.

It is noteworthy that most of our carnivora live chiefly on prey smaller than themselves. The Fox preys on Mice, the Marten on Squirrels, the Badger on Gophers, the Lynx finds its chief support in the White-rabbit. A good Rabbit year is sure to be a good Lynx year, and the disappearance of the Rabbits is followed by a general disappearance of the Lynxes. FOOD

In addition to Rabbits, the Lynx preys on various kinds of grouse—is, in fact, the chief enemy of the Canadian grouse or spruce partridge.

It is a curious fact, as I have often witnessed, that the spruce partridge will allow a man to walk within ten feet and *noose* a member out of a covey, but the moment a dog or anything suggesting a Lynx appears in the distance the whole family take flight in alarm. This may be accepted as evidence that the Lynx, and perhaps the Fox, have for long been the only important enemies of this grouse.

The food of this animal is thus detailed by Audubon and Bachman:[15]

"The food of the Canada Lynx consists of several species of grouse and other birds, the Northern Hare, Gray-rabbit,

[13] Arctic Search Exp., 1851, Vol. I, p. 106.
[14] F. B. A., 1829, Vol. I, p. 101.
[15] Q. N. A., 1849, Vol. I, p. 141.

Chipping-squirrel, and other quadrupeds. It has been mentioned to us that in the territories to the north of the Gulf of St. Lawrence they destroy the Arctic Fox, and make great havoc among the Lemmings (*Georychus*). Hearne informs us (p. 366) that in Hudson Bay they 'seldom leave a place which is frequented by Rabbits, till they have killed nearly all of them.' They are said to pounce on the wild goose at its breeding places, and to destroy many Marmots and Spermophiles by lying in wait for them at their burrows."

We shall probably find in its food list every living creature that it can overcome, which means all smaller than itself, not excluding snakes, frogs, and insects. There are probably but two lesser fellow woodsmen that the Lynx lets alone; these are the Skunk and the Porcupine. Starvation, however, may overcome its fear of these, as is shown by Audubon and Bachman:[16] "At a public house in Canada, we were shown the skin of one of these Lynxes, the animal having been found quite helpless and nearly dead in the woods. It appears that, leaping onto a Porcupine, it had caught a Tartar, as its head was greatly inflamed and it was nearly blind. Its mouth was full of the sharp quills of that well-defended animal, which would in a day or two have occasioned its death."

FOX-KILLER

Most persons are surprised to learn that in the thickly wooded country the Lynx is a deadly enemy of the Fox.

One of my guides in the Kippewa region of Quebec (Archie Miller) tells me that in January, 1904, as he crossed Askoe Lake near Kippewa, he saw a Lynx and a Fox about 80 yards off, fighting on the snow. He watched them for about 15 minutes. The Fox was trying to get away, but the deep, soft snow was against it, and finally it was overtaken and killed by the Lynx. When Miller came up the victor ran off into the woods. In the fight an acre of snow was trampled all over; they must have been at it for an hour. The tracks showed that they began the battle in a woods near by, where there were many Rabbits. The Fox's neck was torn open and its heart

[16] *Ibid.*

pierced in two places, apparently by the claws of its adversary. It was a prime Cross-fox, and brought five dollars.

Similarly, Linklater tells me that once when carrying the mails down from Montreal River, Ont., in January, 1880, he had halted for noon at the edge of a small lake and saw on the ice, a mile or more away, two animals fighting, one either a Fox or a Fisher, the other a Lynx.

After eating his dinner and resting an hour, he travelled on to the place and found the combatants to be a Cross-fox and a Lynx. They had had a long and desperate encounter, but the Fox, as usual, had succumbed to his foe's superior powers, and had been torn into pieces. The head and tail were lying on the ice, but the body had been carried off and buried under snow in the distant woods, where the traveller found it. The tracks showed that the Lynx had attacked the Fox in the woods and chased it round and round on the Rabbit trails for perhaps an hour before driving it onto the ice, where the killing took place. There were plenty of Rabbits, so that hunger was not the excuse. The Fox was at a disadvantage, as the snow was three feet deep and very soft. The Lynx went over the surface on his snow-shoes, the Fox ploughed in deep, and the harder he leaped the deeper he sank.

Both these trappers say they have often heard of Foxes killed by Lynxes and by Fishers.

As soon as these two are trapped out, Foxes increase, but are everywhere scarce in the thick woods.

J. K. McDonald writes me: "I have known of bodies of even full-grown Foxes being found dead, but uneaten, such having been killed by the Lynx."

The latter, however, is not always master of the situation, as proven by the following incident in Nelson's "Alaska":[17]

"Mr. McQuesten, a fur-trader living at Fort Yukon, witnessed one winter day a combat between a Lynx and a Red-fox, which he described to me as follows: 'The Lynx sprang upon the Fox, in comparatively open ground, evidently trying to capture it for food. The Fox instantly made fight,

[17] Nat. Hist. Alaska, 1887, p. 235.

and for a few moments the fur flew right and left. Then a short pause followed and the fight was renewed. A second pause ensued, and after the two had glared at one another for a few moments they slowly withdrew in opposite directions, the hair on each bristling defiance, but each apparently satisfied to close accounts.' This Lynx was probably weakened by hunger, for a vigorous Lynx is certainly more than a match for a Fox."

In this case, I suspect further, that there was very little snow. The Lynx, mounted on his wonderful snow-shoes, has a sovereign advantage when the snow is deep enough to embarrass a Fox. In fact, the Lynx plays crust-hunter, while the Fox flounders helplessly in treacherous drifts.

In my early days about Lindsay, Ont., I several times heard of farmers losing lambs or even small pigs through the attacks of Lynxes; and fawns were believed to be commonly their prey. But the hunters were divided as to whether a Lynx would attack any creature so large as a full-grown Deer.

DEER-KILLER

One of our best naturalists writes:[18] "We have heard one or two accounts of the Canada Lynx having killed a Deer; we are somewhat sceptical in regard to this being a general habit of the species, although when pressed by hunger, which renders all creatures desperate at times, it may occasionally venture to attack a large animal."

Linklater claims that he has conclusive evidence in point. At Green Lake, Algoma (H. B. Post), he once found the remains of a Deer on the ice with no tracks but those of one Lynx about it. The Deer was a two-year-old. He is satisfied that it was killed by the Lynx.

In September, 1901, while camped in the Colorado Mountains, where the Deer and Lynxes both were abundant, my guide, Charles Erickson, told me of another case. About five years before, A. E. Muckey was hunting a band of Blacktails in the snow among the cedar brakes, between the mouth

[18] Bachman, Q. N. A., 1849, Vol. I, p. 141.

of Deep Creek and the Sweetwater, along the Grand River, when the trails of two Bobcats joined on to those of the Deer. A Deer track left the band and one of the Bobcats' tracks disappeared at that place. Muckey turned aside to follow that one Deer. After a short time he came on its carcass with both Bobcats in possession, but they ran way at his approach. He put some poison on the kill, and returned next morning to find both Bobcats dead at their feast.

S. N. Rhoads, after stating[19] "There is nothing in the habits of the Lynx differing from those of a Wild-cat, except what it accomplishes on account of its greater size and agility," adds: "They will not hesitate to fasten themselves on the necks of Deer, trusting to bring them down by sheer exhaustion and blood-letting before the Deer can manage to drag them off by running through brush or branches of thick trees, or by jumping into water. Mr. Seth Nelson on one occasion was trout fishing at a large pool in the woods of Clinton County, when a crashing through the forest made him seize his rifle in time to shoot both a Wild-cat and a doe, which plunged into the pool to free itself of its tormentor. They have been known to seriously wound hunters in their own defence, and even to make an unprovoked attack."

Hearne states[20] that he once saw a Lynx take possession of a Caribou that an Indian had just slain and "suffered itself to be killed before it would relinquish the prize."

In New Brunswick, according to Charles G. D. Roberts, a band of Lynxes do not hesitate to attack even a Caribou; and Linklater relates a similar case that he heard of from an Indian whom he considered quite reliable. It was at a place 35 miles north of Sudbury, Ont. The hunter found the place with all the marks in tree and in snow, showing that the Lynx had been in a tree by the runway and had dropped on a passing Caribou, but the Caribou, by dashing into the thickets, had managed to get rid of its enemy.

From these records we may believe that the Lynx will, if hard pressed, attack Deer and even Caribou, but these

[19] Mam. Penn., 1903, p. 140.

[20] Journal, 1795, p. 372.

are not its usual prey, nor does it care to assail them except when able to do so in force.

NEVER MOLESTS MAN

So far I have not met with an authentic case of the Lynx voluntarily attacking man. It may sneak along the hunter's trail after dark, and close behind him, but it seems to be actuated by curiosity more than anything else, and having come close enough to inspect or wind him, is most likely to glide away in search of its proper prey.

I have several times been followed in this way, but usually did not know it till afterwards, when I happened to come back again to my old track in the snow.

A hunter told me that he once secured a fine specimen through knowledge of this habit.

Chancing on his own track again within a few minutes, he saw the great pad-marks of a Lynx evidently trotting behind him. He crossed an open space into some brush and there sat down to watch. Within five minutes the Lynx came running the trail like a hound and, when within twenty yards, was easily bowled over with a charge of heavy shot.

Several of my companions about Carberry have met with Lynxes among the Sandhills. In most cases the creature walked away, retiring with great dignity, or uttering a deep, defiant growl.

The only account I find of a Lynx facing a man is by Professor H. Y. Hind, and, although he did not witness the affair, he seems to believe it. In this case the Lynx did not attack *voluntarily*, but was at bay. The incident took place on an island opposite Mingan Post on the Gulf of St. Lawrence. Peter Mackenzie, armed but with bow, arrows, knife, and snow-shoes, landed on the island in early spring while yet there was snow, and found the Lynx. After a long pursuit he struck it with two arrows.

"At last he came within twenty yards; the Cat turned round, rose on his hind-legs, snarled, and began to paw the air. Mackenzie discharged another arrow, but at the same moment his snow-shoes tripped him up, and he fell headlong with his face in the snow. The Cat instantly sprung upon him, tearing with one stroke the coat from his back. Mackenzie turned round at once, and caught the Cat by the throat with one hand, and with the other he drew his knife; but as he made a lunge they both rolled over together, and he received some very severe scratches. Still holding on firmly to the throat of the animal, he was not bitten, although he was in danger of having his bowels torn out by the hind-feet of the Cat, who was making a vigorous resistance. A second lunge with the knife was fatal; it passed through the animal's heart, but it left Mackenzie exhausted and bleeding on the snow. He soon recovered, and carried his booty in triumph to the Post."[21]

STORAGE

Canadian hunters and trappers generally credit the Lynx with a well-developed storage habit. When it secures more food than it needs for the present, it carefully hides it in the earth or in the snow for less bounteous times. One case has already been cited.

DISEASES

The Rabbit is the most diseased of our mammals, and the Lynx feeding on the Rabbit should logically inherit the physical troubles of its victim, but I have seen no proof that it does.

As already noted, there is no evidence of epidemic among the wild Lynxes to account for their periodic disappearance.

Those who have the opportunity of conducting *post mortem* examinations on the bodies of Lynxes can render good service by recording in full their condition, as it is probable that the Lynx may be temporary host of a parasite that finds its final and fatal development in the Rabbit.

The specimen from which I made the drawing of feet (Fig. 186) died of sunstroke in the New York Zoölogical Park.

SANITATION

Like all cats, the Lynx is scrupulously clean. Menagerie specimens usually set apart one corner of the cage for the

[21] Labrador Peninsula, 1863, Vol. I, pp. 59-60.

cesspool department, and wild Lynxes are said to bury their dung like cats, but of this I have no conclusive evidence.

CURIOUS PARTNERSHIPS

An interesting kind of commensalism has been noted in the Lynx life. Linklater tells me that he has seen a horned owl and a Lynx together working a Rabbit woods, the owl hovering around the outskirts to pick up the Rabbits as the Lynx routed them out. Of course, in this case we must suppose that the owl was a parasite that the Lynx was helping unwittingly. But the same sort of thing is seen with various hawks; nor do they follow only the four-legged hunters, but will impudently try to utilize the terror spread by the gunner and sometimes, forgetting themselves, come within range, to be added to the common bag.

USE TO MAN

This animal is very easily caught by any of the usual methods of fur-taking, therefore the simplest—the snare—is most in use. In trapping it the half-breeds often use a lure or charm made of beaver-musk, oil of rhodium, asafœtida, and filings from the corn on the inside of a horse's front-leg. How far the practice is founded on mere superstition I cannot say.

This bait is set on a forked stick surrounded by a little fence with one opening. At the opening a noose of wire or cord is set 18 inches from the ground and fast to a short, thick stick. A Lynx coming to sniff the lure is caught in the noose; it tightens as he retires. He tugs till he is strangled, or, climbing a tree to get rid of it, he is hanged through the crosspiece catching in the branches.

The steel trap and deadfall also are used, and in regions where Lynxes are abundant some hunters keep dogs trained to tree and hold them till the gunner can approach and use his fire-arms.

FLESH

Its flesh is a regular article of diet in the North-west. On the occasion when I tried it I found it white and well-flavoured but was debarred—by prejudice, I suppose—from enjoying my meal of cat, in spite of the Hibernian dictum that a Lynx is nothing but an animated Rabbit, anyway.

That the Lynx population fluctuates greatly is well known to all trappers and fur-traders—but does the species migrate? J. K. McDonald, after thirty-five years of service in the fur country as a Hudson Bay trader, writes me: MIGRATION

"It is accepted as a matter beyond cavil by all Hudson Bay hunters, that the Hare, Lynx, and Marten do migrate, and the fluctuation in their numbers is not considered to be caused by epidemics—save in the case of the Hare. Were it so with the Lynx, for instance, *their* bodies would surely be found, yet I have not heard of such a thing. It is always the case that when Lynx and Marten are plentiful, so are the Hares, and I am inclined to think that the former is the cause, at least to a large extent; as they are known to destroy the smaller mammals that prey on the Hares. Thus an influx in Lynxes causes such a decrease in other fur-bearing animals that the fur-traders look upon it as a disaster.

"These periodical waves of increase or decrease cover large tracts of country, and it might be found that where in one tract they were moving east, in another they were going west."

George Linklater and Miles Spencer, northern hunters of life-long experience, reiterate the theory of migration.

The former tells me that at Green Lake, Ont., Lynxes were so plentiful during the winters from 1888 to 1890 that he took in 300 pelts each season at the Hudson Bay post. They then nearly all disappeared, and for three winters he took in only 30 to 35 pelts a season. At the same time they appeared in great numbers at Lake Temagaming, 200 miles away, where they had been very scarce, and for some years several hundred pelts were brought in each winter, instead of 20 or 30. At the time of this change he saw many Lynx tracks pointing eastward from Green Lake towards Temagaming, and one day followed a band of half a dozen for many miles. They were not hunting, but travelling, and so close together that he could not be sure if there were 5 or 6 of them. There was plenty of food at both places, and no disease among the Lynxes, so it was impossible to say why they went, only he was quite certain that *they did go*.

A great deal of evidence of this sort could be presented. The trappers generally agree that the Lynx is migratory and that it follows the White-rabbit. The Rabbit, however, *does not migrate*, so we may understand this to mean that the Lynxes seek out the regions where the White-rabbit abounds. But an unexpected difficulty arises. If the Lynx population merely shifted, the aggregate fur returns of the entire country would not change, for the trappers cover the whole region every year.

FLUCTUATIONS

After spending a life-time as fur-trader, Roderick MacFarlane discusses the question as follows:[22] "This is one of the principal periodic fur-bearing animals which regularly increase and decrease in numbers about every decade. The experience of observers, largely corroborated by the Company's London sales, is pretty much as follows: The catch of Lynxes for each (say) three seasons, when they are least numerous, or rather comparatively scarce, fell sometimes as low as 4,000 or 5,000 skins, as the entire output for the immense extent of territory covered by the Hudson's Bay Company's business operations. The fourth year would double these quantities, the fifth often more than doubled the fourth, the sixth doubled the fifth, while the seventh almost invariably witnessed the maximum trade of skins. The eighth would still be good, while the ninth and tenth would each exhibit a startling decline in the returns, which in quantity would closely correspond with the sixth and fifth years, respectively, in each decade. * * *"

A clear idea of the wax and wane of the Lynx population is found in Alexander Henry's "Journal in the Upper Red River in the Years 1800 to 1808." The old fur-trader thus records[23] the Lynx skins taken in the successive seasons at 20, 67, 194, 167, 38, 0, 4, and 4.

[22] Mam. N. W. Ter., Proc. U. S. N. M., 1905, pp. 691–2.

[23] Journal, 1897, pp. 184, 198, 221, 245, 259, 281, 422, 440.

These, then, are the conclusions presented:

(*a*) The Lynx population rises and falls in cycles of about ten years, and when at its maximum may be as much as tenfold the minimum.

(*b*) There is no evidence that the decrease is due to epidemic disease.

(*c*) There is evidence of local migrations, but not of a kind to explain the great changes.

(*d*) After studying the problem on the ground, I feel no doubt that the decrease is due to starvation through failure of the Rabbits; and the story is grimly and silently told by frequent wasted bodies exposed in the woods when springtime melts the snow, as I myself witnessed on the upper waters of the Mackenzie in the summer of 1907.

The fur of the Lynx is a staple of North-western trade. FUR The number annually exported by the Hudson's Bay Company has long oscillated between 4,000 and 75,000; 1896 was a high-water year, the export being 56,407; 1900 was the low ebb, the number having fallen to 4,473; since then the usual increase has continued, and 1905 must have been near high tide, for in the 1906 sales the Hudson's Bay Company had 58,791 skins. At the previous sale Lampson's disposed of 21,521 skins, or, in round numbers, 80,000 skins were taken in 1905.

Poland's lists,[24] however, show that the other American fur companies collect about 7,000 per annum, with little variation, and that, taking the whole continent, an average of about 30,000 Lynx are killed each year for their fur.

At the London annual fur sales held at C. M. Lampson's in March, 1906, 7,737 Lynx skins were sold. The highest prices realized were 60 shillings ($14.40) each for 4 superb 'blue' skins; 30 shillings to 35 shillings ($7.20 to $8.40) were more usual prices for first-class skins, from which they graded down to about 10 shillings ($2.40) for those of third-class.

[24] Fur-bearing Animals, London, 1892.

XXXI.

Kit-fox or Swift.

Vulpes velox (Say).
(*L. Vulpes*, see ante; *velox*, swift.)

Canis velox SAY, 1823, Long's Exped. Rky. Mts., I, p. 487.
Vulpes velox AUD. & BACH., 1851, Quad. N. A., II, p. 13.
TYPE LOCALITY.—Vicinity of Cherry Creek, Laramie County, Wyo.

FRENCH CANADIAN, *le Re'nard des prairies, ou vif.*
YANKTON SIOUX, *Taghn-kay-ha.*
OGALLALA SIOUX, *Mee-yah'-chah.*

The *Canidæ* or Dog Family comprises digitigrade carnivores of large size; they have blunt, non-retractile claws; bushy tails; pointed muzzles; toes, 5 in front, 4 behind; teeth, 42 or more.

In addition to the characteristics of the genus *Vulpes* (see p. 706), the Kit-fox has:

SIZE

Length, 26 inches (660 mm.); tail, 9 inches (228 mm.); hind-foot, 4 inches (102 mm.).

A full-grown specimen, from Medicine Hat, Sask., weighed 4¼ pounds.

COLOUR

General colour, pale buffy-yellow, becoming a deeper yellowish-brown on the back of the ears across the lower neck, on the outside of fore-legs and back of hind-legs; below, it is nearly white; on each side of the snout is a black spot; the back is covered with a beautiful silvery-gray mantle in which the gray-brown under-fur is peppered and frosted over with conspicuous white and inconspicuous black tips of the long hairs; this silver-tipping is continued onto the snout. The tail is warm yellowish-gray above; strong yellowish below; the

PLATE L.—LIFE STUDIES OF VARIOUS FOXES.
By E. T. Seton.

tail-tip is black. The tail-gland is marked by a black spot, as in the other species. *There is no black on the ears.*

It may be distinguished from its nearest relative, the Big-eared Swift, by its yellower colour and shorter ears.

Two races are recognized:

> *velox* Say, the typical form.
> *hebes* Merriam, larger, paler, and grayer.

LIFE-HISTORY.

This diminutive Fox, no larger than a house cat, is a characteristic native of the Saskatchewan or upper Campestrian region. RANGE

In Manitoba it was formerly found in the Pembina Hills and westward to the Souris. Alexander Henry, trading on the Red River in 1800–8, had one or two Kits brought to him from Pembina Hills, or, as he calls them, Hair Hills, nearly every season; one year, 1804–5, he had 57; of these, 26 were from Pembina Hills and 31 from Salt River.[1] In 1873, Dr. E. Coues found[2] Kit-foxes common along the Souris River at the Boundary Trail.

These are all the Manitoba records I can find, and since then the species seems to have disappeared from the Province, though it still abounds along the Saskatchewan and westward to the mountains.

It is strictly a prairie animal, harbouring in burrows and never venturing far from them, so that it is the most subterranean of our Foxes. ENVIRONMENT

Nothing is known of its mating, beyond the fact that the creature pairs, and that the pair continue together all summer, probably for life, as the male is active in the care of the young. MATING

[1] Journal 1897, p. 259.
[2] United States Geol. Surv., 1878, Vol. IV, Bull. 3, Art. XXV, p. 547.

DENS One of my guides, Lee Hampleman, of Meeker, Colo., tells me that in 1897, when on Pawnee Creek, Colo., he found a Swift's den. It was reached by a tunnel about 9 feet long and was 5 feet from the surface. The chamber was nicely lined with grass and contained 5 young ones. 'Just the cutest, prettiest things he ever saw.'

Fig. 188—Study of Kit-fox in Philadelphia Zoo, August, 1886.

These were taken home to the ranch and easily raised, but they never became tame. Both parents were seen about the den.

Prof. John Macoun relates that in Alberta, June 16, 1895, he saw 2 old Kits and 5 young sitting on a prairie knoll. His dog rushed at them; the young dived into a hole, and the parents busied themselves leading the dog elsewhere.

An interesting account of a pair of Kits that lived near his ranch house, has been given me by R. W. Cowan, of Cochrane, Alta. They were such beautiful and playful creatures that he rather encouraged them until they began to kill chickens, whereupon they fell from favour, and paid the extreme penalty. The family consisted of 2 old ones and 5 young. The latter began to run in the month of May. The old ones were seen

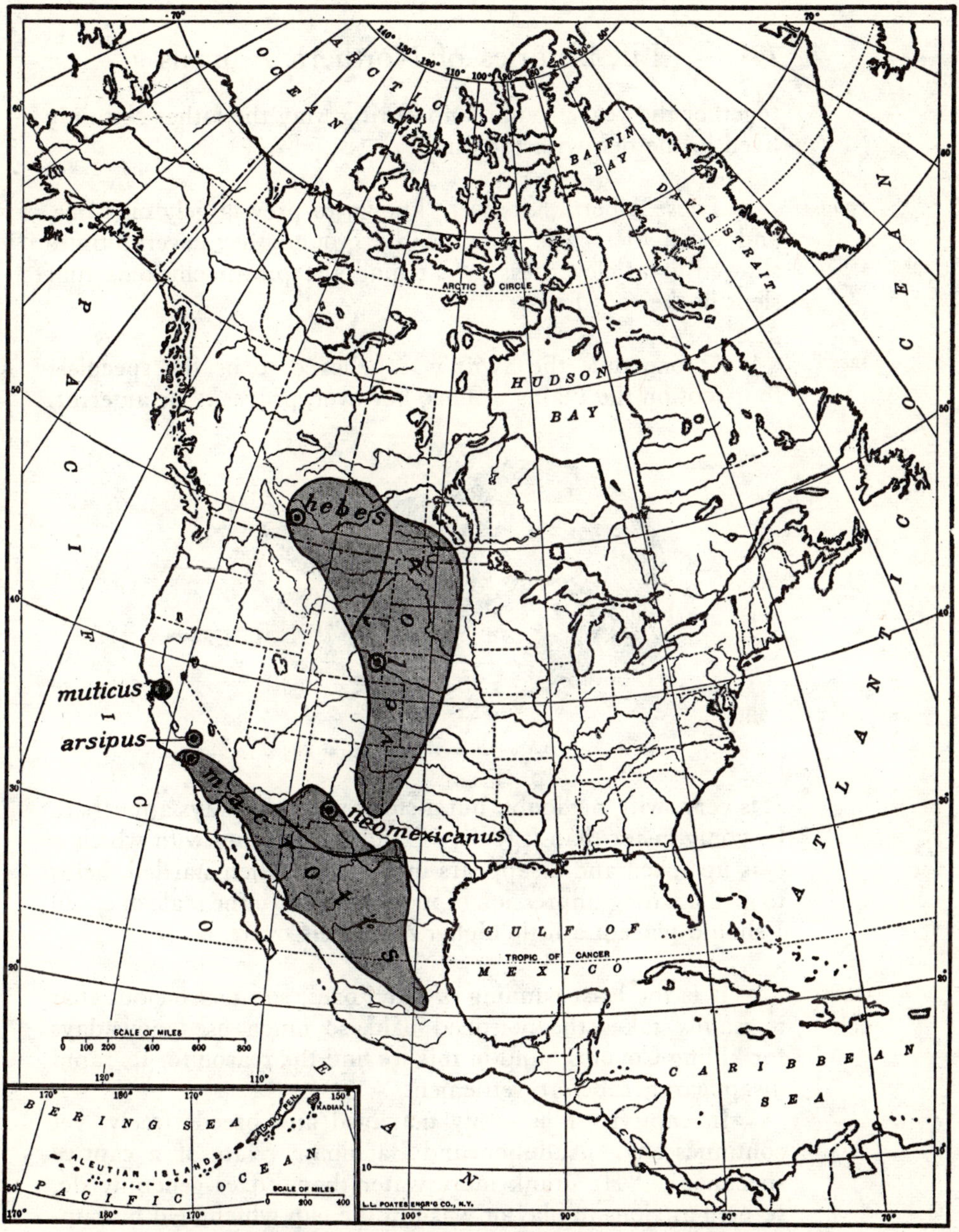

MAP 40—RANGE OF THE NORTH AMERICAN KIT-FOXES.

This map is diagrammatic: the boundaries of the races are theoretical. It is founded chiefly on papers by C. H. Merriam and D. G. Elliot, with records by T. Say, V. Bailey, E. R. Warren, and E. T. Seton, and assistance from E. A. Mearns.

The following species are recognized:

Vulpes velox (Say),
Vulpes macrotis Merriam,
Vulpes muticus Merriam,
Vulpes arsipus Elliot.

together the year around, and during May the father certainly lived in the den with the family.

FOOD

These Alberta Kits were known to prey largely on Mice, and when there was snow on the ground they several times showed great cleverness in catching the prairie-chickens that slept in the soft drifts.

SPEED

At one time the Swift was believed to be the speediest four-foot on the Plains. This, however, proves to be an error.

FIG. 189—Life study of the Algerian Kit-fox.

It is very swift, no doubt, but a small animal always appears to be going faster than it really is, and the rapidity with which it gets up speed and disappears into a hole, when startled, helps to give a wrong impression of its velocity. In the scale of speed I should place it a little higher than the Coyote.

UNSUSPICIOUSNESS

It is the least cunning of our Foxes, so unsuspicious that it readily takes the poisoned baits so much used nowadays for killing Coyotes; and in this we find the reason for its rapid disappearance before settlement.

In captivity it is easily managed and breeds freely, yet continues shy. Audubon and Bachman[3] relate of a captive specimen: "He drank more water than Foxes generally do, seemed anxious to play or wash in the cup which held his sup-

[3] Quad. N. A., 1849, Vol. II, p. 16.

ply, and would frequently turn it over, spilling the water on the floor of his cage."

One of the Indian names of the species is said to mean 'lousy thing,' because it is pestered with lice. The fur is of little commercial value. At Lampson's sales in London, 1905, a total of 5,129 Kits were sold.

In the March sale of 1906, the number fell to 1,404; 5 shillings and 6 pence ($1.32) was the highest price paid, and 1 shilling and 3 pence (30 cents) the lowest.

XXXII.

Royal Fox, Prairie Red-fox or Common Red-fox of Manitoba.

Vulpes regalis Merriam.

(L. *Vulpes*, a fox; *regalis*, royal, because of its superb appearance.)

Vulpes regalis MERRIAM, 1900, Proc. Wash. Acad. Sci., Vol. II, pp. 672–3, December 28.

TYPE LOCALITY.—Elk River, Sherburne County, Minn.

FRENCH CANADIAN, *le Renard royal.*
CREE, *Wah-kus'.*
OJIB. & SAUT., *Wah-gush'.*
CHIPEWYAN, *Nak-ee'-they.*
OGALLALA SIOUX, *Shung-ka-ge'-lah.*
YANKTON SIOUX, *Song-kee-na.*

The genus *Vulpes* (Brisson, 1762) comprises dog-like animals of small size, with long bushy tails (more than half the length of the body); long soft fur, long sharp muzzle, large ears; long, sharp semi-retractile claws; linear eye-pupils; and teeth as follows:

$$\text{Inc. } \frac{3-3}{3-3}; \text{ can. } \frac{1-1}{1-1}; \text{ prem. } \frac{4-4}{4-4}; \text{ mol. } \frac{2-2}{3-3} = 42.$$

The number is the same as in *Canis*, but they are different in detail, as well as much more slender in general.

In addition to these generic characters, the Royal Fox has the following:

Length, about 44 inches (1,118 mm.); tail, 16 inches (406 mm.); hind-foot, 7 inches (177 mm.). The females are about one-tenth smaller. **SIZE**

An adult taken at Carberry, October 27, 1884, weighed 10 pounds. **WEIGHT**

The general colour is golden-yellow, very pale on the hind-quarters, also on the forehead, where it is sprinkled with whitish hairs, and deepening on the back into a reddish-yellow, which extends in a band from shoulder to tail; beginning behind the shoulder, this is sprinkled with whitish hairs, giving a pinkish effect at a short distance. Legs, dark buff; the black on the feet, very limited and mixed with whitish hairs; outer half of ears behind, black; tail, pale brown above, shaded into yellowish below with olive tinge; tip of tail, belly, breast, throat and lower jaw, white; on chin and lower parts is often seen dark or black tinge in the white. **COLOUR**

This is indeed a flat enumeration of its flat tints, but gives no conception of the marvellous colour beauties of its exquisitely blended tawny-pinks, russets, and yellow-browns, set off by the old gold, dull silver, and shining ebony of its extremities.

Black, Silver, and Cross Foxes occur in this species; these forms are mere colour freaks, and may be found in the same brood with those of ordinary colour. **FREAKS**

A notable example of this is given by A. P. Low, in his "Mammals of Labrador," as follows:[1]

"On the Moose River, in 1887, the writer found a litter containing 7 Kits: of these 2 were red, 3 were cross, and the remaining 2 blacks or silver—thus showing that the colour of Foxes no more constitutes varieties than does the difference of colour in a litter of kittens of the common cat. There appears to be a greater proportion of dark-coloured Foxes in the northern region than in the southern."

[1] Mam. Labrador Penin., Can. Geol. Surv., 1896, p. 314 L.

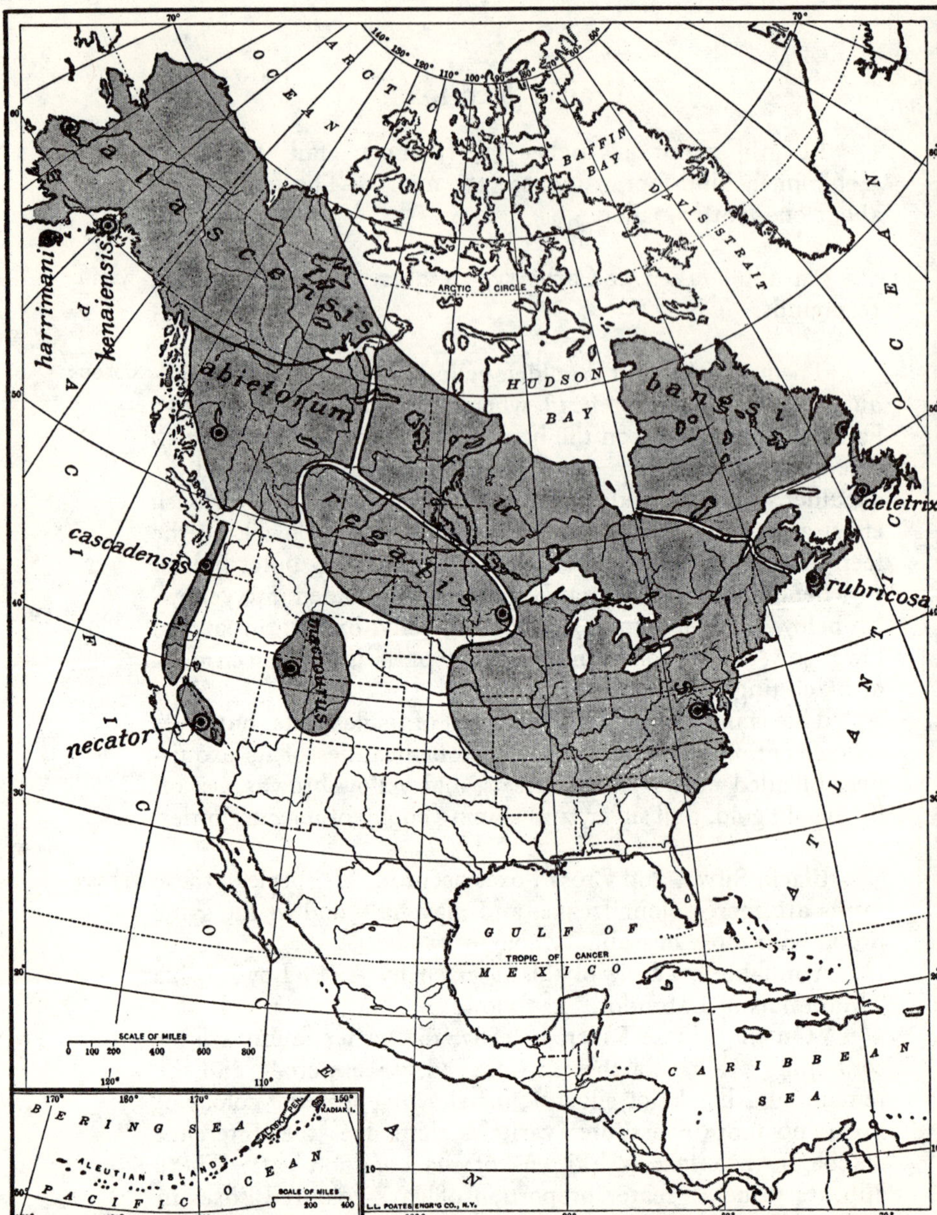

MAP 41—RANGE OF THE NORTH AMERICAN RED-FOXES.

This map is diagrammatic and must be greatly modified by further work, especially in the south and west. Anticosti should have been tinted. It is founded chiefly on Dr. C. Hart Merriam's 'Revision' with additional records by E. W. Nelson, S. F. Baird, J. Fannin, R. MacFarlane, Audubon and Bachman, A. P. Low, V. Bailey, E. A. Preble, O. Bangs, A. E. Verrill.

The following are the species:

Vulpes fulvus (Desmarest),
Vulpes macrourus Baird,
Vulpes necator Merriam,
Vulpes cascadensis Merriam,
Vulpes rubricosa Bangs, with 2 races,
Vulpes deletrix Bangs,
Vulpes alascensis Merriam, with 2 races,
Vulpes kenaiensis Merriam,
Vulpes harrimani Merriam,
Vulpes regalis Merriam.

Another freak is the 'scorched' or 'Samson Fox.' This has no long fur, nothing but wool; the cause of this is not understood, and the pelt is worthless commercially.

When seen running on the prairie, the present species appears a large straw-coloured animal, with black boots, and enormous ears and tail.

All Manitoba specimens hitherto examined belong to the species *regalis*, but it is quite likely that in the north *fulvus* will be found:

Vulpes fulvus (Desm.) may be distinguished when alive by its much smaller size, general deeper and intenser colours, the greater amount of black on the ears and on each leg, and the black spot on the base of the tail above.

Cranially, also, they are well apart; the bullæ of *regalis* being much larger in proportion, etc.

Life-history.

So far as known, this Fox has a very limited range and is confined to the prairie country and adjoining woods. In Manitoba it is found in all the south and west parts, on the open prairie, in the poplar and pond country, and in much of the spruce country. Its favourite localities are the half-open regions—there are, indeed, very few creatures that like the sunless depths of unbroken forests. The great belt of half-timbered country from Roseau River to Dawson Bay probably produces more Foxes than any other part of the Province, and in this they especially affect localities that are broken by low hills and ravines, or that are close to marshes and cover. RANGE

It would be safe to estimate that in the days from 1880 to 1890 there were 2 pairs of Foxes to every township of this great belt, with half as many for the rest of the country. The number of fox-tracks across any section of prairie within three days of fresh snow made this easy to believe. According to these data, there were at least 5,000 Foxes in Manitoba. ABUNDANCE

That this is not an overestimate will appear on setting the facts side by side with those supplied in the 'Old Country.'

After consulting many of my hunting friends, I learn that about 11,000 Foxes are killed each year before the hounds in Great Britain; while the destruction by keepers, etc., at least doubles the number destroyed annually on the 80,000 square miles of the British mainland, and would argue a vulpine population in the autumn of fully 40,000, to compare with the 5,000 that I give as a conservative estimate of the Foxes on Manitoba's 74,000 square miles.

Most observers testify that Foxes are growing less numerous in the Province. This is generally traced directly to the increase of Coyotes; which does not necessarily mean actual conflict of the two, but that the changing conditions have set up new problems of life which the Coyotes have been better able to solve.

The rate of increase among the Blue-foxes of St. George Island, Bering Sea, will help us to gauge the increase of Red-foxes. This island, about 36 square miles, has about 270 pairs of foxes, and, although they are fed and protected and the species has 5 to 12 in a litter, not more than 400 to 500 can be marketed each year without reducing the stock.[2]

The fur returns (given later) show an annual catch of 74,000 Red-foxes with marketable coats. Considering other destruction and their rate of increase, this, I take it, assures a wild stock of at least 500,000, possibly 1,000,000, on the range covered by the Red-fox group.

INDIVIDUAL RANGE

The home-locality of the individual Fox is, I think, not more than 5 miles across. Ordinarily, it does not range so far, but, under unusual stress of famine, will cover even a larger area.

The evidence is—that when pursued by dogs the Fox usually circles at a radius of 2 or 3 miles, differing, of course, with the character of the country; also that a number of well-

[2] See James Judge on Blue-foxes of the Pribilof Islands' Rep. Am. Breeders' Assoc., Vol. V, 1909, p. 338.

known Foxes, such as the 'Mahogany Fox' of Hartford, 'Baldy' of Berkeley, the 'Black Fox' of Blacktail Creek, were known and watched for one or more years and usually found within 3 or 4 miles of their reputed head-quarters.

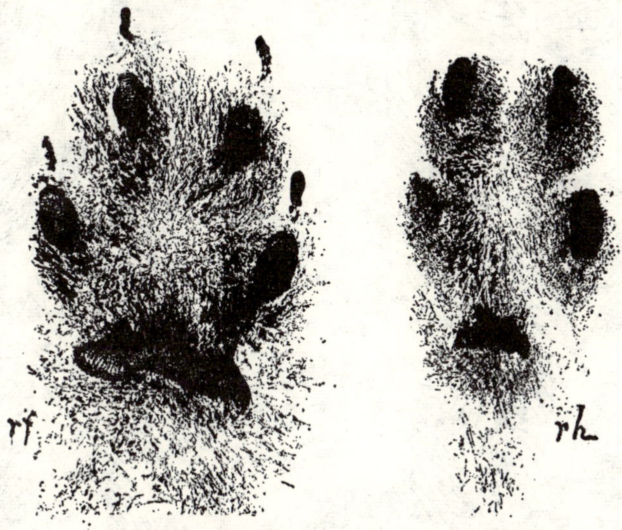

FIG. 100—Impressions of a Fox's feet, from life.
Secured by Mrs. Grace G. Seton. *rf*=right front; *rh*=right hind; both are life size.

Since writing the above, E. Norton, a well-known fox-breeder of Dover, Maine, has given me the following corroborative information.

In March of 1885, he and his brother went out near Dover with their fox-hounds, hoping to run a she Fox to earth and dig her out for a breeder. They soon had one holed up, but were disappointed to find it a dog Fox in half-shed coat; he was not worth skinning, so they cut the top off one ear and slit the other, then turned him loose. Five years afterwards he was killed within 4 miles of where first they had caught him.

At Green Lake, Ont., was a Silver-fox that ranged there for a couple of years. My informant, George Linklater, says it covered 10 miles, or not more than 15, of country, and was there the year round.

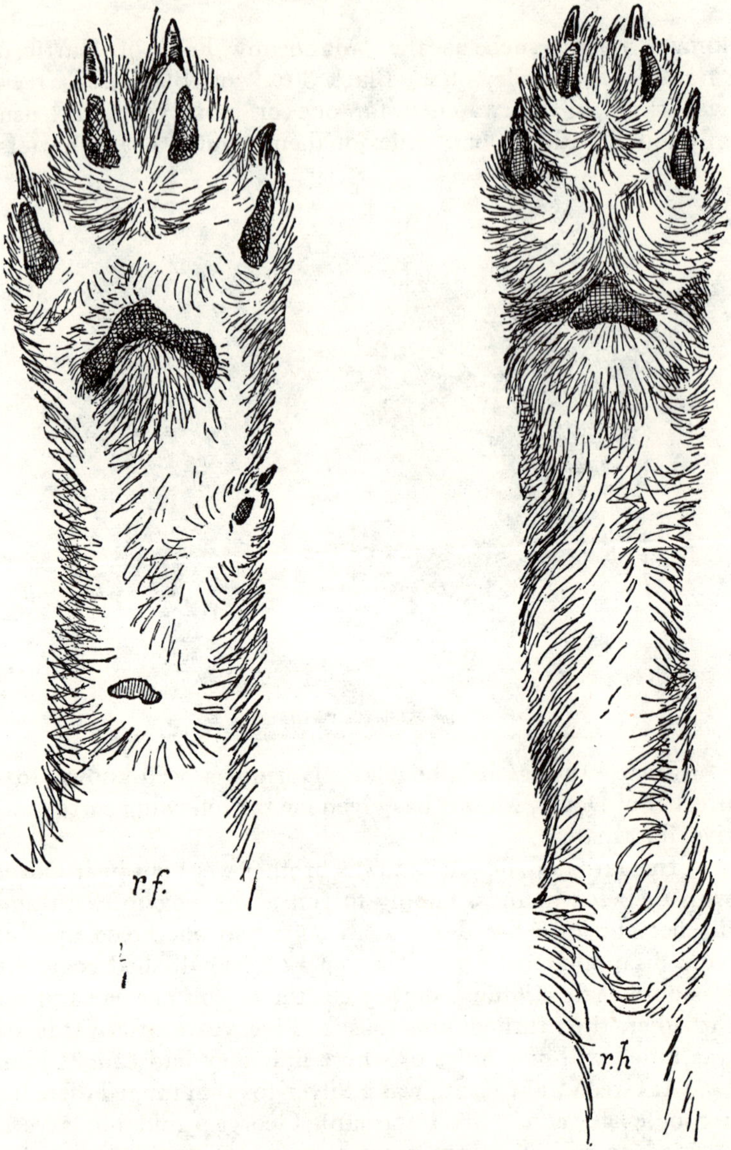

FIG. 191—Feet of Fox (*V. macrourus*).

Sketched in Colorado, Sept. 16, 1902. *rf*=right front; *rh*=right hind; both are life size.

The habits of the English Fox affirm this. All the keepers and hunters that I have consulted believe that a Fox rarely goes more than 4 or 5 miles from home, except when hunted.

But in the winter it roams over an extent of country probably two or three times as large as its summer range. A black Fox that lived on the 'Big Plain,' Manitoba, was seen about Carberry as far north as Petrel, 9 miles, and south to the Sandhills, 4 miles. There is no certainty that it was the same Fox, but when one was killed at Petrel it also disappeared at Carberry, and if it had lived only about Carberry it would have been seen oftener. This seems to give it a winter range of at least a dozen miles in diameter.

SOCIABILITY

Since the relations of mates, or of the young to the parents, are excluded in discussing sociability, we must consider the Fox but slightly a sociable animal. The only detailed cases I have of Foxes working together, refer to two (probably mates) that were combining in a hunt.

AMUSEMENTS

This animal is not much given to social amusements, but Norton and Stevens both tell me that on their fur farms in Maine it is a common thing for the Foxes to gather on moonlight nights and chase each other about with most uproarious barking and churring that do not seem to express anything but good-will and hilarity.

VOICE

This species uses the smell-telephone much less, I think, than the Wolf does. (See Wolf.) Its principal method of intercommunication is doubtless by the voice. It has a short bark followed by a little squall like '*yap-yurr*.' That is the sound oftenest uttered, but it has also a long yell and two or three different yowls or screeches as well as softer *churr-churrs* that doubtless have different meanings to its kind. The voice of the male is notably heavier and coarser than that of his mate.

In the pairing season the she Fox utters a very characteristic shrill squall. The reply of the male is usually two or three short barks. In autumn, I have several times known a Fox come at night around the camp-fire and 'yap' in the manner of a Coyote.

MATING

It seems to be a solitary animal during the winter, but the mating instinct is awakened in late February or early March, and then the track in the snow is often doubled. Those who in New England have followed it for miles at this interesting period, tell me that all the chapters of romance are duly recorded in the snow—the pursuit, the coquetting, the conquest, even the fight between rivals, are fully set forth in the tell-tale white.

These fights I have never witnessed among wild free Foxes, but L. W. Walker, living in the Yellowstone Park, writes[3] that there they are of daily occurrence in late winter, and the part played by the tail is quite important. When the rivals approach each other with hostile intent, they stand sidewise with the tail raised and pointing forward over the back, ready for use as a parry or as a feint; dashing it in the eyes of his foe, the Fox distracts attention or prevents him seeing for a moment, during which time he tries to gain some advantage.

Thomas Anderson, of Fort Smith, Hudson Bay Post, tells me of a curious occurrence that he witnessed at Poplar Lodge River, on the east side of Nipigon Lake, early in March, 1896. As he drove his dog-train around a point he came on a pair of Foxes *accouplé*. Supposing that he could easily secure both, he set his dogs after them, but they turned their heads one way and raced off side by side, allowing no stick or sapling to come between them; and thanks partly to a slight crust, they left the dogs far behind and escaped, without parting company.

PAIRING

There can be little doubt that this Fox truly pairs. I have never seen or heard of *more* than two full-grown Foxes together in Manitoba, but this I have often seen, and have heard of times without number. Many observers, among them W. R. Hine, have found the home in cubbing time, and in each case

[3] Recreation Magazine, May, 1897, p. 339.

PLATE LI.—FOXES FIGHTING.
Drawn from the description of L. W. Walker, who witnessed to the scene many times in Yellowstone Park. The use of the tail as a foil was continually remarked.

both parents were about; it seems quite certain that the father takes an active interest in the young and helps to care for them. All of which tends to prove that our Foxes pair.

The argument of analogy is also in line, for I have observed that in Ontario both parents (*V. fulvus*) take active care of the young. In Maine, E. Norton says, the male Fox has as much to do with raising the family as the mother has.

A veteran fox-hunter (J. H. Whitcomb) writes me from Ayer, Mass., on April 3, 1909: "A day or two ago I saw a fox-den where I think there were young ones, and close by in the sand a Skunk half buried. I suppose it was for madam, to save her from hunting."

In Wales, as T. W. Proger writes me, the male Fox is a faithful partner, bringing food to the female while she is suckling the young, and has a great affection for his offspring, sharing with their mother the labour of feeding and caring for them.

A touching case of this paternal devotion was sent me later by this same naturalist:

"Last spring our old keeper destroyed a litter of young cubs, and stopped them into the earth. Four days afterwards the old dog Fox came right up to the earth and began to open it out again, I concluded, to seek his cubs, but he was caught in a trap which the man had set near the blocked-up entrance.

"I think this is very strong proof that the dog Fox has affection for his young, because he knew very well that the trap was set there on the first night after the deed was done, but his strong desire to find the cubs overcame his habitual caution."

In view of the evidence direct and by analogy, I think we are safe to believe that in life and manners the Manitoba Fox is as good as its near kin.

Granting the pairing, the next question is whether it is for the season or for life. There is much evidence of Foxes consorting in pairs *after* the breeding season, and this points to union for life. The consensus of opinion among hunters and naturalists, according to Dr. Woods Hutchinson,[4] agrees with this conclusion.

[4] Animal Marriage, Cotem. Review, October, 1904, pp. 485–496.

DEN

The den, or 'earth,' is the nursery of the Fox. Although used chiefly while the cubs are nursing, there are cases to show that some Foxes live at home the whole year round.

It is approached by a burrow of 9 to 12 inches calibre, usually on the sunny side of a hill or bank. Sometimes the Foxes make it themselves; sometimes they adapt one that they find.

A fox-den which I examined on a wooded hill, near Toronto, had two or three entrances. This was made like the

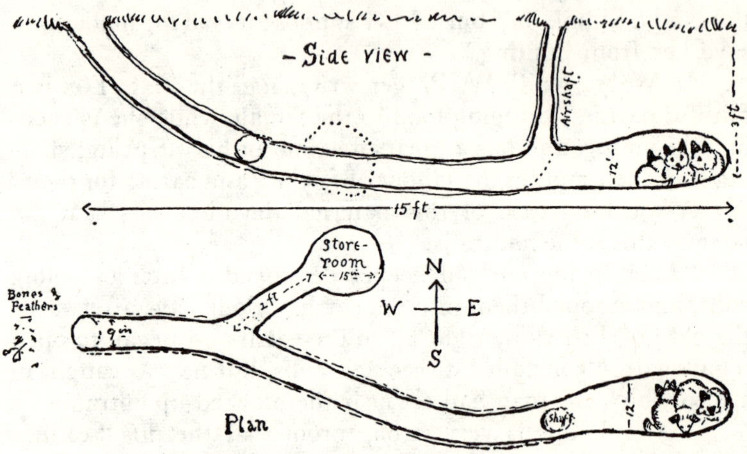

Fig. 192—Side view or elevation, and plan of the fox-den opened by Geo. L. Fordyce.

home of a Chipmunk, that is, all the earth was scratched out of one hole, though there were several doorways. Those chiefly in use had no earth-pile about them to make them conspicuous. Indeed, one might have been within ten feet without suspecting their presence.

I have, however, seen many newly made fox-dens which had no earth-pile whatever, though the tunnel was fully ten feet deep. Evidently the Fox had disposed of the earth by scattering it.

The nest is made in a dry chamber a dozen or more feet from the door, and is sometimes lined with a little dry grass.

George L. Fordyce[5] writes me that in northern New York State he once found a fox-den in a hollow log, another in the base of a hollow standing tree, and yet another in the ground, apparently dug by the old Fox. This was as here shown (Fig. 192). It consisted of two separate parts, the den and a store-room, with a quantity of food.

"I think [he says] this was originally a Woodchuck den, which the old Fox enlarged. I have double-lined the Woodchuck part. Both the store-room and the sleeping room seem to have been made by the Fox.

VENTILATION

"The air-shaft was one of the Woodchuck entrances. The Foxes had only one, but this was the only fox-house I ever knew that did not have two or three different doors.

CLEANLINESS

"There was no bed or lining anywhere, just the clean sandy clay in which the tunnel was made. I noticed in particular also that there was no excrement or offal of any kind anywhere in the den. In fact, everything was decidedly clean and tidy, though, of course, there was the Fox odour. Some bones, feathers, and one or two lamb's legs were found on the pile of dirt that had been thrown out in making the tunnel. There were, however, but few there, and I am inclined to believe that they were perhaps brought out and left there by the young, and would have soon been carried away by the old Foxes. This is a mere guess on my part, but it is very certain that the leavings at the den did not represent more than two or three days' meals. The scarcity of garbage about the den and the remarkable neatness of all indoors, make me believe that the old ones habitually carry away the dung and rubbish."

YOUNG

The period of gestation is now known to be 51 days.

The young are born about the first of April and number from 4 to 9. N. E. Skinner, of Winnipeg, tells me, that at Winnipegosis he found a Black-fox that had young so early

[5] Of Youngstown, Ohio. Letter, April 14, 1905.

in the season that the mother's pelt was prime, bringing $75, though the young were big enough to rear by hand.

The cubs are clad in lead-coloured fur and look as much like kittens as Foxes when they come. They are blind till 8 or 9 days old. They do not venture out of doors till they are three or four weeks old, and the den continues to be their only home for 3 months.

Fordyce sends me also the following interesting notes on the Fox family that he observed near Auburn, N. Y.: "One morning in the end of March, 1877, a man came to the store in the country-village where I lived and said that while coming across the fields of a neighbour, an old Fox had jumped out of a hole in a tree near which he was passing and run away barking. I overheard what he was saying, and as it was the same wood-lot in which I had found the two young Foxes the previous June, I at once thought there might be some cubs in the tree. I started out across the fields and 'back-tracked' the man, there being snow on the ground, until I came to the place where he had stopped. The tree was a standing basswood about four feet in diameter. The Fox-track led from the hole at the base of the tree, but no track came to it. As it had snowed the night before, this indicated that the old one had gone in before the snow-storm, and remained there until the man had frightened her away. I reached into the hole and found that the hollow diameter of the tree was about two feet. I felt a warm bunch of little creatures, one of which I pulled out, looked at, and, as I had expected, it was a young Fox. I then removed my coat, laid it on the snow beside the tree, to put them on a warm spot, and took out the others, 9 in all. The little creatures did not look unlike new-born kittens, but were about twice as large. Their eyes were not open, and I do not think, from their condition, they were more than one or two days old. I selected one of the lot and took it home to raise, but, having no way to feed it properly, it died within a few days. I then followed the track which the old Fox had made when the man had frightened her away from the tree. She had run directly to a hilltop, nearly a quarter of a mile away, and there stopped to watch him. I followed her track farther and found where she had stopped many other places, and, finally, where she had stood on a high stump and probably watched me while I was at the tree looking at her young. In fact, I believe she had watched me from this stump while I had

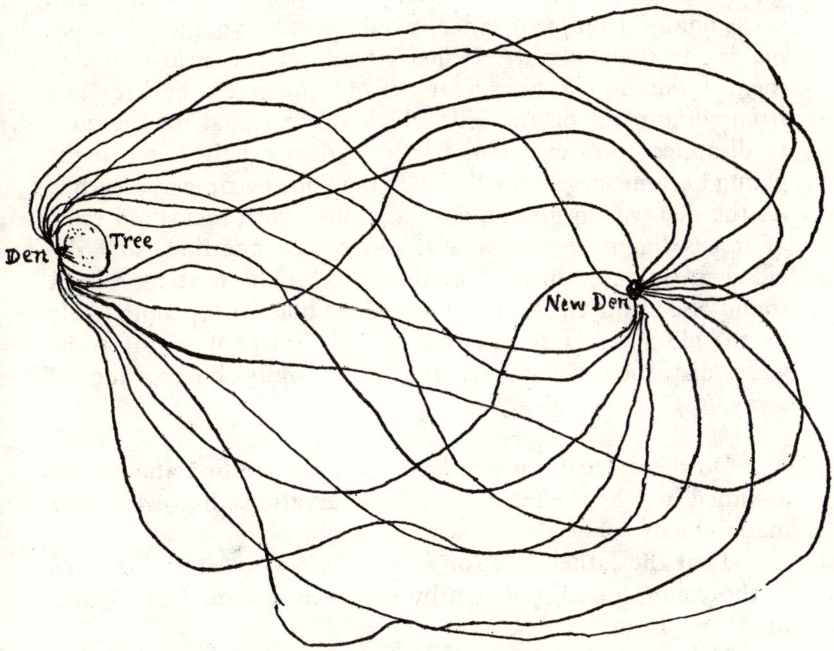

FIG. 193—Diagram of fox-tracks (by G. L. Fordyce), showing approximately the tracks left by the mother Fox in moving her brood from the hollow tree to the new den.

been following her trail, because her tracks showed that she had run with greater leaps from the stump straight away into the open country.

"The next morning I went to the tree soon after daylight and the young were gone. There was a maze of tracks coming and going to the tree. I started out to follow them, settling down to one which I followed with considerable difficulty, on account of other tracks crossing it, but I finally succeeded in locating the den into which she had carried them. It was an old burrow in a hillside about a quarter of a mile from the tree. In going back and forth she had gone over a different route each time, sometimes far out from a direct line on either side, and sometimes far beyond. The number of tracks showed that she had carried the young one at a time. For the next two or three months I went to the locality of the den frequently, and occasionally, from a distance, would see the young Foxes out playing in the sunlight. When I attempted to steal nearer to them, I would invariably hear the old one give a bark or two from some place on the hill, which was a signal for the cubs to disappear into one of the holes. About July 1, when the young became larger, my desire to take one became irresistible. As the den was in ground that was filled with rocks and roots of trees, digging for them was out of the question, so I decided to set traps in each of the holes. The next morning I found one of the traps pulled out of the hole and sprung, likely by the old Fox. I set this trap again, but none of the traps were disturbed thereafter, the Fox family having moved away."

On these occasions Fordyce saw only one Fox about, and assumed it to be the female. His observations, however, were made chiefly while the cubs were small.

FATHER

That the father Fox (of *Vulpes vulpes*) is never forgetful of the young brood is shown by evidence sent me from Wales by T. W. Proger:

"When the cubs are very young, he will frequently bring food right up to the earth, but as they get older he does not do this, but drops it at a distance from the earth, a hundred yards or more. This may be done to teach the cubs to hunt for themselves, as the rabbit or bird, as the case may be, is often lightly covered with leaves and mould, and it would certainly be very good practice for the youngsters."

The place around a fox-den is usually littered with bones and feathers of their prey, but they also have an indoor banqueting hall. As already noted, Fordyce dug out a fox-den early in May, 1878. "During the winter of 1878," he says, "I had tracked a Fox into this den. In early May it was reported that Foxes were killing lambs in the neighbourhood. I went at once to the place and found about the hole abundance of bones, wool, and feathers, indicating that it was the home of a Fox family. The next morning two other boys went with me to dig them out. We followed the main hole, which went down with a gradual slope for about 3 feet from the surface of the ground. It continued nearly level for about 20 feet, where we found 3 young Foxes grown to about the size of an ordinary house cat. After getting them into a bag we went back to a branch from the main hole that we had passed in digging. Following it for about 4 feet, we found an enlarged dug-out space which was used as a store-room." (See plan, page 716.)

Concerning one which he kept captive, my informant writes: "One day in June, 1876, my cousin and I chopped 3 young Foxes out of a long hollow log. They were about one-third grown and quite savage.

"We each took one to our homes. After keeping mine for several months, until well grown, it was killed by a neighbour, on account of catching chickens. It seems to me this Fox did some 'thinking' in its method of taking the chickens. I had dug a trench in the ground, gradually sloping it down until about 2 feet deep and 6 feet long. At the bottom of the trench I placed a wooden box with a wooden arch leading down to it. When the dirt was filled in, this made rather a good den underground. The Fox was fastened with a chain attached to an overhanging pole, which gave it free access in and out of this den with a radius of perhaps 20 feet on the outside. Our neighbour's chickens were running about the yard more or less, and the Fox began catching them. I watched to see how it was done, and found that when food was given the Fox, it would, instead of eating it, place it almost as far away from the opening of the den as its chain could reach. The Fox would then back down into the hole and wait until the chickens came for the food, and when one got inside the radius of its chain, it would have chicken for dinner instead of the food I had given it. One thing I noticed in particular was that it seldom missed getting a chicken when it jumped for one; in fact, it never made a jump except the chicken was inside the chain limit."

During June and July I often visited the Ontario fox-den referred to above, sometimes going at night, and usually taking my hound with me. And on each occasion, long before nearing the place, we were met by one of the old Foxes, who would deliberately cross our path or bark at us from a hillside, tempting the dog away in pursuit. The latter would dash off at full cry, and I could tell by his tonguing that the trail led to a distant part of the country, either along the railway track or down the river, where the Fox easily got rid of him by some trick. One of these tricks I witnessed in daylight. The Fox led the dog down the river, then, retracing his steps for forty or fifty yards, he scrambled along a steep bank of sand that edged the stream. The sand apparently carried no scent; the hound could not follow it at all.

I saw the Fox do this several times at the same place. Indeed, he was shot here by one of my friends, in the act of repeating the performance. This turned out to be the male Fox. Evidently he was actively interested in the care of the young.

One day (July 15) I saw the mother Fox carrying a live hen from our barnyard across the river and up towards the den. I believe she was keeping it alive with intent to let the young ones have the practice of killing it, just as a cat will bring live game to her kittens.

As soon as the young are large enough to come out of doors, they romp and play about in a delightful fashion, combining the elegant suppleness of the Otter with the frolicsome ways of kittens, but are ever ready to fly home at the slightest alarm. Usually the alarm is given by the ever-watchful parents.

There is, of course, only one brood to the season. The young are nearly full grown by the end of August, but are still in the old home with their parents. Such quantities of game, dead and alive, have been brought to them during the summer that their front door is now dangerously marked with the bones and feathers of the victims.

The young probably scatter voluntarily before winter, but it is interesting to remember that in England a systematic effort, called cub-hunting, is found necessary each year, in October, to disperse the families and improve the hunting by equalizing the distribution of Foxes.

I have never seen the young accompanying their mother; indeed, 2 Foxes are the most I have known together.

The Eastern Fox seems to hold its own very well, wherever there is rough country for final retreat. I suspect that there are even more Foxes in New England and Ontario to-day than in the early times, for the reason that food is more plentiful in winter now, and at no time did the Fox prefer the deep forest. I remember very well once in June, 1885, near Coboconk, Ont., seeing a Fox trot out of the woods ahead of me for fifty yards along the path, then disappear without knowing I was near. I mentioned this at a camp I came to in the evening. Two lumbermen were its total population, one had been 16 years in the Muskoka woods, the other a little less, but both said that they had never yet seen a live wild Fox in the country. HABITS

The world-famed cunning of the European Fox is due in part, no doubt, to the ceaseless persecution it has suffered so long. Yet our American Foxes are not unworthy of their trans-Atlantic cousins. From birth they have a deep-laid fear of every strange or peculiar object, and they early acquire a horror of anything that bears the taint of man. Their mode of life is, moreover, a constant sharpener of their wits. And the quickness with which they learn to distinguish and distrust the latest devices of the trappers, is wonderful evidence of their cunning, perhaps also of their power to communicate certain ideas. MENTALITY

When caught by the foot a Fox will struggle violently, twisting and tearing at the foot, sometimes till it is torn off.

Some trappers say it will amputate the leg if need be. I have never seen this, but I have known the Fox to bite off the imprisoned toes *below* the trap. This, whether design or accident, is the best thing it can do, as it can then jerk free with the least possible loss and pain.

Many hunters and farmers in England have told me that a Fox never kills near home. It has no wish for trouble with the near neighbours. The barnyard next its den is perfectly safe so far as this pair is concerned. I am not sure that this is the case in Manitoba as yet.

There is a device that I have several times known the Ontario Fox to resort to when pressed by the hounds, that is, run along the railway ahead of a train, and cross a high trestle bridge. On one occasion I knew of a hound being thrown by the locomotive from the trestle into the river below, minus his tail, but otherwise unhurt. I was told, however, that all were not so fortunate, as some hounds had been killed at the same place in a similar way. It is very hard to say how much was intentional on the part of the Fox. The fox-hunters who know the animal, say *it was intentional throughout*. Some maintain that it was entirely accidental. It certainly was not necessary for the Fox to know anything about train times, as he could *hear* the train coming miles away. The track is a notoriously bad place for scent to lie, the trestle was a place of difficult footing, like a sloping tree, which often furnishes refuge, or the steep sand bank already noted, where I several times saw the Fox baffle the hounds. He might run to the train, just as I have known a Deer or Hare run to a wagon or sleigh when flying for its life, preferring the unknown terror to the certain death. Add to this the element of luck when first the Fox made the attempt; success that time would lead him to try again.

I have several times been told by hunters, of Fox mothers poisoning their captive cubs because they could not free them. I am more sceptical now than I was formerly of these accounts, not because, as some have illogically asserted, this would postulate a knowledge of the nature of poison and of death on the part of the Fox—it would be just as true to claim that a Cougar has a comprehension of locomotor paralysis because it aims to disrupt the spinal cord of its victim—but because the evidence was faulty. The possibility of murder under such circumstances is proved by the facts that cattle will often kill one of their own kind that is in dire extremity; a crow in trouble is sometimes destroyed by his friends; a Mouse in a trap is often devoured by its own companions. E. Hofer reports a case in the Yellowstone of a little Bear cub that on the first night of its captivity was killed and eaten by one of its adult wild kinsmen. In menageries many carnivorous mothers, including Foxes, kill a large proportion of their own young, especially when they learn that man has tampered with them. And, finally, Wolves and Foxes have certainly grasped the idea that poison is a thing of danger. These various facts bring us much nearer to acceptance of the hunter's tradition, without, however, being conclusive. They at least remove it from the category of the wildly impossible.

It is well known that the English Fox will unite with another, probably its mate, to catch a Hare, by the old stratagem of drive and ambush. W. R. Hine reports a similar subtlety on the part of the Manitoba species.

Near Morris, in 1885, he once saw 2 Foxes working together to stalk some Canada geese that were feeding on the open prairie. One Fox was lying in wait in some slight cover; the other was approaching from the opposite direction, one hundred yards away. It crawled as close as possible, then, seeing that the ever-watchful geese were alarmed, it began to roll about on its back in plain view, and tumble over, looking much like a bundle of dry grass that is being blown about by the wind. Each move brought it nearer to the geese, who, knowing it well for an enemy, kept moving away as they grazed, and thus drifted towards the Fox in ambush. Hine was satisfied that the two were working together, but did not see it out. When he had got within about fifty yards he 'collected' the Fox and a goose with 'right and left.'

NON-MIGRATORY

I have not seen any good evidence of migration among Manitoba Foxes. The fact that they are much more often seen in fall is due, of course, to the fact that the Fox population is then at its maximum; the families are breaking up, and the young are running about in search of the best hunting grounds. Their habits change but little, however, with the changing season.

SPEED

The best speed of an average Fox for 1 mile is at the rate of about 26 miles an hour. This is faster than a Coyote, but slower than a Jack-rabbit.

A. S. Barton, of Boissevain, sends me an interesting item on this head. "Once," he says, "while mounted on a fast saddle-horse I ran a Fox for half a mile, both of us doing our best, but it was an even race all the way. I should say the horse was doing two-minute time, as his record was 1.51 for the mile." Therefore this Fox was running at the rate of 30 miles an hour. Doubtless it was his highest speed, and he must have been an exceptional Fox.

USE OF THE TAIL

No one can long watch a caged Fox in winter time without discerning the use to which it puts its great bushy tail. Its nose and pads are the only exposed parts, and these might easily be frost-bitten when it sleeps during severe weather. But it is always careful on lying down to draw these together, then curl the brush around them; it acts both as wrap and respirator. I have many times seen wild ones do this same thing, and am satisfied that the tail is a necessary of life to the Fox, as well as to the Squirrel and Wolf. I believe a Fox or Coyote would die before spring if turned out in the autumn without a tail.

The brush is large in proportion to the coldness of the climate. In Foxes from the Southern States it is a very meagre thing, but on the Saskatchewan and further north, it is enormous, looking at a distance almost as large as the Fox's body, and, of course, it reaches its greatest size in the depth of winter. The coat in general is developed by cold, but not apparently to the same extent as the tail.

PLATE LII.—THE FOX SPRINGING ON THE PINTAILED GROUSE THAT HE LOCATED BY SMELL AS IT SLEPT UNDER THE SNOW.

As already noted, this useful member serves further as a fender in fighting, but it has also its disadvantage. Dunham Wheeler, of New York, tells me that once while hunting a Fox in the Adirondacks, during early spring, when the snow was deepest and wet, he saw the creature coming toward him; it stopped and seemed to worry its tail; again it did this when nearer. He shot the Fox and found that its tail was heavy with water, and, when the crafty one stopped, it had been to wring or stamp out the water with its front paws and so reduce the heavy burden of the water-logged brush.

In the early part of the winter the Fox hunts chiefly at dusk, but the growing scarcity of food increases the need for diligence, and in February and March it may be seen abroad at all hours. HUNTING

I have often followed a fox-track for miles to learn this hunter's methods. He goes on a general up-wind course, but turns aside to examine every promising thicket and sedgy hollow. He goes to all the places where he remembers having good luck on previous hunts; he calls and sniffs at all the signal posts as described in the Wolf account, though to a less degree. He adds his own record to those already inscribed. He trots along ridge after ridge, he seeks out a bare knoll on which he has voided dung before now, and, finding the spot, endeavours to repeat the act. He stops at the slightest click of leaf or twig, freezes to a statue in an instant, holding one foot up in a pose of wonderful grace. Sometimes he stands on hind-legs to overlook the grass or bounds aloft for an observation hop, after the manner of a Jack-rabbit. He searches the wind with his nose, he trots on by the hour, missing nothing, and passing from cover to cover, in a somewhat zigzag line, but with a general up-wind course. He sneaks by settlers' homes, looking for luck, and is not above feasting on offal. He looks out sharply for the dog, and, if pursued, easily leaves his foe behind in a few hundred yards, then will sometimes turn and bark in defiance, tempting the dog to further pursuit. He runs across the fresh track of a Rabbit, follows this for a time, and may even succeed in springing on the crouching Bunny; but the

FIG. 194—Fox-tracks in snow.

A, showing where the Fox, entering at top, came down and unearthed a snake, which he killed. *B*, the second part, showing where the Fox stalked two Grouse hidden in the snow and secured one. The various pauses as well as the touches of the tail are clearly shown.

latter is as alert as the Fox, and has the advantage of awaiting approach. Usually it gets into the brush, where the hunter must give up the hunt.

In following the trail of a hunting Fox once, I saw where he had dug out a torpid garter-snake, bitten it nearly in two, and left it lying on the snow, intending, it seemed, to come back for it if he found nothing better.

But, farther on, the track recorded how the prowler had scented two prairie-chickens asleep in a drift of soft snow, had stalked them with nose worthy of a pointer and step worthy of a cat, had come just above them before they awoke to their danger, and when they burst out of the drift he had sprung and secured the nearest. Having now abundance of this finest food, he was not compelled to go back for the cold snake, which is never good eating, and on a cold day would have been a very cold lunch indeed.

When satisfied or tired, he lies down for a nap, not usually in a hollow, but on some exposed place, the top of a bank, a boulder, a log, or a stump. Here he curls up in a ball, his blanket is on his back, and his travelling rug is his tail, his big black ears, sticking a little above his tail, are the only things that break the rounded yellow of the ball.

He looks like a yellow stone, and seems to know it. Once while travelling on the Souris in 1882, my brother and myself noticed a yellow boulder, among others, on a ridge. He said: "Look at that; doesn't it look like a Fox?" I said: "No, I see nothing but a yellow boulder." We marched within thirty paces, ourselves, our wagon, and oxen. When twenty yards past, a puff of wind seemed to cause a crack in the boulder. My brother stopped and said: "I'm sure that's no boulder; it looks to me like a Fox." He turned aside, took one step towards it, and at once the Fox sprang up and ran for dear life. He skurried across a stretch of burnt black prairie, then, reaching a bit of unburnt yellow grass three hundred yards away, crouched down in this and watched us again, not, I suspect, because he knew the grass to be a good match with his own colour, but simply because it was cover.

I do not think he was asleep when we passed him, because we had a heavy wagon and the oxen were driven with the usual noise. I believe he was watching over or through his tail, and would have lain still, trusting to escape notice, had not my brother alarmed him by leaving the trail and stepping towards him.

The Fox does not have its regular hours of sleeping any more than of eating, except that it prefers to sleep in sunlight,

Fig. 195—The Fox playing at 'boulder'.

but its sleep may be broken into a dozen naps, for it curls up when it feels tired and has satisfied its hunger.

Foxes have little use for a den in the winter. At one time I thought they kept entirely clear of them while the snow was on the ground, but in following one that had gone off with a trap on his foot, I found that he went into every den and Badger hole that he came to, apparently in hopes of leaving the trap behind.

W. R. Hine tells me that he also has tracked Foxes into dens when there was snow.

I have several times seen a Fox mobbed by birds, usually crows, but once by a lot of kill-deers. These noisy plovers seemed to be actuated by fear for their young, recognizing very clearly that the Fox was an enemy, but the crows could not

PLATE LIII.—THE FIRST WHIFF.

PLATE LIV.—THE LAST GLIMPSE.

From paintings of the Manitoba Fox made by E. T. Seton in 1886.

have been inspired by such a thought; in one case, indeed, it was mid-July, but the Fox had already secured a fowl, and the crows were mobbing him because, knowing his dislike of 'a scene,' they hoped he might abandon his plunder to get rid of them, and so they would profit by his success.

This animal is popularly supposed to subsist chiefly on poultry, Rabbits, and game birds. I have known it to kill each of these, but I suspect that Mice form the largest part of its diet. FOOD

The Fox spends so much time catching Mice that he is often seen in the act. Many times, by means of a telescope, I havc observed one in broad daylight, while he secured his easy prey. Selecting some well-known mouse-haunt, usually a grassy hollow, he advances quietly, looking this way and that for the slightest rustle, alert to the finest sound, tiptoeing, even standing on his hind-legs to see more clearly over the grass. A squeak, or perhaps the movement of the grass-tops, catches his eye, and he springs for the root of the long vibrating spear, slaying with a nip the Mouse that he probably does not see, then separates it later from the grass, to chew and swallow the morsel in a few seconds. His movements are full of elegance and his habits of graceful poses. I know of no prettier sight than a Royal Fox, red and rich in his sleek new coat with its black velvet facings and its trimmings of silver and gold, as he hunts for Mice among the rank foliage and flowers of a prairie-hollow in Manitoba. MOUSE HUNT ING

A similar scene has been described to me by W. R. Hine. In this case, however, the Fox was not a common yellow one, but a superb Silver-black.

In the October of 1887, while out shooting grouse on the Emerson Trail, two miles south of Winnipeg, he saw a large Black-fox on the open prairie, some five hundred yards away. It was catching Mice, and paid little heed to him as he drove by in a rig with his father and brother. Hine whistled heedlessly and passed the mouser at one hundred yard distance, taking care not to go straight towards it. At length its suspicions were aroused and it crouched; although the grass was but six inches high, it sunk so low that he could see nothing but its black ears. The gunner drove in a circle ever nearer without alarming it, except that the Fox crouched yet lower, and at fifty yards shot him with a charge of heavy shot. Although it was October, the pelt was already prime.

The squeak of a Mouse has such a charm for the Fox that even a poor imitation will bring him at a run towards the squeaker. Even when pursued by the hunter he will jump at the sound of a mouse-squeak and, if the dogs be not too close, will turn for an instant to a statue, then try to locate that sound of sweetest promise.

STORAGE HABIT

Most animals of the Dog Family store up food when they have more than they need. T. W. Proger writes me concerning the Fox (*Vulpes vulpes*) in Wales, that "it usually buries surplus food. These caches he returns to infallibly. I think it probable that the Vixen stores up food as her time draws near, so as to have plenty to eat while she is unable to hunt. I do not think one Fox would touch a cache belonging to another, unless hard pressed. They certainly never forget the place, though I do not believe the story that they mark it well by urination.

"A cock pheasant killed by a Fox and cached for a week in cool beech leaves is considered by poachers the finest eating on earth."

The Ontario Fox is said to hide food in this way, but I have no evidence for the Manitoba species. Observations on these points are much desired.

On the fur farm at Dover, Maine, the Foxes (*V. fulvus*) habitually bury food. They watch near the place and are ready to fight any other Fox trying to appropriate the store. If it is interfered with by man they bury it elsewhere. They return to it as soon as hungry, and if there is more than they need, they re-cache the remainder. The Red-fox has not been seen to urinate on its cache, but the Blue-fox does.

In the den already described by G. L. Fordyce was a large separate apartment for stores. In it he found two lambs, one partly eaten, a ruffed grouse, a Cottontail Rabbit, and a Muskrat, all perfectly fresh, with the exception of one lamb, they had not been eaten at all. The Muskrat was not injured in any way by the old Fox, the only marks on the body being where she had crushed its back in killing it. He found this out in removing the skin, which he afterwards sold.

B. R. Ross credits the northern Fox with the highest kind of storage. He says:[6] "When a Fox finds a piece of meat or a fish, he almost invariably hides it, and returns to eat it at some future time. I have remarked this trait even in cubs which I have reared in confinement, and which used previous to eating, to dig holes in the snow, to bury their food, pushing the snow with their noses to cover it. During the commencement of summer he will lay up a store of the eggs of the wild-fowl, for his winter consumption. These he deposits in holes dug in the sand bars of the river, or in beds of moss, and at the expiration of several months will, when hard pressed by want, visit his caches. Even when there are several feet of snow, he will readily distinguish the place by scenting his urine with which a Fox usually sprinkles in a liberal manner all his secret hoards."

This storage habit is asserted also by Thomas Anderson, of Fort Smith. He says that it is well known in that country that the Fox stores eggs for time of famine.

Fox-dung is not an attractive study material at first, but after a few weeks' exposure to rain and sun it is dried, bleached, and purified. All fœcal matter is gone, though it retains its shape. Now it is odorless and closely resembles owl-pellets. DUNG

The undigested remains of Mice, birds, etc., their skulls, feet, hair, and feathers are easily separated and distinguished. A valuable chapter on Fox food can be gathered from such examinations. Those that I have conducted go to prove that

[6] Fur-bearing Animals, Mack. R. Dist., Can. Nat., 1861, p. 17.

by far the largest proportion of this food is Mice and Gophers, so that, on the whole, our Manitoba Fox must be considered a creature beneficial to agriculture.

HYBRIDITY

On January 28, 1902, at the Cincinnati Zoo, I was shown a curious creature supposed to be a cross between a Fox and a dog. The Director said it had been found in the woods near Cincinnati when it was perhaps two months old. It was lying in a hollow stump with two others of the same kind. The boy who found them carried them home without waiting to see of what sort the parents were. Two died, but this one was successfully reared on a bottle. Its appearance suggested Coyote rather than Fox, I thought, but there were no Coyotes within some hundreds of miles. The creature was good-natured and friendly and had much the appearance of a slim, yellowish sheep-dog.

ENEMIES

On September 16, 1901, while camped in Colorado, on the South Fork of William's River, the boys of my outfit found a sick Fox (*V. macrourus*) in a willow thicket. It was very thin and weak, and its hind-legs were paralyzed; it could hardly walk. One of the men killed it with his riding-whip and brought it to me for examination. Its mouth was full of Porcupine quills, doubtless its inside also was suffering from the same, but its condition was such that I did not risk an autopsy. Though an adult male, it weighed only 6 pounds 10 ounces.

There can be no doubt that in the northern wilderness Foxes are preyed on by Wolves, Lynxes, and Fishers, while their young are destroyed by every evil beast that can find them, as well as by most of the larger birds of prey.

The killing of a Fox by an eagle is described in great detail by a correspondent of *Forest and Stream*.[7] The eagle was the aggressor, and the Fox, being on the open prairie, had no chance to seek cover. As the tragedy took place near Estevan, Sask., the Fox was probably of the present species.

[7] W. M., Forest and Stream, February 9, 1896.

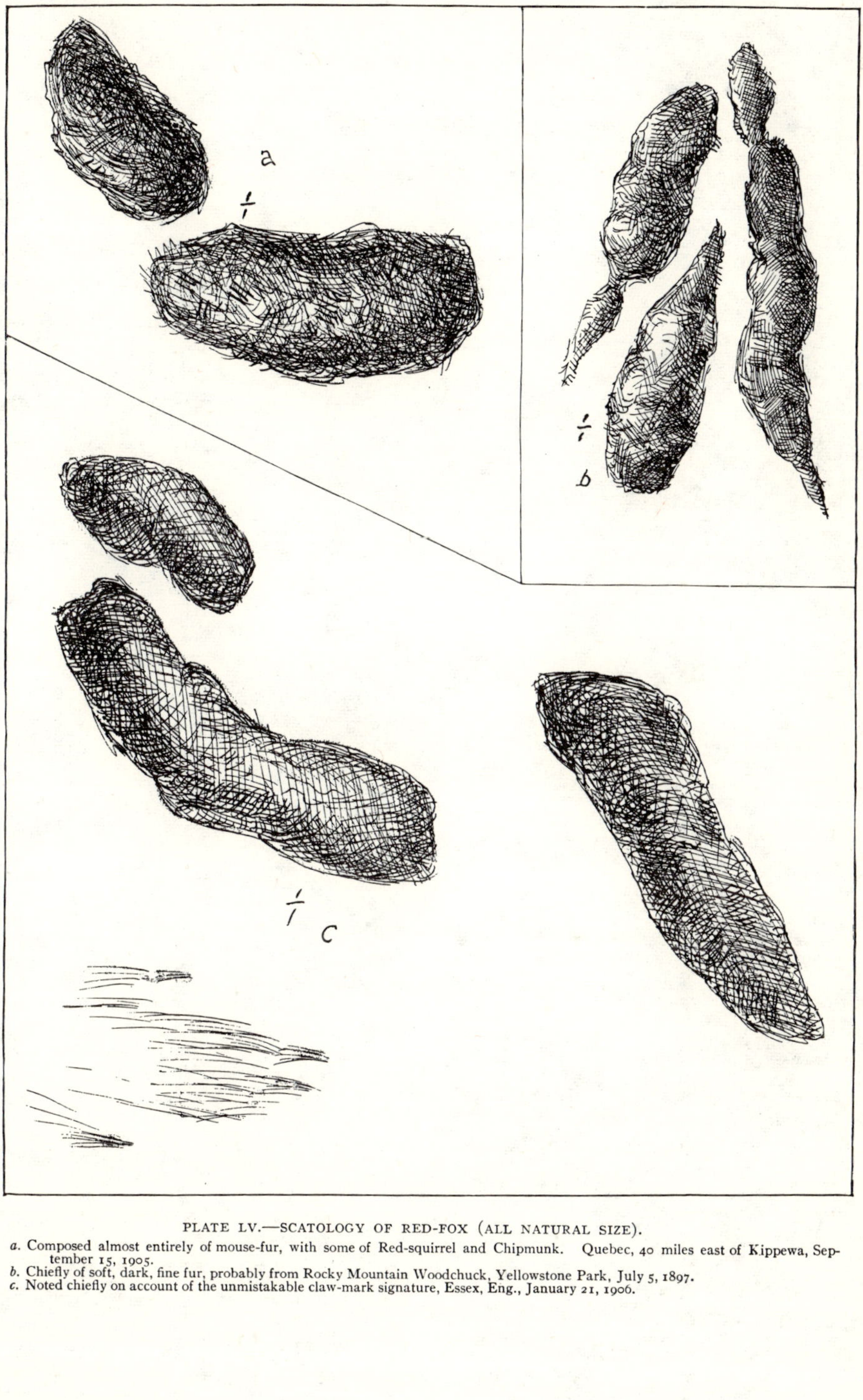

PLATE LV.—SCATOLOGY OF RED-FOX (ALL NATURAL SIZE).

a. Composed almost entirely of mouse-fur, with some of Red-squirrel and Chipmunk. Quebec, 40 miles east of Kippewa, September 15, 1905.
b. Chiefly of soft, dark, fine fur, probably from Rocky Mountain Woodchuck, Yellowstone Park, July 5, 1897.
c. Noted chiefly on account of the unmistakable claw-mark signature, Essex, Eng., January 21, 1906.

There are several records of rabies among Foxes, but none of other epidemics, so far as I know. DISEASES, ETC.

The following affords important light on the age attained by this animal. AGE

Christian Sanderson, of Chadd's Ford, Pa., tells me (October 30, 1905) that in March, 1897, an old dog Fox,

FIG. 196—Life-study of the Fox that attacked the Porcupine, Colorado.

locally famous as a runner, was trapped near the Ford. A silver collar with inscription was put on his neck, and he was released on the one-hundredth anniversary of Bayard Taylor's fox-hunt in "Kennett." He lived and doubtless was hunted many times until 1905, when, after a good run, he was killed at a point 60 miles due south of Kennett. He was evidently aged when killed. Thus he had run for 8 years after he was fully adult. This agrees with the belief that a Fox is old at 10 years, and rarely reaches 15.

A corroborative note is supplied me by Dr. J. W. Walker, of Wakefield, Eng. He knew of a Fox that ran before the local hounds three or four times each year *for 7 years*, before it was killed.

STRANGE INSTANCES

An interesting case of a Fox cub that knew how to take care of himself is vouched for by Dunham Wheeler, of New York. He had five young Foxes in a cage; one of these had the ill-luck to break his hind-leg. The others plagued him so that their captor gave him a little box in which was room only for one. Here he at once ensconced himself, snarling savagely and threatening, with back-turned ears, whenever any of the others approached; and he stayed there until the broken leg healed, when he leaped out as sound as ever.

Possibly connected with the instinct for rolling on any strange strong scent, is the following related by L. R. Gridley, of Appleton, Wis. His wife's father, a trapper in Wisconsin, found a certain trap sprung again and again, but nothing in it except long Fox hairs. At length he sat up to watch. At four in the morning the Fox came and rolled over the trap. It sprung at once, but could not grip on his broad body; he ate the bait in comfort and went his way. But the trapper now set one of those abominations called a clawed otter-trap, and next night Reynard was caught by the back.

FUR

During the 85 years, 1821[8] to 1905 inclusive, the Hudson's Bay Company collected 1,536,420 skins of this species; an average of 18,075 for each year. The lowest was 2,757 in 1826, the highest 52,693 in 1876. The average for the 10 years, 1895 to 1905, was 22,671.

Poland's lists[9] show that during the 71 years, 1821 to 1891 inclusive, 3,831,516 skins were taken by the other American

[8] 1821 was the first year when the Hudson's Bay Company's operations extended without opposition over all the fur country of British North America.

[9] In using these lists one must remember that he gives year of marketing, whereas the furs were taken the year before; also certain returns had two years' catch represented, others but half a year's catch.

companies, an average of 53,965 each year. So that the average annual catch of American Red-foxes for fur is about 74,000.

FUR OF COMMON-FOX

At the London annual fur sale held by Lampson's, March, 1906, there were 25,496 Red-fox skins. The highest price reached was 41 shillings ($9.84) each for 288 dark skins. First-class skins brought usually 15 shillings to 30 shillings ($3.60 to $7.20), but inferior skins sold as low as 1 shilling and 2 shillings (24 cents to 48 cents).

CROSS-FOX

The Cross-fox is the half melanism or partly black freak. At the above sales 3,697 Cross-foxes were sold. The highest price realized was 75 shillings ($18.00) each for 26 first-class dark skins, but 30 shillings to 35 shillings ($7.20 to $8.40) may be considered ruling prices for first-class, from which they graded down to 20 shillings ($4.80) for second-class, and 7 shillings ($1.68) to 15 shillings ($3.60) for third-class.

THE SILVER-FOX

The most valuable fur in the world is doubtless that of the rare and wonderful Sea-otter. A prime skin of this brings from $500 to $800. Next to the Sea-otter comes the Black or Silver-fox. This is, of course, simply a superb melanism of the common Red-fox. It is intense black with more or less silver tipping of the hairs on head and rump, the less tipping the higher the value.

Miller Christy sends me the following interesting item: "At the Hudson's Bay Company annual fur sale, held in London, March, 1900, 601 Silver-foxes were sold, bringing an average price each of £50 16*s.* 1*d.* ($247), and 3 especially fine pure black skins brought, respectively, £310 ($1,507), £340 ($1,652), and £400 ($1,944); that is, £1,050 or $5,103 for the 3.

The record price for a Black-fox is, according to D. A. Boscowitz, £540 ($2,625), given by Grunwaldt, of Paris, in 1889, at C. M. Lampson & Co.'s sale. Such a purchase, however, can scarcely be considered a representative commercial transaction.

At the London annual fur sales, held by Lampson's, March, 1906, 992 Silver-fox skins were sold.

The highest prices realized were £320, £310, £310, £280, £220, £210, £190 (that is, $1,555, $1,506, $1,506, $1,360, $1,070, $1,020, $883). These were for skins of unusual size and marvellous beauty. The ordinary run of first-class skins brought only £50 to £80 ($243 to $388), and many Silver-foxes of inferior quality brought only £1 or £2 ($4.80 to $9.60). Lest the frontier trapper who sees the above be led into undue appreciation of his fur, it is well to remember that the fur-dealer has three great risks to face, damage in transport, damage in storage, and, above all, damage through the freaks of fashion. The intrinsic value of Silver-fox as an article of clothing is little more than that of the Red-fox, say $5 or $10 a skin. But Silver-fox is beautiful, rare, and fashionable, hence the fancy prices paid. It is always within the range of possibility that the fashion may suddenly change and the price of choicest skins drop to a fraction of the last ruling figures. The prices paid at the trading posts to-day have been fairly adjusted by keen competition; they certainly are not too low.

The lustre, fulness, and beauty of the Silver-fox fur are unique and inimitable. Its market value is so high that successful attempts are being made to breed Foxes for their pelts. As the subject is of wide interest, I reproduce an article which I wrote for *Country Life*, in 1905.

Fox-Farming for Fur

The rapid disappearance of certain wild Fur-bearers, combined with the steady demand and ever-rising prices commanded by good furs, has led many to look for means of artificially supplying the want.

It is years since the idea of breeding for fur was first discussed, but never were the times so ripe as now, and it is the object of this article to set forth the important branches of the new industry for the benefit of those who wish to embark in it.

This kind of farming offers two distinct fields. First: the production of a new variety of some already domesticated and easily multiplied animal—as cat, dog, goat, rabbit, or cow—with a coat of such quality as to have a new value as fur.

The second, the breeding, under protection, of certain wild animals whose fur has already an established market value. In this class are Beaver, Mink, Otter, Skunk, Marten, Fisher, and Fox.

In other words, one makes a fur-bearer of an animal already domestic; the other makes domestic an animal already a fur-bearer.

The second is the only department that will be treated herein. There are two ways of dealing with this; we may call them the wholesale and the retail.

The first is the instinctive choice of the beginner. He usually plans to get possession of an island, a mountain valley, or at least a couple of hundred acres of wild land with a high fence around it. This he expects to stock with fur-bearers that will increase speedily to thousands, after which he has nothing to do but shovel in a few tons of offal weekly and draw off a few thousand of the choicest pelts yearly. This is what I call the wholesale method. It has never yet succeeded with Red-foxes, nor indeed with any creature that I know of, except, perhaps to some extent, the Blue-fox on the islands of Alaska.

Success in breeding any domestic animal turns on personal care that can be directed and adapted to each individual, if need be; which is, obliquely, a reason why the would-be fur-farmer is better off with five acres than with five hundred. This individual method is what I call the retail plan; it has been proved a success many times.

Fur-farming is a good chance for small capital. A man of experience may put in $1,000 and get a remarkable percentage as soon as well started. But any one who thinks he can put in $10,000 or $20,000 and do the same with little experience and labour is certainly going to end in disaster.

There is no object in breeding cheap furs. A Muskrat

with its fifteen-cent pelt is almost as much trouble to raise as a $300 Silver-fox, therefore only the high-class fur should be considered.

What is the most valuable fur of all? No doubt the Sea-otter. But the animal is so rare that a large fortune would be exhausted in getting the stock, and nothing is known of the method necessary to its propagation.

Next on the list is the Silver-fox. The Black or Silver-fox is nothing but a black phase or freak of the Common-fox, just as the black sheep is a colour freak of the common sheep. A pair of pure Red-foxes may have a Black-fox in their litter, and that Black-fox may grow up to be the parent of nothing but Red-foxes, but a Red-fox will bring only a dollar or two, and the Silver-fox a hundred times as much.

The thoughts of the fur-farmer, then, are likely to turn at once to the Silver-fox. The first objection usually made to it is its sterility in captivity. At one time, indeed, it was said that the Fox never breeds in confinement. This, however, is far from the truth. Experience proves that the Fox is as fertile in captivity as any other carnivore when properly managed. Another popular error that is wide-spread, even in books of good repute, is the idea that a Fox cannot be tamed. It is highly probable that some individuals will always continue wild and treacherous in captivity, but most of them respond to judicious treatment, and some of them, as I have seen, become as tame as cats.

I first saw Foxes successfully managed by N. E. Skinner, of Bangor, Maine. He began a fur-farm in Winnipeg in 1899. But the best working out of fox-farming as a paying commercial enterprise that I have seen, is at Dover, Maine, where I had the privilege of inspecting the farms of E. Norton and M. F. Stevens, in July, 1905.

Stevens's enclosures were 30 feet each way and surrounded by a mesh-wire fence 10 feet high with an 18-inch overhang at the top, and sunk 3 feet into the ground. Six feet would have been high enough, but at Dover they have to reckon on snow-drifts 4 feet high. The overhang above

PLATE LVI.—MAURICE F. STEVENS AND ONE OF HIS BREEDERS.

PLATE LVII.—YOUNG AT MEAL TIME.

Foxes on fur farm.

From photographs by E. T. Seton.

is necessary, as the Foxes climb up the wires like cats. It is quite a common thing to see one or more of them clinging to the cage at a height of 10 feet from the ground, and there they stay for minutes at a time trying to get out by forcing the wires.

The Norton yards at Dover are considerably larger than the Stevens, his smallest being 50 by 25 feet, and the large runs 230 by 64 feet. The fences are alike.

After inspecting these two farms, and hearing of the various problems to be met, I should thus plan a small fox-yard —and no wise man will go into a large one until after he has had enough experience to avoid the inevitable blunders that on a large scale would be fatal (Fig. 197).

The whole space is 120 by 230 feet, a little over half an acre. All the fences should be 10 feet high, of 1½ inch mesh, No. 16 galvanized wire, as a Kit, that is a weanling in his first summer, can go through a 3-inch mesh, and a full-grown Fox can squeeze out of a 4 by 4. The netting should go 3 feet below ground, or down to hard-pan, and would be still safer with a turn in of 18 inches at the bottom, although a line of flat stones laid with the fence, inside, is usually enough to prevent any digging out. At the top it should have an overhang of 18 inches. This is simply the ordinary mesh wire supported on crosspieces nailed to the posts.

The outer lane is 15 feet wide, that is, wide enough for a wagon, but it can be reduced to barrow width, if space is an object. In each of two corners is a kennel for a watch-dog that patrols the lane; for among the plagues of the fox-farmer are the sneak thief and the malicious neighbour, who might open the gate by night and thus cause the loss of all the stock. At each of the other corners is a long refuge box with a 5 by 9 inch hole at each end. In these any Fox getting out of his pen will take refuge from the dog and remain till put back where he belongs.

The small cages are designed each for a breeding female. If more than 5 were on hand, these cages might be halved without serious cramping. All their doors are 2 by 4 feet, and open on the central alley. Their sill sides should be raised a foot to allow for snow.

In each, at its driest point, a hole about 3 feet wide and 2 feet deep should be dug and roofed over with something to turn the rain. A 9-inch hole, angling down to this at one side, completes the den. In the Norton yards these dens were shingled, as it is very important to have the nursery well-drained and dry. No lining is needed. The mother Fox can add it at will, or leave it out, as she mostly does.

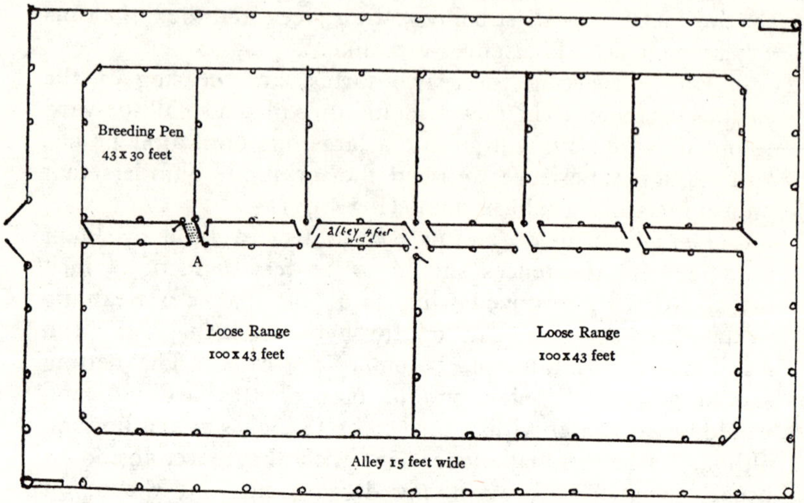

FIG. 197—A model Fox-yard. Scale 50 feet to one inch.

The large cages are loose ranges for any or all that are not nursing mothers or very young. All open into the central lane, as an additional safeguard against escape. This lane is 4 feet wide. It is furnished with a movable alley-cage about 5½ feet long by 2 feet wide and 4 high; it is made on the bias and provided with a drop door at each end. Both top and bottom should be of light boards, as sometimes one is floor, sometimes the other. This is put across the lane from door to door when it is intended to change an animal from one cage to another. At A it is shown in position to allow of the Fox being driven from the pen. By this means a Fox can be transferred from any one pen to any other without handling.

Most beginners will ask themselves—or other competent authority—what does the Fox feed on in a state of nature, and then decide that that is his proper food in captivity. Curiously, and happily, this decision is not backed by experience. Those who feed the Foxes on fat pullets, Rabbits, and Mice will soon find their charges a lot of worm-eaten dyspeptics. A much better answer is given by Norton and endorsed by Stevens, "Feed your Foxes the same as your dogs." Bread, table scraps and a very little meat, is their diet on the farms. Norton feeds butcher's scraps, offal of animals, and in the winter he often gives them the flesh of a horse, taking great care that it be not one that died of disease. FOOD

Stevens feeds his chiefly a cake made of the following recipe: One quart sour milk, 1½ teaspoons of soda, enough corn-meal or "Daisy flour" to make it stiff, spread half an inch thick on pans, and bake. No sugar, salt, or eggs are needed. This will keep for two weeks, and is eagerly eaten by the Foxes. The young ones that I saw came running and shouldering like a lot of little pigs to get at it in the trough when the evening meal was announced by a familiar whistle. A piece 2 by 3 inches makes a meal, and they get two meals a day—morning and night. Each Fox also gets daily a pint of skim milk, and once a week about half a pound of raw meat. They are extremely partial to Woodchuck. One fat Woodchuck is enough to make a week's allowance for eight Foxes.

Norton occasionally shoots a crow and throws it in. The Foxes do not touch it for two or three days, then, as it gets gamy, they devour it with relish. Neither farmer has tried dog-biscuits. If a Fox secures more food than he can eat, he buries it till he is hungry, keeping a watch on the place lest any fellow captive should steal his hoard.

Stevens has been in the business for two years; he has 20 or 30 Foxes, and reports none lost at any time by disease.

Norton has had five years' experience; he carries 30 to 40 Foxes, and reports that one or two die each year from a disease that affects the head. It causes a running of the nose and fills the ears with scab. A careful comparison of their systems of diet shows that the Norton Foxes get more meat than the Stevens Foxes.

BREEDING In a state of nature the Fox is a monogamous animal; is believed to pair for life. In captivity the breeders encourage polygamy, letting one very choice male serve several females. The embarrassingly high moral standard of the Fox is one of the difficulties of the breeder, but there is good reason to believe that human care and influence will undermine their awkward scruples exactly as with the dog and other animals that have submitted to domestic life. I am told, on the other hand, that polyandry is bad. If a breeding female is allowed to join with two males in the same season, the result is said to be sterility. I do not know what this is grounded on.

Over-fat animals do not breed. It is a great mistake to over-feed. In this we probably find one reason for the infecundity of Foxes in most menageries. An ordinary Fox weighs 8 or 9 pounds. The largest and fattest Norton ever had was 16¾ pounds. The breeders should be kept down to about 10 pounds. Stevens claims that of those that mate in his yard half are fertile; Norton claims only one-third. As also noted, Stevens feeds them less, but something may also be due to the fact that the Stevens farm is on a quiet hillside in the country, while the Norton farm is in the busy town of Dover.

The wild Foxes mate in late January or early February, the captives are four or five weeks later. When the time comes the female utters her peculiar squeal and the male answers

PLATE LVIII.—TEN FEET FROM THE GROUND.

PLATE LIX.—FORCING A MESH.

PLATE LX.—CLIMBING UPRIGHT WIRE.

PLATE LXI.—IN THE YARD.

YOUNG FOXES ON FUR FARM.

To illustrate especially the climbing propensities of the young.

From photographs by E. T. Seton.

with a deeper, coarser bark. Both show a restlessness and a good deal of excitement.

The period of gestation is 51 days without 12 hours' variation; of this I am assured by both breeders.

The wild dog Fox is a model father and faithfully helps to provide for the young, but in captivity he is not needed, and it is best to keep the mother in the breeding pen by herself. The quieter she is the better. The young should not be approached, much less handled. Strangers should be forbidden the range till the young are able to run. If disturbed or frightened the mother is liable to carry the young about in her mouth, seeking a safer place for them, till they are worried to death.

The young number from 3 to 9; 6 or 7 are the usual litter. They are born blind; their eyes open about the seventh or eighth day. When they are a month old they begin to come out and play together like kittens. They are now considered past the critical period. They begin to eat solid food, and, when three months old, the mother weans them, and they may be taken away. At midsummer they are half grown, by winter apparently full grown, and in the following February or March, while yet less than a year old, they also breed. The young are much wilder than the old ones at first, but good management soon convinces them that all their fears are ill-founded, and they soon get tame and gentle.

Foxes are much less quarrelsome than most carnivores. Norton has lost only 2 through fighting. Both were killed by an abnormally vicious old dog Fox that bullied the others. Such instances are rare, and a Fox of that temper should be considered a freak, and not allowed to breed.

Every effort should be made to 'gentle' the Foxes. First, by selecting the most docile to breed from; second, by giving them no cause for alarm.

Cleanliness is of the utmost importance. Clean drinking water and clean pens are essential. The Foxes are naturally clean. They do not bury their dung with cat-like scrupulousness, but they leave it at one end of the run. Stevens's pens, though smaller, have escaped all diseases because, he says, each year he digs over the soil to sweeten it. The pens must have shady places and sunny places for the varying weather. It is well also to caution the beginners against giving the Foxes a pile of new earth to dig in. This is sure to cave in and cause trouble.

Fur is prime in November, but should be left until late December, as it continues to get fuller, and the fading process does not set in until the end of winter. The fur is improved by cold weather and by plenty of food in the fall, especially oily foods, such as fish heads, etc., and oil cake might prove effectual. Castration, to make a larger, finer robe, has not been tried.

COST OF STOCK

In getting stock the fox-farmer must remember that the farther north it comes from the better, and that the Gray-fox of Virginia and the South is not wanted at any price. The best way to get Foxes is by digging out the young in May and June.

Wild Red-foxes can often be bought for one or two dollars each. The breeder gets \$8 to \$15 a pair for good cage-reared seasoned specimens of the common Red-fox. But we wish to raise Silver-foxes, not common Reds. As already stated, the Silver, Black, and Cross-foxes are mere colour freaks of the Red-fox. They are found wild in every shade and intergrade. The ideal and inimitable Silver-fox is glossy jet black with a silvery tip to each of the long hairs, giving a frosted finish of exquisite beauty. A pair of Silver-foxes may produce a litter of Red-fox young. But these things run in families and a pair of Silvers or Blacks are almost sure to produce some of their own colour. By selecting the dark ones for breeding, the desired type can soon be fixed. In five generations, that is, five years, Norton tells me he found it possible to breed out all the 'Red' and have a strain of pure 'Silver-foxes.' From a pair of fairly good Blacks, valued at \$50 each as pelts, he last year raised 7 perfect Silvers worth \$200 each.

There are two sources of profit to the fur-breeder. First, the sale of fur; apparently this is his only one. The Winnipeg *Commercial* gives the following quotations for Fox pelts, March, 1904: PROFIT

Prime	Common Red . .	$1.50 to $4.00
"	Cross	$5.00 to $15.00
"	Silvers	$50.00 to $200.00

These may be considered as conservative figures. W. F. Sheard, the fur-dealer at Tacoma, tells me that he once sold three perfectly matched Silver-foxes for $1,200. A man, therefore, with an acre of ground "under Foxes" might raise 20 to 40 Silvers a year with an average market value of $150 each, *i. e.*, $3,000 to $6,000 a year gross, and the expenses would be very low.

How long the price would keep up to this high figure it is impossible to say, but it has kept up for over a hundred years and there is no present sign of decline, rather the contrary.

My own impression is that twenty years from now there will be a great many fur-farmers furnishing first-class Silver-foxes, more than ever before were put on the market. This will eventually lower the prices, but it will be largely offset by the increasing market, and it is safe to say that so beautiful a fur will always fetch a figure sufficient to make it remunerative to produce.

The second profit of the dealer, and perhaps his best, is the sale of his fine stock to other breeders. A Fox whose pelt is valued at $100 is worth $300 as a breeder. Messrs. Norton and Stevens hold their breeders at: Reds, $15 per pair; Cross, $75 per pair; Silvers, $300 to $400 per pair, and even at that figure they tell me that they cannot keep pace with the demand. This would naturally decrease, but to judge from experience in other fine stock it is more permanent than it would seem. It will probably last quite long enough to enable the prudent to gather a very handsome return and perhaps a comfortable fortune.

From this we can see that the Fox breeding does not differ essentially from the breeding of any other high-class stock. Judgment, economy, cleanliness, and thrift will make it a success. I am satisfied that any man who has made a success of hens can make a success of Foxes, with this advantage for the latter—a Fox requires no more space or care than a hen, but is worth twenty times as much, so gives a chance for returns twenty times as large.

XXXIII.

Gray-wolf, Buffalo-wolf or Buffalo-runner.

Canis occidentalis Richardson.

(L. *Canis*, a dog; *occidentalis*, of the west.)

Canis occidentalis RICHARDSON, 1829. F. B. A., Pt. I., p. 60.
TYPE LOCALITY.—Probably Plains of Saskatchewan.

FRENCH CANADIAN, *le Loup gris, la Louve grise.*
CREE, *May-hee'-gan.*
SAUT., *My-in'-gan.*
OJIB., *My-in'-gan*, or *Kit'-chi My-in'-gan.*
YANKTON SIOUX, *Song-toke-chà Tung-ka.*
OGALLALA SIOUX, *Shunk'-ah Mah-nee'-tu.*

GENERIC CHARACTERS

The genus *Canis* is composed of the true Dogs. They have long, pointed muzzles, long legs, long bushy tails, and pointed ears; have 4 toes on each hind-foot, and 5 on each front, but the innermost of the 5 toes is very short and small, and raised so that it does not touch the ground; the claws are blunt and non-retractile.

The teeth: Inc. $\frac{3-3}{3-3}$; can. $\frac{1-1}{1-1}$; prem. $\frac{4-4}{4-4}$; mol. $\frac{2-2}{3-3} = 42$

Until the Wolf group is thoroughly reviewed by competent authority, with abundance of material, it is impossible to do more than guess at the name that properly belongs to the Wolf of the Manitoban prairies.

The oldest name for the American Wolves appears to be *Canis mexicanus* of Linnæus, 1766. Next of those tenable for the Gray-wolf probably is *Canis occidentalis* of Richardson, 1829. No special type locality was given, but the interior of the continent.

For the present it seems well to apply this name to the big Gray-wolf, or Buffalo-wolf, of the plains.

SIZE OF MALE

A good-sized male Gray-wolf that I caught in Colfax County, New Mexico, December 13, 1893, was 5 feet 2 inches (1,575 mm.) from nose-tip to tail-bone tip; of this, its tail was 16 inches (406 mm.); shoulders, 27 inches (686 mm.); girth of neck, 18 inches (457 mm.); girth of chest, 28½ inches (724 mm.); girth of forearm, 8¼ inches (209 mm.). Its weight was 102 pounds; other males caught in the region weighed 90 and 78 pounds. W. R. Hine weighed the Winnipeg Wolf (a male) at 104 pounds.

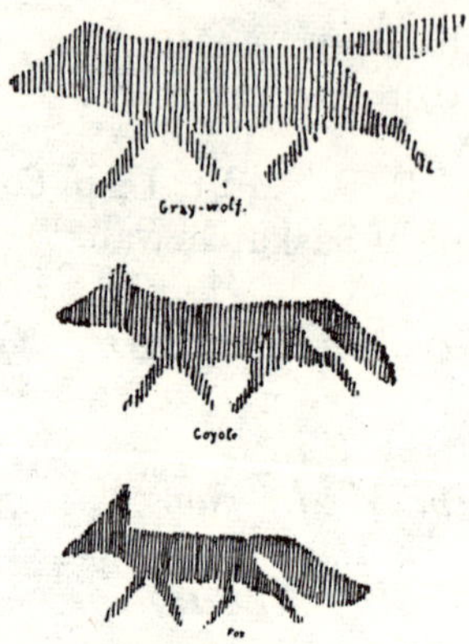

FIG. 198—Distant views and characteristic outlines of :
Gray-wolf,
Coyote,
Fox.

T. P. James, of Clayton, New Mexico, assured me that in the fall of 1892 he killed a huge Wolf that turned a standard scales at 150 pounds. This, however, is extreme, and the weights given above more nearly represent the normal male.

SIZE OF FEMALE

A female taken at the same place, December 29, 1893, was 4 feet 7½ inches long (1,410 mm.); tail, 12 inches (305 mm.), but imperfect; hind-foot, 10 inches (254 mm.); height at shoulders, 25 inches (635 mm.); weight, 75 pounds. Another female weighed 80 pounds, and a third, a poor one, only 55 pounds.

COLOUR

The skin of the first-mentioned male is now before me. It is, in general, a dull, yellowish-white, becoming nearly pure

PLATE LXII.—GRAY-WOLF.

Drawn by E. T. Seton, in 1892, for *Forest and Stream*, and reproduced by courtesy of that publication.

white on cheeks, chest, and inside of hind-legs. The upper part of the muzzle, crown, and outer side of each limb and the entire plantar surface of each foot is tinged a clear pale sienna. On the backs of the ears the sienna is much deeper and stronger. Beginning on the muzzle between the eyes are many black-tipped hairs, which increase in length and number and continue over head, upper neck, shoulder, and back to the basal third of the tail, where they end in a black spot an inch wide and two inches long. After this the tail hairs are faintly tipped brownish-black; the tail itself ending in a dark tip of blackish hairs, with a few white ones interspersed.

The under-fur is brownish gray on the under parts, becoming much darker on the limbs and much browner and darker on upper parts generally.

The dark spot on the tail near its base is formed by a curious tuft of black-tipped hairs, below which there is no wool or under-fur, but evidently a skin odour-gland.

The claws are dark horn-colour.

Compared with a number of Coyote skins taken at the same place, there is no absolute difference. The Coyotes are more strongly tinged with sienna above, and more nearly pure white below. Also, their under-fur on the back is a rich brown instead of dark gray-brown. But they have the tail-gland, and there is little but size to distinguish them when living.

The above colour description of the Gray-wolf agrees exactly with Merriam's colour description of the Coyote quoted on pages 790-791. So far as can be told from outside characters, the Winnipeg Wolf killed near Winnipeg (see later) was a gray Buffalo-wolf. But E. W. Darbey had 6 skins from Riding Mountain that are puzzling; 1 is pure white, 5 are nearly black, and yet all belong to one pack, probably one family; so that colour seems to count for little.

The New Mexican specimen here detailed represents the prevalent colour.

But individuals are found of any shade, from white to deep yellow and almost black. Its size, short tail, short, wide-spread ears, and pale, straw-coloured eyes are characteristic. Unfortunately for those who propose a friendly study of the living Wolf, the skull furnishes the most reliable means of identification.

Life-history.

RANGE

The Wolf most common in Manitoba is probably the Gray-wolf or Buffalo-runner. But it seems likely that in the timbered country we have also the Timber-wolf or *Canis nubilus* Say. At present it is impossible to decide any of these points with certainty.

Map No. 42 shows what little is known of their ranges. It is quite possible that all these forms are races of two or even more species. The type localities are given for the principal forms that have been recognized.

INDIVIDUAL RANGE

The home-region usually corresponds somewhat with the size of the animal. It is probable that the Wolf's home area is larger than that of any other of our non-migratory animals, because it is a large animal—therefore compelled to find much food—a flesh-eater, whose food supply is notoriously uncertain, and a swift-footed animal that can travel great distances.

Dr. James R. Walker and others of Pine Ridge, S. Dak., told me that in 1902 an enormous white Wolf had been living around there for three years. It was usually seen within 15 miles of Pine Ridge.

In New Mexico, several Wolves were well known by their individual marks and believed to be permanent residents of a region about 30 miles across. In Dakota, near Medora, was a well-known Wolf, called Mountain Billy; he was so named because he was always found near a certain mountain called Sentinel Butte. This would limit his range to a radius of a dozen miles.

An extraordinary story of Wolf endurance is related by Archbishop Taché, of St. Boniface, Man.[1] A large Wolf had

[1] *Esquisse sur le Nord-ouest, de l'Amerique*, 1901 (original edition, 1868), pp. 120–1.

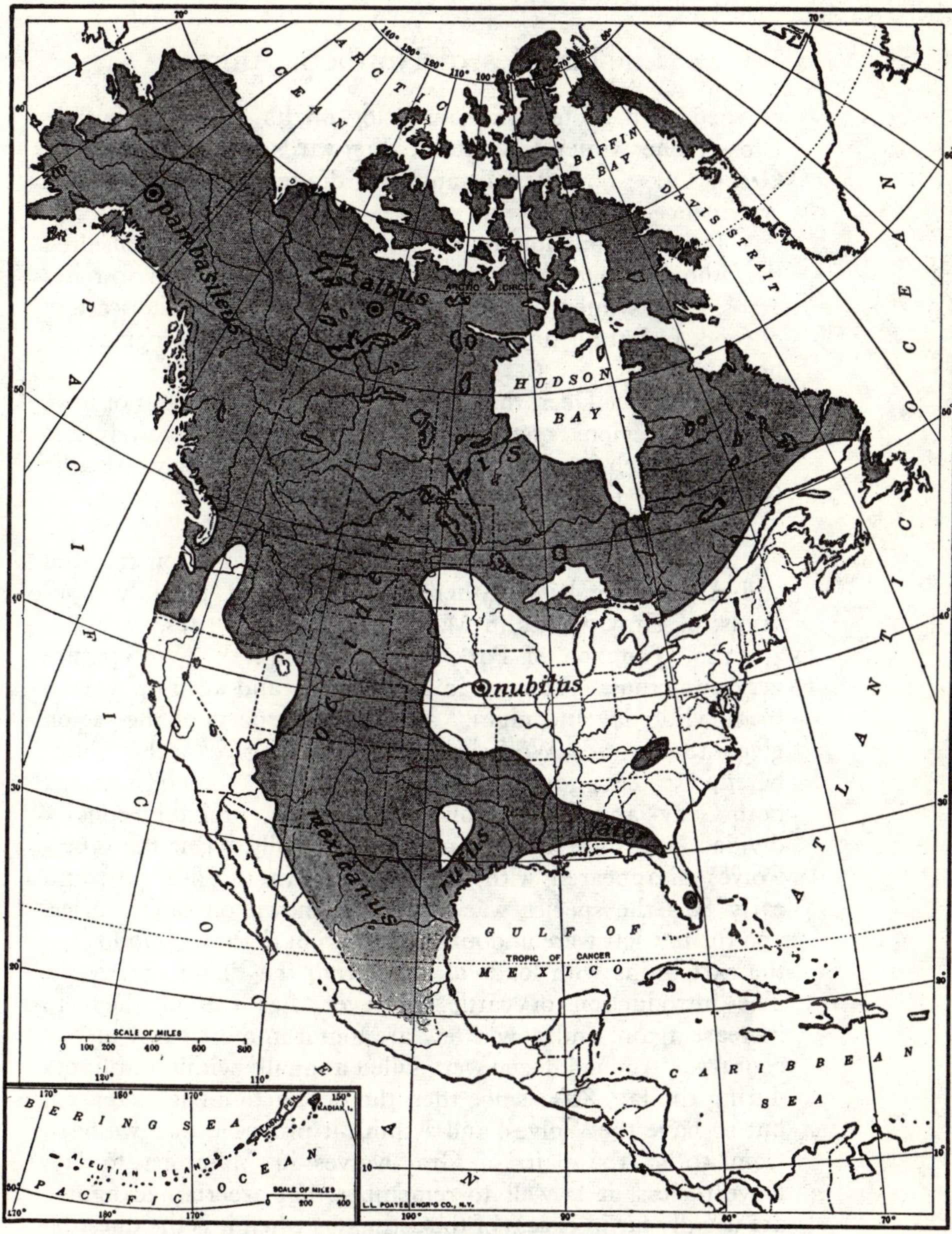

MAP 42—RANGE OF NORTH AMERICAN WOLVES.

Canis occidentalis **Richardson.**

It is impossible in the present state of knowledge to map the species, therefore I give:

(*a*) Those parts of North America where Wolves are found to-day, taking as authority, in the United States, the maps recently published by the Bioglogical Survey, corrected for Florida by F. M. Chapman.

And in Canada, published records by Dr. Robert Bell, John Fannin, A. P. Low, E. A. Preble, E. W. Nelson, O. Bangs, with many original observations and much additional information supplied by Dr. Robert Bell: Prof. John Macoun for the north country and Labrador, Mrs. H. Durant Cheever, W. C. Roberts for New Brunswick, H. C. Pierce for Gaspé and Quebec, Dr. C. L. Bagg for Newfoundland.

(*b*) The names of species that have been described are entered with type locality indicated when known. Concerning distribution in Mexico I have no information. Originally there were Wolves in all the region east of the Rockies, but never, apparently, in the blank area of the South-west.

escaped with a steel trap and clog on its foot, at Isle a la Crosse, one winter. A month afterwards it was killed near Green Lake, 90 miles distant, still dragging the trap. This is evidence of a very wide range.

It is the opinion of all hunters that I have consulted that the summer range of a Gray-wolf is less than 50 miles across. In winter, however, it may be doubled by the scarcity of food, but at all times there is a region that it recognizes as home.

Unlike the Deer, the Wolf, so far as known, does not have two home-regions, one for summer and another, entirely distinct, to which it migrates for the winter; in other words, the Gray-wolf is a wide ranger, but non-migratory.

ABUNDANCE

In the early days when the Buffalo swarmed on the Red and Assiniboine Rivers, there were hundreds, possibly thousands, of Gray-wolves in Manitoba. Henry writes,[2] October 4, 1799 [Junction of Park and Red Rivers], "Wolves are very numerous. They go in large droves and keep up a terrible howling day and night," and his fur reports of the region give: 1800–1, 204 Wolf skins; the following years they numbered, 256, 801, 360, 690, 862, 420, and 68. These included both Gray- and Prairie-wolves or Coyotes, probably equally divided, and most were from Pembina Hills. But the Gray-wolves disappeared with the Buffalo. In the late 70's and early 80's the species was almost unknown on our prairies, and the few left were undoubtedly brought in through following the cart trains with loads of meat from the Plains each year. The introduction of cattle, however, has caused them to increase again, and now a few are found in most parts of our country. About a dozen were killed annually within our limits during the late 80's; since then the destruction has increased, but so have the Wolves, and I think it probable that we have from 50 to 100 pairs of Gray-wolves in Manitoba to-day. Nevertheless, it is well to remember that ascertained figures are usually far in excess of the estimates when it is the question

[2] Journal, 1897, p. 112.

PLATE LXIII.—LIFE STUDIES OF WOLVES.

By E. T. Seton. That in the centre and the two toward the right-hand upper corner are Gray-wolves from the Buffalo Plains. The rest are from France.

of animal population. Vernon Bailey's recent investigations[3] in central Wyoming have shed much light on the habits and number of Wolves. In 100 square miles of cattle country in Wind River, where Wolves are fairly numerous, he found in March, 1906, 20 breeding dens of Gray-wolves.

The State of Wyoming has paid in eleven years for the killing of 20,819 Wolves. As a good third of those shot or poisoned are never found, we are safe to believe that 30,000 Wolves have been killed in that time, or 2,600 each year, and yet these numbers are rather increasing, from which I should infer that there are between 5,000 and 10,000 Gray-wolves in Wyoming alone, and that they are in like proportion over all of the cattle country from the Gulf of Mexico to the Saskatchewan. Taking the lower figures as safer, they would show a total of 500,000 Gray-wolves still roaming the West, although their numbers are unquestionably much less than in primitive times.

SOCIABILITY

The Wolves are the most sociable of beasts of prey. Not only do they gather in bands, but they arrange to render each other assistance, which is the most important test of sociability. The most Gray-wolves I ever saw in a band was 5. This was in northern New Mexico, January, 1894. The most I ever heard of in a band was 32 that were seen in the same region. These packs are apparently formed in winter only. I think, further, that this species is not gregarious in the sense that the Antelope and Wapiti are. The packs are probably temporary associations of personal acquaintances, for some temporary purpose, or passing reason, such as food-question or mating instinct. As soon as this is settled they scatter. No doubt these same individuals are ready to reunite as soon as a new occasion requires it, and would resent the presence of a total stranger. This I take to be true sociability.

An instance in point was related to me by Gordon M. Wright, of Carberry, Man. During the winter of 1865 he was logging at Sturgeon Lake, Ont. One Sunday he and some

[3] Forest Service Bull. 72, U. S. Dep. Agri., Wolves, 1907.

companions strolled out on the ice of the lake to look at the logs there. They heard the hunting cry of Wolves, then a Deer (a female) darted from the woods to the open ice. Her sides were heaving, her tongue out, and her legs cut with the slight crust on the snow. Evidently she was hard pressed. She was coming towards them, but one of the men gave a shout which caused her to sheer off. A minute later six Timber-wolves appeared galloping on her trail, heads low, tails horizontal, and howling continuously. They were uttering their hunting cry, but as soon as they saw the Deer they broke into a louder, different note, left the trail, and made straight for her. Five of the Wolves were abreast and one that seemed much darker was behind. Within half a mile they overtook her and pulled her down, all seemed to seize her at once. For a few moments she bleated like a sheep in distress; after that the only sound was the snarling and crunching of the Wolves as they feasted. Within fifteen minutes nothing was left of the Deer but hair and some of the larger bones, and the Wolves fighting among themselves for even these. Then they scattered, each going a quarter of a mile or so, no two in the same direction, and those that remained in view curled up there on the open lake to sleep. This happened about ten o'clock in the morning within three hundred yards of several witnesses.

MATING The mating season of Gray-wolves begins about the last week of January and may last into the first week of March, differing according to the region; the colder it is, the later.

PAIRING Does the Gray-wolf pair? This is so important in the natural history of monogamy that I give evidence at length.

Ordinary dogs, we know, are promiscuous, but domesticity is notoriously bad for the morals of animals; here, therefore, the argument of analogy would be unsafe. Dr. Woods Hutchinson, in an important article on "Animal Marriage,"[4] points out the promiscuity of the dog as anomalous and maintains the superiority of monogamy as an institution. "A

[4] Contemporary Review, October, 1904.

monogamous race," he says, "will, in the long run, defeat a polygamous," and then he claims that monogamy is the rule in all the higher animals. Hearne says[5] of the Wolves in the Barren Grounds: "They couple in spring and generally keep in pairs all summer."

Miles Spencer, an observant fur-trader, at Fort George, Hudson Bay, is thoroughly conversant with the Wolves of that region and maintains that the male assists the female in caring for the young.[6] The Wolf hunters in New Mexico tell me that when they find a Wolf's den, *two* old ones are sure to be hovering about, and in January I saw there at least one case of a male being deeply devoted to a certain female. A good father is a good husband among animals. Bailey says:[7] "Men who have made a business of hunting Wolves for the bounty assert that they are usually able to shoot one or both of the old Wolves at the den by watching the trails, or hiding near the den early in the morning before the Wolves return from the night's round. These statements are fully corroborated by my own experience. While watching dens in Wyoming I could easily have shot the male who was doing sentinel duty; for although he watched from a high point, from which he could see a man long before being himself seen, still in his anxiety to decoy me away he often came within rifle range.

* * * * * * *

"It is now positively known that Wolves always pair, at least for the breeding season."

How long does this ideal condition last? For that season or for life? LIFE-LONG UNION

Probably for life. I have several times seen a male and female Wolf together at a time when the sexual passion was dormant; and yet the male showed the female more attention than he would have done had she been simply a smaller male. This points to permanent partnership.

[5] Journey, 1795, p. 362.
[6] A. P. Low, Expl. James Bay, Can. Geol. Surv., 1888, App. III, p. 76 J.
[7] *Op. cit.*, Note 3, pp. 22–3.

In the London Zoo, at present (December, 1904),[8] is a pair of Wolves, officially called 'Lobo' and 'Blanca.' The male is from western Texas, the female from Arizona; they are good, typical examples of the Gray- or Buffalo-wolf of those high plains.

They have been there for three years and are supposed to be five years old. They bred last year, producing 9 cubs, and this year 8. The father has always been kept away from the young, so it is not known whether he has any parental feelings.

These two old Wolves live in harmony except when the keepers come to the cage; both are fond of being noticed, and eager to monopolize all attention; each strives jealously to be next the bars, pushing the other away, barking and growling meanwhile, with bristling mane and evident temper.

Lobo often springs at his mate as though to bite her, but is always restrained at the last moment *by something*. What is it, if not *a feeling akin to chivalry?*

In these quarrels, if Blanca sees that she has gone too far, she apologizes by licking Lobo's face in a conciliatory manner, always effectual.

The fact that the male shows chivalrous feeling, and that the pair continue as mates in the autumn and winter, when the sexual instinct is dormant, are partial evidences that Wolves pair for life.

D. A. Thornbury, Superintendent of Schools, Grinnell, Iowa, writes me thus:

"In the latter part of October or early November, 1886, in Mitchell County, Iowa, while we were hulling corn, my brother and myself saw two Gray-wolves come out of the woods. One of them was carrying a dead Cottontail in his mouth. They passed within fifty yards of us, and watched us as they passed. They seemed to know that we had no gun."

The fact that two Gray-wolves *unhungry* should be travelling together in fall shows a friendly alliance most easily explained by a life attachment between the pair.

[8] In 1909 this same relationship continues.

There is, however, some evidence for the other view. The Alaskan sled-dogs are known to be domesticated Wolves; all are much mixed with wild Wolf blood, some even are Wolves captured when young. Captain Dick Craine, who spent nine years among them, owning and handling in that time about 200, tells me that he has several times known a pair of half-wolf train-dogs to mate and remain together as mates *until the pups were well grown*, after which they parted. Two very marked cases happened at the same time. In these the father took an interest in the pups and the mother allowed him to approach them, but warned all others away. He never saw the father feed the pups, but the mother often did so by disgorging.

From these cases he infers that the Wolves mate for one season but not for life.

Dr. Woods Hutchinson, in his paper on "Animal Marriage," says:[9]

"There is a general impression among field-naturalists and trappers that many of these animals, having once paired, come together again in the succeeding seasons, although they may be widely separated during the intervening part of the year. Indeed, some of them positively declare that the union among Wolves, Foxes, Panthers, Lions, and Seals, is practically for life. In a few species, however, does it terminate until the expiration of the period required for the young to become able to shift for themselves. In some of these, like certain Wolves, the father practically disappears during the nest period of the young, but joins the family again when the cubs are able to run."

A degree of this latter custom is seen in Foxes, Coyotes, and several other carnivores. It seems that in all these creatures there is a deep-laid instinct to leave the mother quite alone during parturition, and until the young are some days, or in some cases weeks, old, after which the father is allowed to join the family. This, it will be remembered, is closely paralleled by the practice of many tribes of human savages.

[9] See Note 4.

As a general argument, mating customs are so deeply rooted as to be very uniform in a given family; thus all the true pigeons pair, all the true Deer are polygamous; and every evidence, direct or collateral, I can find on any of the true Dogs, except the case of the Alaskan team dogs (in domestication), points to perfect and permanent *monogamy as the rule.*

DEN The nursery den is either a natural cave, a hollow log or stump, or a hole in the ground, dug out by the parents themselves. Sometimes they enlarge a Badger hole, and in any case the bed is not far from the entrance.

The Gray-wolf apparently does not line its nest. Roland D. Carson, of the Philadelphia Zoo, writes me concerning those that bred in the gardens:

"The females dug a hole in the earth but made no attempt at lining the nests, and when hay and other materials were put in it to form a bed they were promptly thrown out."

As the lining habit is instinctive in the kinds that practice it, we note with interest that many animals are tormented with parasites which harbour and breed among the nest material, so that lining is a dangerous comfort.

GESTATION According to all observers the period of gestation is 63 days in the Gray-wolf, as in most, if not all, of the true Dogs.

YOUNG The young number 3 to 13, but are usually 6 or 7. When born they are blind and almost naked, and, like young dogs, their eyes are not opened until the ninth day. Carson says that the only litter of the Wolf pups that he was able to watch closely "did not get their eyes opened till the thirteenth day." Possibly these were prematurely born.

W. H. Blackburn reports that he has watched several litters in the National Zoo at Washington and found their eyes opened a little on the seventh day and fully opened on the ninth. Litters have been born there on March 23, 27, 29, and April 4.

Those in the London Zoo (parents from Texas) were born March 28; all the four litters bred in the Philadelphia

PLATE LXIV.—GRAY-WOLF SCRATCHING HIMSELF.
Life studies by E. T. Seton.

Zoo were born in March and April, the earliest March 18, the latest April 19. Even in the Red River Valley they are born about the same time, as the following from Henry's Journal attests:[10]

[Park River Post on Red River] "April 7, 1801. One of my men brought in 3 Wolves of this year which he had found in a hole in the ground; they sometimes have their young in a hollow log or stump. * * * Another of my men brought in 6 young Wolves he had found in one hole."

MATERNAL INSTINCT

"Shortly after birth and long before their eyes were open the mother-wolf [in the Philadelphia Zoo] would come to the front of the enclosure with one of her pups in her mouth, sometimes returning for another one or two, but all were never brought out at one time; one was carried about for a while and then returned. This was generally done when the keeper was in the back passage or adjoining cage. Nervousness from fear of harm to her babies seems, in most cases at least, to have been the cause of this habit."

The maternal instincts of the she-wolf are of a high order. Carson, above quoted, remarks:

"We have no instance of a Gray-wolf killing or eating her young, but Prairie-wolves in our Zoo have not only killed their young, but eaten them when they have died from other causes."

When I was at Sidney, Ohio, March, 1902, I met an old hunter who related a curious and interesting story that illustrates the motherliness of the she-wolf. About twenty years before, when he lived in Wisconsin, a bounty of ten dollars each was put on Gray-wolves, and he spent a good deal of time hunting them. One day he saw a Wolf come to the river to drink. He shot and killed it, then found that it was a female suckling young. He searched many days for the nest and could not find it.

Two weeks later he shot another female Wolf coming out of a hollow log. She also was suckling young. He crawled into the log and found 13 Wolf pups of two different sizes—

[10] Journal, 1897, pp. 174–5.

6 very small ones, 7 much larger. She had but 6 teats in commission, so he concluded that this she-wolf had rescued the young of the first female he had shot.

GROWTH OF YOUNG

The young ones of the Philadelphia Zoo would whine like puppies as soon as they were born, and the mother would answer and call them in the same manner.

As soon as they could see they began to play together like the young of the domestic dog.

At from three to four weeks, according to their vigour, all would come out daily, never at night, and sit or play in the sun about the door of the den, but were ever ready to skurry in again at the slightest alarm. According to Lee Hampleman, my Rocky Mountain guide, the young ones in Colorado first began to follow the mother for short distances from the den in June when they were about three months old.

FEEDING YOUNG

The question how they are fed has been much discussed; and sifting down the evidence of many observers, it may be considered sure that they are simply suckled for about six weeks, during which time the father has little to do with them. Now the mother begins to *disgorge* solid food for them. Many observers say that they have watched closely but never saw anything of this. Carson, however, states positively and conclusively:[11]

"At five or six weeks they began to eat the food disgorged for them by the mother, and later, when they wanted food of this kind, they would jump at her mouth, sometimes several at once, until she supplied them."

The father soon becomes active, not disgorging, so far as we know, but bringing fresh game to the den. On this point D. A. Thornbury, Superintendent of Schools, Grinnell, Iowa, also writes me: "My father has many times found in the mouth of the hole in which he took some young Wolves, bodies of Rabbits and chickens, and in one instance he took from the hole the partly eaten body of a lamb."

[11] In personal letter.

Of the above-mentioned brood of 8 young Wolves in the London Zoo, 4 were raised by a collie foster-mother, and they became very tame as well as unusually fine specimens. The rest were left with the mother; 2 died. The others grew up, but were weaklings and very shy and wild. The collie was fed on dog-biscuit, the mother Wolf on raw meat, the only food she would touch.

"After the collie-raised 4 were put in a paddock by themselves they also grew wild and timid, except 1, which for some unknown cause remained tractable and dog-like. This is a most interesting case of individual variation in temperament and suggests how breeds of domestic dogs have been brought to their present condition of tameness by breeding from stock artificially selected for that attribute." (*R. I. Pocock in letter.*)

"One was sold to a member of the Society, who lives in the country, where the young Wolf has all the liberty of a dog. It follows a carriage, attends garden parties, and is a favourite with children. It is larger than either of its parents." (*Dr. J. D. Drewitt in letter.*)

The following year Dr. Drewitt wrote me, June 25, 1905: "Blanca has had another litter of puppies, and, as usual, those that were brought up by a collie wet-nurse were tame, and those which were nursed by the mother, wild. All the puppies had lead-coloured iris.

Miles Spencer says[12] that the young are suckled for two months after birth, in the region about Hudson Bay.

Carson, however, writes me: "It is probable they would nurse for five or six months, but as a rule we took them from the mother before they were entirely weaned. The weaning was very gradual, and towards the last the mother would snap when they attempted to suckle her."

At this time the only enemies that the young Wolves have to fear are eagles, man, insects, and disease. Eagles very often pick up the youngsters, as they play around the den door; man ENEMIES

[12] See Note 6.

destroys the mother, and digs out the pups, if he can locate the den; parasites, insects, and disease also are to be dreaded.

Hearne makes some remarks which show the Northern Indians and the Wolves on a very friendly footing.

The Wolves, he says,[13] "always burrow underground to bring forth their young; and though it is natural to suppose them very fierce at those times, yet I have frequently seen the Indians go to their dens and take out the young ones and play with them. I never knew a Northern Indian hurt one of them; on the contrary, they always put them carefully into the den again; and I have sometimes seen them paint the faces of the young Wolves with vermilion or red ochre."

In August the young are so far grown that they begin to follow the mother about in her hunting expeditions, and the den is abandoned. By this time the doorway is littered with the bones, fur, and feathers of the game brought home by the parents.

EDUCATION

The education of the cubs now begins in earnest. The chief means is example. Whether consciously or not, on the part of the teacher or of the class, there can be no doubt that it is by seeing the mother do, or not do, that the little ones learn much that is necessary to their success in life. Thus she inspires them with terror of a trap, by showing her own terror of it; no matter whether conscious or unconscious, *this is teaching*. The same is true, I suppose, of all the ideas that modern Wolves have; that is, the ideas so recent that they have not yet had time to become ingrained as instinct.

"They are still puppy-like at one year, and hardly full grown until eighteen months old; even then they did not have the fully adult look. The females in the Philadelphia Zoo did not come in heat till they were two years old." (*Carson.*)

This agrees with observations made on the Wolves in the London Zoo. It is likely, therefore, that the Gray-wolf is not mature until its third year, thenceforth the female breeds once each year while in vigour; that is, probably, till her ninth or tenth year.

[13] Journey, 1795, pp. 362–3.

The range of the Gray-wolf has a known history. When the Buffalo swarmed over Western America from the Alleghanies to the Rockies, and from Great Slave Lake to Central Mexico, their herds were followed by troops of Buffalo-wolves that preyed on the weak and helpless. As the Buffalo disappeared the Wolves were harder put for a living. When the last great Buffalo herds were destroyed and the Wolves were left without their usual support, they naturally turned their attention to the cattle on the ranges. HISTORY

The ranchmen declared vigorous war against them: traps and poison were imported in vast quantities, a bounty was offered for each Wolf scalp, and every inducement held out to wolf-hunters.

In those days the Wolves were comparatively unsuspicious, and it was easy to trap or poison them. The result was that enormous numbers were killed in the early days of 1880 to 1888 or 1889; so many, indeed, that the species seemed on the verge of extinction. The remnant of the race continued on the foothills of the Rockies or the Badlands, but they were so rare as to be no longer a factor in the cattle question. Then new knowledge, a better comprehension of the modern dangers, seemed to spread among the Wolves. They learned how to detect and defy the traps and poison, and in some way the knowledge was passed from one to another, till all Wolves were fully possessed of the information. How this is done is not easy to say. It is easier to prove that it *is* done. Few Wolves ever get into a trap, fewer still get into a trap and out again, and thus learn that a steel-trap is a thing to be feared. And yet all Wolves have that knowledge, as every trapper knows, and since they could not get it at first-hand, they must have got it second-hand; that is, the information was communicated to them by others of their kind.

It is well known among hunters that a piece of iron is enough to protect any carcass from the Wolves. If a Deer or Antelope has been shot and is to be left out over night, all that is needed for its protection is an old horseshoe, a spur, or even any part of the hunter's dress. No wolf will go near such suspicious looking or human-tainted things; he will starve rather than approach the carcass so guarded.

With poison, a similar change has come about. Strychnine was considered infallible, when first it was introduced. It did vast destruction for a time, then the Wolves seemed to discover the danger associated with that particular smell, and will no longer take the poisoned bait, as I know from numberless experiences.

It is thoroughly well known among the cattle men now that the only chance of poisoning Wolves is in the late summer and early autumn, when the young are beginning to run with the mother. She cannot watch over all of them, the whole time, and there is a chance of some of them finding the bait and taking it before they have been taught to let that sort of smell-thing alone.

The result is that the Wolves are on the increase, have been, indeed, since the late 80's. They have returned to many of their old hunting-grounds in the cattle countries, and each year they seem to be more numerous and more widely spread, thanks to their mastery of the new problems forced upon them by civilization.

HABITS

The Gray-wolf is one of the shyest of wild animals. I have talked with men who have lived their whole lives in regions where Gray-wolves were far from rare, and yet they have never seen one. They hear them at night, they see their trails and their work in the morning, but never see the animals themselves until after they have been trapped or poisoned. Their extreme shyness is partly a modern development, as also is the respect for man, which now fully possesses every Gray-wolf in the cattle country. There are many records that show the Wolf to have been a continual danger to mankind in the bow-and-arrow days. There can be no doubt that then man was considered a fair prey, a difficult and wide-awake one, no doubt, but still a creature to be eaten in times of scarcity. Consequently, each winter in America, as in Europe, a number of human beings were killed and devoured by hungry Wolves.

During the last twenty years, however, I cannot find a reliable instance of Western Wolves, or especially Manitoba Wolves, killing or even attacking human beings. NEVER ATTACK MAN

The following, related by George Fraser, of Winnipeg, aptly illustrates the disposition of Wolves to-day: In 1886, he was travelling near Whitewater Lake, in southern Manitoba. He came on a Swede who was drawing a long box wagon in which were three or four quarters of beef. Sometimes on the load and sometimes running around were two large Gray-wolves, feeding on the beef in spite of the Swede's efforts to keep them off with a pitchfork. The driver and the Wolves dodged around the wagon for some time before the man heard Fraser's shouts to stand aside; when he did, Fraser shot both Wolves. The Swede said these two had been a pest for some time, killing his sheep and one colt. They had never offered violence to man.

I have seen many recent newspaper clippings that recorded harrowing tales of men, women and children devoured by grewsome packs, but each and all have crumbled into newspaper stories when fully investigated. The question then arises, are the old records wrong, or are the modern Wolves of different species? The answer is, the modern Wolves are the same as the old ones, except in one particular, viz., that they have been *educated by fear* to let man alone. Man with the modern gun is a different creature from man with the bow and arrow. The Wolves have learned this, and are now no more a menace to human life than are the Prairie-wolves or Coyotes. Not only do they abstain from harming man, but they have learned that they are likely to be harmed by him, unless they keep out of sight in the daytime. This, I think, is why Wolves are so rarely seen, even when comparatively common.

In accounting for these changes it is not necessary to attribute human intelligence to this animal. Evidently much hard luck and many unpleasant surprises have engendered in it a deep and general distrust of all strange things, as well as a well-founded fear of anything that bears the taint of a human being. This distrust, combined with its exquisite sense of smell, may explain much that looks like profound sagacity in this animal. Nevertheless, this will not explain all, as I have had very good reason to remark again and again, when I have endeavoured to trap or poison Wolves on the cattle ranges.

And even ascribing much to mere shyness does not remove it from the sphere of intelligence, though doubtless ranking it lower in that department, making it a vague fear of the unknown, in place of a dread of danger well comprehended.

FISHING One of the most curious instances, I find, is given by B. R. Ross. The evidence is purely circumstantial and not complete at that, but Ross was a good naturalist and evidently believed the case proven:

"In the month of May," he says,[14] "when the holes cut in the ice do not freeze up, the fisherman at Fort Resolution on visiting his trout lines, set at some distance from the Fort, discovered that several had been visited; the lines and hooks were lying on the ice, as well as the remains of a partly eaten trout, and a Wolf's track was observed about the place. The fact was the Wolf had hauled up the lines and helped himself to what fish he required. This occurred again and then ceased, the animal having been probably driven away by the dogs of the Post."

FOOD The diet of this species includes every kind of animal food, that he can secure, from Mice to Moose. Throughout the summer Mice and such 'small deer' are doubtless the staples. The coming of winter makes a radical change. First, it puts the small game beyond reach; second, it robs the Moose and Deer of the safe refuge afforded by the lakes and rivers, and thus brings these great ruminants into the dietary of the Wolves.

[14] Fur-bearing Anim., Mack. R., Can. Nat., January, 1861, p. 10.

Writing of the northern species, R. MacFarlane says:[15] MOOSE-KILLER

"These Wolves yearly succeed in killing as prey quite a large number of Reindeer and not a few Moose. On one occasion, while travelling upon the ice between Forts Liard and Nelson, in the Mackenzie River District, we came across a big patch of hard-packed snow on the Liard River where a large buck Moose had evidently been surrounded and no doubt overpowered, after a most gallant fight for life, by perhaps a score of ferocious and cowardly Wolves. A few well-picked bones and the skull were the only relics left. At a short distance, however, we perceived a full-grown Gray-wolf, which was at once shot. It had one of its hind-legs shattered by a kick from the Moose, which so disabled it that it could scarcely crawl. Had its companions not been fully gorged, they would doubtless have fallen upon and eaten it, too."

The havoc wrought by Wolves during winter among the Whitetailed Deer is well known, but at all times they prefer an easier prey, the easier the better; even carrion is always acceptable food, and I have several times heard of Wolves hard pressed in winter, filling their bellies with horse dung gathered on the highway.

The habit of burying surplus food seems to be common to all the Wolf tribe. Roland D. Carson writes me of the Wolves in the Philadelphia Zoo: "Our males and females often bury surplus food, but the females have not been observed to do so more than usual just previous to the birth of the young." STORAGE

Captain Craine's half-wolf train-dogs, if not hungry, would bury their food, and water on the place or even on the food. This latter performance is explained in the Wolverine chapter.

These train-dogs afford much light on the ways of their wild kinsmen. One of them will watch his cache all day and in its defence fearlessly attack another that ordinarily he was afraid of. The big dog rarely presses the point under these PROPERTY INSTINCT

[15] Mam. N. W. Ter., Proc. U. S. N. M., 1905, p. 692.

circumstances, but acts as though he knew his cause was weak. This beginning of property law is of cardinal interest.

These hoards may be of vital service to the Wolf, but the instinct, as at present developed, is very crude, and scarcely to be compared with the fine providence of Beaver and Squirrel.

DOPING Wolves, as well as dogs, have a singular habit of rolling in carrion, or 'doping,' as it is called. They seem delighted with the opportunity of making themselves reek with stench of the foulest flesh or fish they can find. Although the dogs are without the personal nicety of cats, they do have some habits of cleanliness, and spend a certain amount of time in dressing the fur. Who has not seen a dog bite the burrs out of his coat, or the ice balls from his legs? A dog or a Wolf which is wounded or bleeding will take the trouble to remove the stains from his fur, and the mystery of his rolling in carrion is unexplained. It cannot be a pleasant smell to him, one would think, because oftentimes it is stuff he will not eat.

The suggestion that it is based on the sexual instinct does not seem to hold, as female dogs, as well as males, will do it at any time. I know of no satisfactory explanation.

VOICE The usual cry of the Wolf is a long smooth howl. It is quite musical, though decidedly eerie when heard in the woods at night. I cannot distinguish it from the howl of a large dog. Its beginning is also much like the hoot of a horned owl. This is usually the 'muster' or 'rallying cry'—the intimation of the Wolf to his friends that he has found game too strong for him to manage alone. It is the call usually heard at night about the settlers' huts. A second sound is a higher pitched howl, vibrating on two notes. This may be styled the 'hunting song'; it corresponds exactly with the full cry of a pack of hounds on the hot scent. A third is a combination of a short bark and a howl. It seems to mean the 'closing in' for a finish. There are several others that I have often heard, but cannot comprehend. Some of my hunting friends claim that they can discriminate the calls of the she-wolf to her mate and

PLATE LXV.—GRAY-WOLF APPROACHING TO ATTACK.
Life study by E. T. Seton.

her young; the call of the young to their mothers, etc. I doubt not these signals are used, just as surely as dogs use corresponding sounds among themselves, but I have not been able to distinguish them. The whining used by the young while still in the nest has already been spoken of, as well as the mother's similar response.

Besides these sounds as a means of intercommunication, Wolves use example as already set forth, and scents. INTERCOMMUNICATION

The scent method of communicating ideas I made the subject of an article in *Forest and Stream*, January 23, 1897. I reproduce the substance of it here.

It is well known that not only each species of animal but that each individual has its own peculiar smell, conclusive evidence of which is found in the fact that a good dog has no difficulty in following his master through a crowd, or keeping to the track of the animal he is hunting, though it be crossed by the tracks of many others. SMELL-POWER

It is further known that, even though it always retain its individuality, this personal odour varies with the condition of the animal. Thus a horse smells strong after exercise; Canada grouse and Snow-shoe Hares smell of spruce or cedar when they feed on these; a Mink smells differently when angry; dogs in ill-health become malodorous; Deer in rut become offensively strong-smelling; a female animal in rut is recognized afar by the scent.

In many species additional effect is given to the body scent by the development of special glands which secrete a strong odour. These glands are usually situated in a part which is habitually brought in contact with the ground or the vegetation. Thus, in a Musk-deer they are on the side of the belly; in the Peccary, on the back; in our common Deer on the tarsus, between the toes, and in the lachrymal fossa. In some animals, however, the contact with the ground is secured in a different way. The glands are situated within the anal ODOUR-GLANDS and preputial orifices, so that the natural excretions *in transitu* bear with them the taint which reveals so much to the next passer-by of the same species.

WOLF TELEPHONES

In order that this second animal may find the depot of intelligence quickly, it is necessary that his discovery of the place be not left to chance; and, incredible as it may seem at first sight, there is abundant proof that the whole of a region inhabited by Wolves is laid out in signal stations or intelligence depots. Usually there is one at each mile or less, varying much with the nature of the ground. The marks of these depots, or odour-posts, are various; a stone, a tree, a bush, a Buffalo skull, a post, a mound, or any similar object serves, provided only that it is conspicuous on account of its colour or position; usually it is more or less isolated, or else prominent by being at the crossing of two trails.

Now, a man returning to town goes at once to his hotel or club, glances over the last three or four names on the register, adds his own, then makes a more thorough inspection. And the behaviour of an animal arrived at an odour-post is precisely the same. It approaches, hastily sniffs the post, adds its own odour, then makes a more thorough investigation. The attention that dogs pay to lamp-posts in town is precisely the same habit, a trifle over-developed through idleness, etc., but it will serve to illustrate. I have many times seen a dog approach the post, sniff, then growl, register, growl again, and, with bristling mane and glowing eyes, scratch fiercely with his hind-feet, and walk off very stiffly, glancing back from time to time. Again, it is common to see a dog, after the preliminaries, become keenly interested, trot about the vicinity, and come back again and again to make his own record more evident. At other times one sees the animal, suddenly aroused by the news, take up a recent trail or fly to the next signal post, and so continue in pursuit of whatever it was that was sensed.

REGISTERING

Wolves do precisely the same, but I believe they carry it to a higher pitch, and there can be no doubt that a newly arrived Wolf is quickly aware of the visit that has recently been paid to the signal post—by a personal friend or foe, by a female in search of a mate, a young or old, sick or well, hungry, hunted, or gorged beast. From the trail he learns further the direction whence it came and whither it went. Thus the main items of news essential to his life are obtained by the system of signal posts.

The Wolf, as well as the dog, has further a habit of urinating or defecating on certain things that appeal to his nose, without arousing his appetite. He usually follows this action by vigorously scratching the dust with his hind-feet over and around the object defiled. This treatment is commonly accorded to poisoned baits as well as to traps. I am inclined to think that Wolves have been taken in some of my hidden traps while thus serving them with a Wolf's contempt, and heedlessly going too near while doing so. On other occasions, stones raked into the trap by this scratching have sprung them, and thus they have been fully revealed. EXPRESSION OF SCORN

I am satisfied that scorn—that is, hate with superiority—is among the feelings thus expressed, for I once saw the victor in a fight between two female Coyotes urinate gleefully again and again on her fallen foe as she crouched in a corner of the cage.

A scent-gland that has long been overlooked is on the base of the tail above; its exact place is marked by a dark spot on most of the dogs. In the Gray-wolf this spot is black, the hairs composing the spot are bristly, and there is no underfur at the place. EXPRESSION OF ANGER

When a dog or Wolf, ready to do battle, approaches a stranger, the tail is raised at base and drooped beyond, so that this gland is at the highest point and the hairs on it are raised. This doubtless allows the escape of more of this scent.

In play, in battle, or in abject fear, there is no suggestion of this pose of tail. (See Plate LXV.)

SOME REMARKABLE WOLVES

Just as there are geniuses and heroes among men, so there are wonderful individuals among Wolves. These have always interested me, and I have endeavoured to make records of their lives. One of the first of them that I met was the Winnipeg Wolf. In March, 1882, while coming to Winnipeg from St. Paul, I saw a sight that stirred my blood. As the train flashed through an opening of the poplar woods south of St. Boniface, there stood a big Gray-wolf, erect and defiant, surrounded by a motley pack of town dogs, big and small. He was holding all at bay. A small dog was lying in the snow near him, and a big dog was bounding about doing some splendid barking, but keeping his safe distance. The train passed and I saw no more.

A dog-driver was killed next winter on the ice of the Red River while bound for Fort Alexander. The team were big fierce Huskies, and he was a strange driver. It was thought that he had struck at one of them with the whip, it had snapped back, and he, in retreating, had fallen, whereupon the four savage creatures had set on him and ended by devouring him. The counter theory was that he had been killed by a Wolf or Wolves, of which the dogs are notoriously afraid. The latter explanation found favour only with the dogs' owner, for the reason, people said, that he did not wish to lose his valuable team.

A large Wolf was seen several times afterwards about the city, and at length was killed near the slaughter-house, some said, by poison, dogs, guns, or all three. This was a male and weighed 104 pounds. It was mounted by W. R. Hine, the taxidermist, and shown at the Chicago Exposition of 1893. This interesting relic was one of the valuable specimens lost in the Mulvey Grammar School when the building was destroyed by fire in 1896.

I have, of course, no evidence that in each case it was throughout the same Wolf, but in writing the story of "The Winnipeg Wolf" I took a writer's liberty in making them so. The other adventures ascribed to him really belonged to other Wolves in distant regions.

In the story of "Lobo," I assumed a similar freedom. I as-

PLATE LXVI.—BLANCO IN THE TRAP.
From photograph by E. T. Seton.

PLATE LXVII.—LOBO IN THE TRAP.
Currumpaw, New Mexico, January 31, 1893.
From a photograph by E. T. Seton.

cribed to one Wolf the adventures of several, and I selected for him the most heroic exterior I could find in fact. But the final chapter recording his capture and death is given exactly as it happened, and was indeed the inspiring motive of the story.

The following Wolves also became known by name in various parts of the Province of Manitoba:

At Carberry, in 1897–8, a huge black Wolf appeared. He killed many sheep and calves and spread terror among the parents that had children going to school, but he never even threatened a human being. He was known as the Black Buffalo-runner. He was killed by Alexander Langmuir.

Another, the Virden Wolf, was killed at that place after a short but exciting career, by F. S. Baird, February 20, 1898. The photograph of this shows it to have been an ordinary Gray-wolf of medium size.

While at Pine Ridge, S. Dak., in August, 1902, I was told by Dr. James R. Walker and many others, that during the past three years the country between here and the Badlands (15 miles) had been frequented by an enormous white Wolf. The Wolves, in this region, were increasing and becoming so troublesome that a twenty-five dollar bounty was paid for each scalp, but double was offered for that of the white Wolf. It was a female, as it was once seen with seven cubs. One of them was caught and staked out for a decoy, but the mother came by night, eluded the watchers, pulled up the stake, and bore off her offspring in triumph. She is flourishing yet.

COURAGE OF WOLVES

It is often said that Wolves are cowards, but this sweeping statement seems not well-founded. They never voluntarily attack mankind, for the fear of man has been widely spread among them; yet a Wolf will attack and kill almost any dog. A Wolf has often been known to face a whole pack of dogs and carry off one of them in spite of the others about.

Richardson says:[16] "During our residence at Cumberland House, in 1820, a Wolf, which had been prowling, and was wounded by a musket ball and driven off, returned after

[16] F. B. A., 1829, I, p. 64.

it became dark, whilst the blood was still flowing from its wound, and carried off a dog, from amongst fifty others, that howled piteously, but had not courage to unite in an attack on their enemy."

A full-grown Wolf will indeed fight any number of dogs in self-defence, and will die without a thought of surrender. This is not cowardice.

Nevertheless, individuals vary so much, in this highly specialized animal, that we may expect to find some that are downright cowards, as well as others of heroic bravery. The one described by Richardson may have been a noted desperado of his tribe.

Exceptions to the rule may be accounted for precisely as in man; bodily well-being is an essential of physical courage. Richardson says[17] of the Barren-ground Wolves:

"When reduced by famine they are very abject and unresisting. Mr. Bell once, while residing on Mackenzie's River, caught a full-grown, but famished Wolf in a marten-trap tied to a small log which it had not the strength to carry away. He went to the Fort for a line to lead it home, and the children who accompanied him back assisted in bringing it in by pushing it on from behind. It made no resistance and suffered itself to be tied quietly to the stockades of the Fort. The experiment of taming it was not, however, made, and after the curiosity of the people was satisfied it was killed."

CHIVALRY

Chivalry in its simplest aspect may be defined as consideration by a male for a female, on account of her sex, when the sexual passion is dormant. In this light it is fair to say that there is much chivalry among Wolves. Richardson records many instances of such kindly consideration; indeed, I have heard the question raised as to whether male dogs or Wolves will at any time attack female dogs or Wolves, and *vice versa*. I have no personal evidence to give that they *will attack*, but I have some evidence to show that they will *refrain* from attacking. A case has already been noticed in the chapter on pairing.

[17] Arc. Search Exped. (of 1848), 1851, Vol. II, p. 87.

FIG. 199—Tracks of large Gray-wolf. (Life size.)

SPEED The speed of the Wolf is often exaggerated. My impression is that 21 or 22 miles an hour would represent the *highest rate* of an average individual *for one mile*. This is much less than the speed of the Coyote, Jack-rabbit, Deer, Antelope, greyhound, or even foxhound; but the Wolf can keep it up longer than most animals. A comparative scale is given on page 233.

TRACK The track of a Wolf cannot be distinguished with certainty from that of a large dog. (See Fig. 199.)

STRENGTH Although we must be cautious about receiving accounts of the Gray-wolf's ferocity, we are sure to be surprised by facts about its strength. I have known a young Gray-wolf, scarcely six months old, drag off a 100-pound bar of iron, to which it was chained, taking it 200 or 300 yards without stopping, and a quarter of a mile before discovered. This same cub could almost hold its own against an ordinary man pulling at its chain. I have several times seen a Gray-wolf in a trap go off with a drag that weighed considerably over 100 pounds; and on one occasion I saw an 80-pound female that was trapped drag a 52-pound beef-head over rough ground faster than I could follow on foot, and keep up the flight for one and a half miles.

I have known a Gray-wolf go off carrying the head of an ox in his jaws, and take it so far that I gave up following his trail in the dust. I did not weigh the ox-head, but found that a small cow-head weighed over 50 pounds, so that it must have been at least 75 pounds.

The Wolf's great strength, indeed, is in his jaws. It is doubtful whether any dog, of truly domesticated race, has such powerful jaws as the Wolf. It is generally believed by the hunters that for this reason no dog has yet been found which, single-handed, could conquer a full-grown Gray-wolf.

The rope used for lassoes on the Plains is half-inch manilla, and yet has often been cut through by a single clip of the Wolf's jaws when he has been lassoed.

PLATE LXVIII.—THE GREYHOUND THAT FOLLOWED TOO FAR.
Wolf study by E. T. Seton.

The Hon. Theodore Roosevelt gives thus an instance[18] of a Gray-wolf killing a horse. "With a few savage snaps the Wolf hamstrung and partially disembowelled it." Many similar cases could be cited.

The strength of its jaws is, doubtless, a cardinal factor in the Wolf's life-problems, not only putting it beyond danger from other carnivora, but also leaving all herbivora at its mercy.

Doubtless its hold on environment is largely due also to endurance. A Wolf can live on one full meal a week; that is, a dozen meals at equal intervals would carry it through the winter.

The Wolf that Archbishop Taché tells of[19] roamed for a month in deep winter, at Isle à la Crosse, with a heavy trap and clog on his hind-foot. It is hard to see how he could have got a meal in all that time, and, though emaciated, he was very lively indeed when found.

The species is credited by most hunters with cunning enough to hunt by combined drive and ambush, exactly as described in the chapter on the Coyote, but this I have not personally witnessed.

SWIMMING

The Wolf of Ontario is known to be a good swimmer. W. Lewis Fraser once described to me the antics of a family of Gray-wolves that he saw playing in the water like a lot of water spaniels. This was in Muskoka, during the month of September, and they therefore were probably a family.

SOCIAL AMUSEMENTS

The dogs, and especially train-dogs, howl much on moonlight nights in winter, and in a less degree at other seasons. They do not sit around in a circle as has been stated, nor have any accompanying ceremonies been seen. They howl usually when some loud noise or one of themselves begins it. These remarks apply in a measure to Wolves. Unless this nightly chorus belongs to the class, I do not know of any social amusements among these animals. A hint, however, is supplied by

[18] Wilderness Hunter, 1897, p. 394.

[19] See Note 1.

a Saxon name still used in Teesdale, England. A certain place there is called "Wolf Lake," although there is not, and never has been, any water near, but my friend, James Backhouse, informs me that it was originally "Wolf lek," that is, the place where the Wolves were supposed to play. (Anglo-Saxon, *laeken*, to lark or play.)

SANITATION

In sanitation Wolves have the habits of ordinary dogs. They do not bury their dung, but they keep their dens clear of it.

HYBRIDITY

The Eskimo or Husky dog is understood to be simply a domesticated Wolf, mixed with a strain of some other dog stock. The readiness of the Wolf and the Husky dog to cross is noted by all writers on the subject. Henry, in his famous "Journal on Red River,"[20] refers to this as a regular thing and gives a very graphic account of the way in which the female dogs were unwittingly made to play Delilah and betray the he-wolves into the merciless hands of their human enemies.

A similar account is given by Richardson for the female Wolves about Cumberland House.[21]

Two large dogs, supposed to be the offspring of a Wolf and a Husky, lived about Kildonan, Man., and terrorized the district for about a year in the early 80's. One was gray, one red or liver colour. No one owned them; they lived wild. George Fraser, of Winnipeg, my informant, fired at them several times with a shot-gun, without visible effect. One day he got a close chance at the red one with No. 5 shot; the beast got away, but never was seen again; probably it died.

AS TRAIN-DOGS

W. F. White, the taxidermist, of Winnipeg, informed me, not long ago, that he had no difficulty in selling living male Wolves, as they could be utilized to cross with and improve the train-dogs.

Henry also speaks of saving young Wolf cubs to be used for the trains.[22]

[20] Journal, A. Henry, 1799–1814, pub. 1897, p. 166.

[21] Franklin's First Journey, 1823, p. 90.

[22] Journal, 1897, p. 175.

Captain Dick Craine, of Petoskey, Mich., tells me that he spent 9 years among the train-dogs in Alaska and Yukon; owning and handling in that time 200 dogs. Among these he had 3 full-blooded Wolves also used as train animals. Many half-breeds of course were among the dogs, and all are more or less of Wolf blood.

The latter, he says, is not so good as a train-dog. It is strong enough but always more or less shy, watching its driver as though cowed, and shrinking from the touch of the hand.

The only tangible difference between a Husky dog and a wild Wolf is in the tail. A Wolf's tail is rarely above level or curled up; a Husky dog's is always excessively curled. Why? Perhaps it is a result of the harness toil. In hauling, unusual energy is forced into all the extremities; that in the tail is not specially directed, and therefore causes the tail to curl up, obedient to the strongest muscles, just as a man's teeth clinch under violent effort of the limbs. I doubt not, if the *flexor* muscles of the tail were strongest, instead of the *levators*, the train-dog's tail would be permanently curled between his legs.

Corroboration of this is found in a fact that I have several times observed. A train of half-bred Wolves may set off in the harness with tails down, but the moment they come to a bad place, where they must strain at the traces, their tails fly up into curl.

The Husky dog's ears are frequently drooped. A wild Wolf's ears are erect, but, according to Captain Craine, the train Wolf at the age of nine or ten is apt to droop his ears.

Many observers attest the tamableness and dogginess of this animal. Ross says:[23] "A full-grown Wolf became, during the months of July and August, 1857, quite domesticated at Fort Resolution. Though rather shy of the people, it lived in great harmony with the dogs, playing and sleeping with them, and sharing their food. Around the smoke made to keep off the myriads of noxious flies from the cattle, it reposed with the other animals, and, although there was a small calf DOGGINESS

[23] Fur-bearing Anim. Mack. R., Can. Nat., January, 1861, p. 11.

in the band, it never attempted mischief. It was shot at by an Indian and never seen after."

The Rev. J. A. McLaughlin,[24] of Berens River, Lake Winnipeg, wrote me, March 9, 1893:

"Wolves are quite numerous here this winter, but do not seem to band together, to any extent, and are not, so far as I know, dangerous. Last week one of the Indians on going to his cache, where he had been doing his fall fishing, found a splendid black Wolf in a trap. He tied his mouth with a line, took him out of the trap, hitched him up to his dog train, and made him help haul in the load of fish. The Hudson's Bay Company officer here had him chained up at the Fort, and intends trying to make a cross with one of his dogs. I have seen a number of Wolves, but none like this specimen. The fur is exactly like a Silver-fox in colour, thick and beautiful. The ears are much more rounded than usual, giving the head more of the appearance of a Bear than a Wolf. I have handled it, but there is never any sign of crossness, and no attempt at biting."

D. T. Hanbury on his journey from Selkirk to Norway House, February 26, 1899, says:[25] "At this place [Berens River], I had the novel experience of riding in a sleigh drawn by a team which included a Wolf. This animal was muzzled and, though rather savage, worked well. I was told that a pure Wolf does not retain its stamina in captivity, but a half or quarter cross makes a most useful animal."

It is hard to understand why the train-dogs should so fear the Wolves, if they are such near kin. Probably the wild ones are larger and the train-dogs more or less cowed by their life.

LATENT FEROCITY

The savage nature of the Wolf, however, is apt to break out at times in the train-dog, as already noted in my account of the Winnipeg Wolf on an earlier page. Another tragic incident of the kind took place recently on the Saskatchewan.

[24] Mr. McLaughlin was drowned in Lake Winnipeg, September 12, 1903, while faithfully doing his work.

[25] Northland, Canada, 1904, p. 6.

It was reported to me by Dr. D. A. Stewart, of Winnipeg. A half-breed dog-driver was taking his team and his little boy to a distant post. He left the boy in charge of the team while he went after a Deer. On his return he found the dogs curled up asleep and nothing left of his son, except fragments of his clothes. The half-breed was a devout Catholic; he drove the dogs to the Trading Post, shot the four brutes, and gave them Christian burial.

The diseases that have been observed to torment the Gray-wolf are mange, scab, and rabies. I have several times heard of mange removing all of a Wolf's hair except a ridge along the spine, and in consequence have arisen many rumours of strange beasts in the land. DISEASES

Warburton Pike says:[26] "There was some sort of disease resembling mange among them [Gray-wolves] in the winter of 1889–90, which had the effect of taking off all their hair, and judging from the number of dead that were lying about, must have considerably thinned their numbers."

Henry in his Journal makes frequent mention of scab. Thus:[27]

"March 3. A large Wolf came into my tent three times, and always escaped a shot. Next day while hunting I found him dead about a mile from the Fort; he was very lean and covered with scabs."

Rabies or hydrophobia seems to break out among them at times. Although Wolves do not ordinarily attack man in America, there are one or two recent cases on record, from the western United States, but there is also evidence that in each case the Wolf was rabid.

Even as early as 1800 it appears to have been considered evidence of madness for a Wolf to attack a man, as Henry thus makes record at Park River:[28]

November 2. "Last night the Wolves were very troublesome; they kept up a terrible howling about the Fort, and

[26] Barren Grounds, N. Canada, 1892, p. 53.

[27] Journal, A. Henry, 1779–1814, pub. 1897, p. 194.

[28] *Ibid.*, p. 133.

even attempted to enter Maymiutch's tent. A large white one came boldly into the door and was advancing towards a young child when he was shot dead. Some of them are very audacious. I have known them to follow people for several days, attempt to seize a person or dog, and to be kept off only by fire-arms. It does not appear that hunger makes them so ferocious, as they have been known to pass carcasses of animals, which they might have eaten to their fill, but they would not touch flesh; their object seeming to be that of biting. The Canadians swear that these are mad Wolves and are much afraid of them."

And again:[29]

"April 18, 1810 [on North Saskatchewan]. Another mare was bitten in the nose by a mad Wolf and died the day after, foaming at the mouth and running around distracted."

WOLF-KILLING

Wolves are so rarely seen that shooting is not to be relied on as a means of keeping them down.

Hunting with dogs has been carried on with fair success, but it requires a composite pack of running, tracking, and fighting dogs, as well as the best of horses, so that it is somewhat expensive.

In the early days the Indians captured many Wolves in pitfalls. The following, from Henry's Journal, bears on this:[30]

"We had now [south of Turtle Mountain] a well-beaten path, but were several times in danger of breaking our necks in deep pits which the natives had dug in the path to catch Wolves and Foxes in winter. Some of them are 10 feet deep, hollowed out to a space about 30 feet in circumference, whilst the entrance is no wider than the foot-path and about 5 feet in length. These holes are covered with dry grass at the season when Wolves are good, and every morning are found to contain some of these animals. In summer the grass grows strong and high about the mouths, entirely concealing them, until one arrives upon the very brink, and is in danger of tumbling in headlong."

[29] *Ibid.*, p. 594.

[30] *Ibid.*, p. 322.

POISONING

Poisoning, once quite easy, is now very hard to practise, since the Wolves have learned the smell and dangers of strychnine. One method is to bore an auger hole into a post that the Wolves use as a 'calling station,' fill it with a mixture of strychnine and tallow, and then over the outside spread a coat of pure tallow or butter. The Wolves will lick and gnaw at this till the poison has time to work, at least in former times they did so; now it seems to be losing its charm for them.

In early days I have had some success in poisoning with a drag. To do this I would take a lump of meat, or a bunch of Jack-rabbits, and drag it behind my horse for ten miles around the camp. At intervals of a quarter of a mile I dropped a carefully prepared poisoned bait, two grains of strychnine in a gelatine capsule hidden in a piece of liver about two inches square. These baits are carried in a rawhide bag, are lifted out with a pair of wooden pinchers, and are never touched with iron or the human hand. It is well to mark in some way the place of each bait for future reference.

The Wolves will follow the drag out of curiosity, even if not hungry, and, coming to the juicy bait, they will take it, or at least in olden days they used to take it. Then, again, the drag does good service, the poison is not likely to act before the Wolf travels a quarter of a mile, and he may go a mile, but he follows the drag still, and is picked up later *on the line*, instead of going off to die in some hollow where he cannot be found.

Of late, however, the Wolves seem to have got a comprehension of the device, and are no longer to be easily beguiled. Though they yet follow the drag, they commonly urinate on the baits and pass on. It still answers for the Coyote, but incidentally gathers in many of the neighbours' dogs. This breeds inharmony.

TRAPPING

Steel traps are more successful because they call for less initiative on the part of the Wolf. One way to employ these is on a drag as though for poison bait. Then at some spot on the line, preferably where two or more trails meet, bury a lump of meat and around that, three or four feet away, several traps, carefully rubbed with blood or cow-dung or smoked, and never touched with the bare hands. Each trap is fast to a separate drag; that is, a log or rock weighing forty or fifty pounds. Drag and all must be carefully concealed. The trap is sunk in the ground till the pan is exactly on a level with the surface, then the space under the pan is filled with dry grass, Coyote fur, or, best of all, cotton-wool. Dry dust is now sprinkled on everything, the trap logs and bait are completely concealed, a few weeds thrown about, and, last of all, the foot of a Coyote or a Wolf is used to make a few reassuring tracks at the place. The foot of a female dog at the mating season lends an especially helpful touch. This plan seems to play on the Wolf's habit of burying surplus food. If the bait were in plain view he might find some suspicious taint, enough to make him keep his distance, but when it has to be dug out before examination, he has time to pass all around and to get into one or more of the traps.

Sometimes the traps are set on the trails used by the Wolves in crossing cañons or going to water. The disadvantage of this is that a great many cattle get into them and it is an awkward job getting the trap off the foot of a range steer. He is not so grateful as he should be. If, however, the traps be not too large, they slip off the hard hoofs of the cattle when they happen to tread in them.

An excellent plan is to put a bait up three or four feet from the ground in a rough place; then set traps in the open places that a Wolf would naturally walk in as he circled about suspiciously to inspect the bait.

Yet another plan is to put the trap under water. A thin piece of stone is laid on the pan, and the trap sunk so that only this stone is above water. This is set eighteen inches from the dry bank, then a foot beyond the trap a bait is put on another stone.

The Wolf reaching out to sniff the bait, naturally sets a foot on the dry stone between him and the meat, and is caught. The water in this method assists greatly in disguising the smell of the iron. This plan answers also for most other carnivorous animals.

PLATE LXIX.—BLOOD ON THE TRAIL.
Wolf study by E. T. Seton.

Wolf-hunters sometimes throw a marrow bone in the fire at sundown; this smoulders all night and makes an attractive smell that the Wolves can detect and are drawn by, though miles away.

When seized by the trap a Wolf bounds off with all his strength. If the trap be held solid, something is likely to break under the violence of the struggle, but fastened to a drag, which yields to each jerk, the Wolf is securely held. His efforts merely tire him out, and he is usually found in the nearest cover or hollow within a few hundred yards of the bait.

As to the humanity of setting out such devices for catching wild animals there is little to be said. Nevertheless, it is not so much injury of the steel as the days of struggling and starvation that have caused the chief suffering, and this every trapper aims to avoid by going at very short intervals to the traps. As a rule, the less the animal has suffered the better the pelt. The ranchman puts the matter briefly: We do not trap and poison for fun, but because the Wolves would soon ruin every man in the cattle business if we did not keep them down. And we kill that way because there is no other way of doing it.

Since the above was written, Vernon Bailey, of the United States Biological Survey, has shed unexpected light on the Wolf question. He proves by actual experiment that, since the young Wolves are born in March, when the snow is on the ground, it is easy to track the parents home and exterminate the family. An energetic repetition of the process soon rids a region of Wolves. The details of his method are published in Bulletin 72, Forest Service, United States Department of Agriculture, 1906.

The skin of the Gray-wolf is split open flat like that of a Beaver, while Coyote skins are cased like Fox and Mink. The fur is rich, full, and beautiful. It makes a fine robe, but is not very durable as a rug. It is prime from November 15 to April 15, and brings from $1.00 to $10.00, according to quality. FUR

At the London annual fur sales held at Lampson's, March, 1906, there were 15,843 Wolf skins disposed of. The highest price realized was 64 shillings ($15.36) each, for two unusually fine blue skins, but 32 shillings each ($7.68) was considered a high price for 5 extra large fine skins of ordinary colour, and first-class skins varied, according to size, from 2 shillings (48 cents) to 30 shillings ($7.20).

XXXIV.

Northern Coyote, Big Coyote, Prairie-wolf or Brush-wolf.

Canis latrans Say.

(L. *Canis*, a dog; L. *latrans*, barking; because it is more of a barker than is any other wild dog.)

Canis latrans Say, 1823. Long's Exped. Rky. Mts., I, p. 168.
Type Locality.—Council Bluffs, Iowa.

French Canadian, *le Coyote.*
Cree, *Mes-cha-cha-gan-is'.*
Saut., *Mes-cha-chag'-an-is.*
Ojib., *Mes-ta-cha'-gan-es.*
Yankton Sioux, *Song-toke-cha.*
Ogallala Sioux, *Mee-yah-slay'-cha-lah.*

'Cased Wolf' is the old trade name of the Coyote, because its skin was cased like that of a Muskrat, while the Gray-wolf's pelt was spread out flat like that of a Beaver.

The generic characters are as in the preceding, but the Northern Coyote may be distinguished from the Gray-wolf first and chiefly by its much smaller size, slender build, and almost fox-like muzzle and ears; second, its general warmer sienna colour; and from the numerous other Coyotes of the far Southwest by its larger size, paler colours, and teeth of which the premolars and carnassials are "very large and greatly swollen."

A very fat male killed at Touchwood Hills, Sask., by Ed. Hollis, in the winter of 1901–2, measured: SIZE

Head and body	2 feet $9\frac{1}{2}$ inches	(850 mm.)
Tail	13 "	(330 mm.)
Hind-foot	$7\frac{3}{4}$ "	(198 mm.)
Ear	$4\frac{5}{16}$ "	(110 mm.)

This was not unusually large, but it was the heaviest he ever saw, weighing 42 pounds on standard scales.

A still heavier is reported to me by George L. Rimington, of Penrith, England. He spent the winter of 1907-8 hunting Coyotes near Calgary, Alta.; out of 85 killed, the heaviest was a very large male, which weighed 46 pounds on standard scales. As the date was early November, it was of course at its fattest time.

A Winnipeg specimen (female) collected in the October of 1908 by E. W. Darbey was:

Length of snout to tail-bone tip . .	4 feet 1 inch	(1245 mm.)
Tail	1 foot 2½ inches	(369 mm.)
Hind-foot	8 "	(203 mm.)
Height at shoulders	1 " 9 "	(534 mm.)

Weight 25 pounds.

Richardson gives[1] 3 feet (915 mm.) as the length of the head and body of a specimen he took on the Saskatchewan. In New Mexico, among a score of Coyotes of the local form (*lestes*), I found many of the above dimensions, but the heaviest, a male, weighed 31 pounds; the ordinary males were but 28 pounds and the females 24 pounds.

COLOUR

"Muzzle dull and rather pale fulvous, finely sprinkled with gray hairs (chiefly above) and with black hairs (chiefly on cheeks); top of head from front of eyes to ears grizzled gray, the pale fulvous zone of under-fur showing through, but the gray predominating; ears deep, rich fulvous, sparingly sprinkled with black hairs; upper parts from ears to tail coarsely mixed buffy-gray and black; under parts and upper lip whitish; long hairs of throat sparingly tipped with blackish, giving the broad collar a grizzled appearance; fore-legs and feet dirty whitish, becoming dull clay colour on outer side of leg; hind-legs and feet dull fulvous on outer side, white on inner side and on dorsal surface of feet, the change from fulvous to white rather abrupt; tail narrowly tipped with

[1] F. B. A., 1829, I, p. 74.

PLATE LXX.—COYOTE HEAD.
Life study made by E. T. Seton, in Jackson's Hole, Wyo., September 1, 1898.

black; its under side whitish basally, becoming pale fulvous on distal half and tipped and edged with black."[2] (*Merriam.*[3])

The first of the above colour descriptions fits word for word and hair for hair to the typical male Gray-wolf described in the Wolf chapter (pp. 750-1). The only difference I find on comparing many skins of Gray-wolf and Coyote is in the under-fur of the back, which usually is gray-brown in the former, and sienna-brown in the latter. There are many exceptions, however, so that we must look to the size of the animal, with its cranial and dental character, for reliable diagnosis.

Four races of the large Coyote are recognized:

latrans Say, the typical form.
nebracensis Merriam, similar but "everywhere paler; backs of ears buff instead of fulvous; skull and teeth smaller."
texensis Bailey, like *nebracensis*, "but darker, brighter-coloured, and with lighter dentition; smaller, brighter, and more fulvous than *latrans*."
lestes Merriam, very similar to *latrans* in size and colour, "cranial characters as in *nebracensis*, but skull and teeth averaging somewhat larger."

Life-history.

While we speak broadly of the Coyote as though it were one species, it should be remembered that scientists recognize at least a dozen kinds that are closely akin and yet have their own peculiarities and habitat. But they agree in their general style and character; the Coyotes everywhere are sons of the

[2] The Winnipeg specimen measured above agrees fairly well in colour with this description, but has the lips fulvous, a black spot on the forepart of each fore-leg, and a large black spot on the base of the tail over the gland. The under-fur everywhere is plumbeous, except on the throat, where it is pure white.

[3] Revision of Coyotes, Proc. Bio. Soc., Washington, Vol. XI, pp. 19–33, March 15, 1897.

desert, Ishmaelites living by their wits. Further, they are alike in their vocal gifts—our Ishmaelite is also a Troubadour.

The first of the Coyotes to be discovered was of course the one that is found farthest East. It happens also to be the largest. This is *latrans*, 'the barker,' so called by Say, because it was the only known species of wild Dog that habitually barked.

RANGE

Its range, so far as known, is set forth on Map No. 43, though I suspect it goes much farther north-westward than the lines would indicate.

The spot near the Alaskan Boundary marks a new record. In 1907, Madison Grant secured a complete skin and skeleton of a Coyote killed near Whitehorse on Alsek River, Alaska, in February of that year. Dr. J. A. Allen examined the specimen and found it closely related to *lestes*.[4]

In general, it prefers the untimbered portions of the north temperate regions, but the Prairie-wolf is far from confining itself to the open country. The woods from Pembina to Riding Mountain, as well as immediately east of Winnipeg, is well supplied with the species. I found it abundant about Lake Winnipegosis even on the east side of the water, even where fully 100 miles in direct line from open country. And north-westerly its range extends into the forest 500 miles to Great Slave Lake. It is never found, so far as I can learn, in the north-eastern or coniferous region of Manitoba, but it is more or less plentiful in all the south-western half of the Province.

HOME-RANGE

So much for the range of the species. The range of the individual is less easy to establish. How large is the home-range of a Coyote, or rather, a pair of Coyotes? For we shall see that this interesting little brute is highly moral as well as clever. I should think, notwithstanding the popular notion of the Coyote as a world-wanderer, that its home-range is much less than 10 miles across. After consulting many hunters and

[4] Bull. Am. Mus. N. H., Vol. XXIV, pp. 584–6, September 11, 1908.

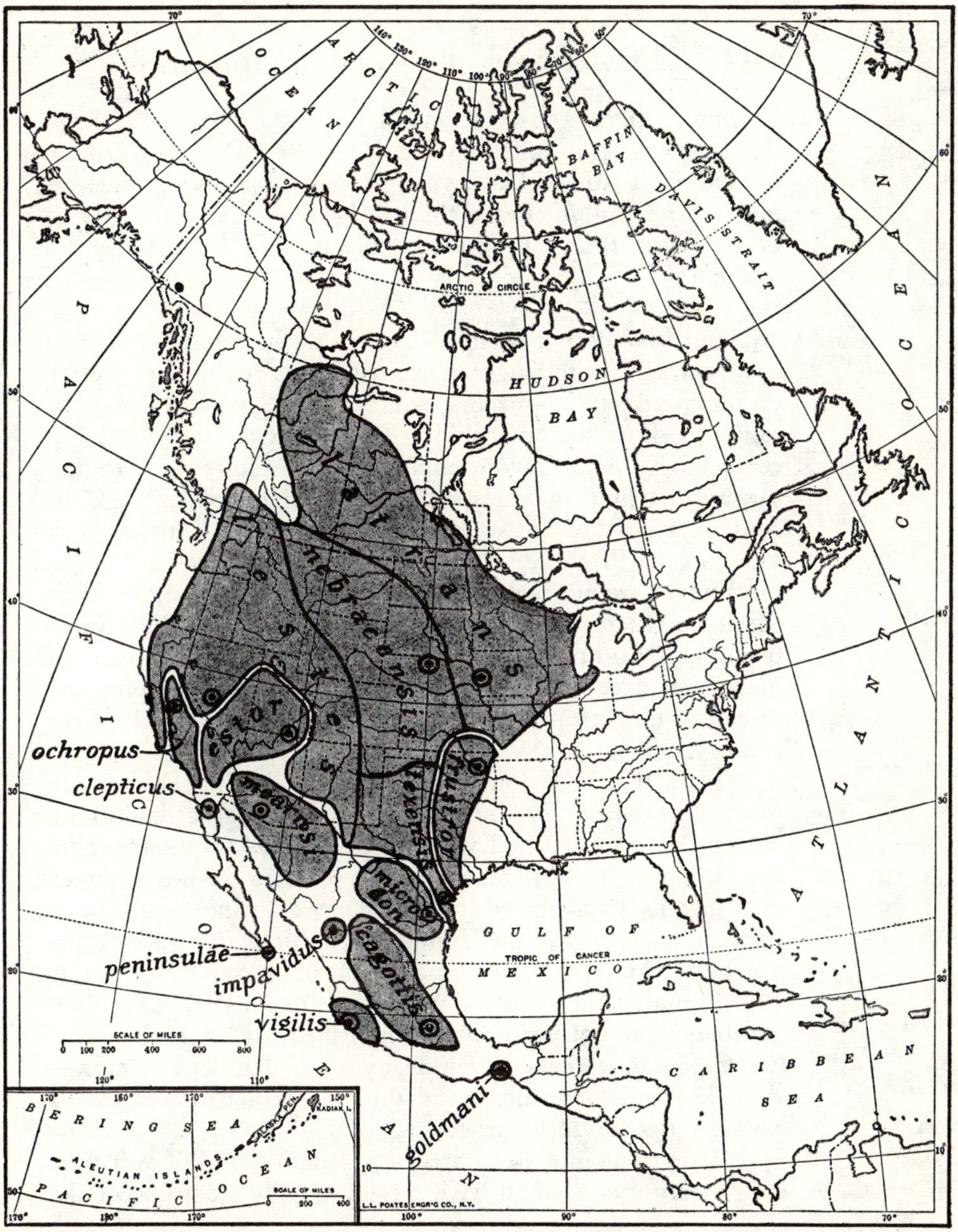

MAP 43—RANGE OF THE COYOTES.

This chart is purely diagrammatic; many of the forms undoubtedly overlap or intergrade. It is drawn up chiefly from Dr. C. Hart Merriam's Revision of the Coyotes, 1897; the Biological Survey map (Doc. 132, Senate 1907), also D. G. Elliot, Vernon Bailey, E. A. Preble, H. A. Allen, and from my own notes in many parts of the West. Full investigation must greatly change the boundaries between the many forms and all the boundaries in Mexico.

The following are recognized:

Canis latrans Say. Common, with 4 races,
Canis frustror Woodhouse. Woodland Coyote,
Canis cagottis (H. Smith). Red Coyote,
Canis peninsulae Merriam. South California Coyote,
Canis microdon Merriam. Rio Grande Coyote,
Canis mearnsi Merriam. Mearns's Coyote,
Canis estor Merriam. Desert Coyote,
Canis ochropus Eschshoetz. California Coyote,
Canis vigilis Merriam. Colima Coyote,
Canis goldmani Merriam,
Canis clepticus Elliot,
Canis impavidus Allen.

making numerous observations, I conclude that in the summer a township (6 miles square) is more than ample hunting-ground for a pair of Coyotes. In winter, perhaps, twice as much is needed, in the north, and beyond this they never go of their own free will; outside this limit is foreign country to them.

ABUNDANCE But they do not occupy any area to the exclusion of their kind; probably the ranges of at least half a dozen pairs overlap on the same hunting-ground, which assumes a general population of 10 to the township. These calculations would, if correct, give us a Coyote population in Manitoba of 12,000. Or, approaching the question from another view, in 1904 the Manitoba government paid bounties on 4,541 Prairie-wolves killed in the Province. The testimony of all observers is that the Wolves are increasing in spite of this destruction, therefore the annual increase is greater than the annual kill. This would presuppose an original population of at least 5,000, which we may safely accept as a minimum of Coyotes in Manitoba, and of course they are condensed in the south-western half of the country.

That this is a low estimate the following shows: In *Shields' Magazine* for April, 1904,[5] Jack Comegys describes a recent Coyote drive at Evans, Colo.; about 20 square miles (half a township) were included, and some 40 Coyotes rounded up; that is, 80 to the township, or say 2 to the square mile. Further, according to the United States Biological Survey,[6] the State of Kansas (81,700 square miles), in the year ending July 1, 1904, paid bounties on 20,000 Coyote scalps, but their numbers were not perceptibly diminished; at least as many—the Colorado evidence would say even double as many—were left, which would make the population above 1 to 2 square miles, or 20 to the township.

If anything like these rates of population prevail over their entire territory, we shall have a total of fully 1,000,000 of the

[5] P. 215.

[6] D. E. Lantz, Bull. No. 20, Biol. Surv. U. S. Dep. Agr., 1905, p. 9.

large Coyotes on the 2,000,000 square miles over which they are found.

SOCIABILITY

The species is but slightly gregarious. The most I have ever seen in one day were 8, and the most at one spot, were 3. They were gathered at a dead calf and scattered immediately after their feast. The most I have heard of together were 12, also attracted by a carcass. W. R. Hine tells me that he has seen 5 together, never more; these were at a dead animal; 3 are the most I ever saw travelling in company, and the most he ever met with in one day was a dozen during a 60-mile drive along Red River, in the autumn.

Six Coyotes were seen by Lew Wilmot,[7] as they were chasing a Deer, in the spring of the year at Oroville, Wash.

Eight Coyotes were seen together in August by Henry W. Wende, of Sunnyside, Wash. They were near a drinking place in the Yakima Valley and may have been a family, but they looked fully grown.

A dozen are the most in one band that I can learn of. These were seen and heard near Humboldt by Professor John Macoun on October 15, during his journey from Fort Carleton to Winnipeg in 1875. Early one morning, as he sat by the fire after his wagon had gone, they gathered about him and sitting on their haunches, some 75 yards off, all howled their loudest personal information.

As will be shown later, the Coyotes frequently combine their efforts for the common good, although they do not habitually go or live in bands. I should therefore say the species was sociable, though but slightly gregarious.

INTERCOMMUNICATION

Intercommunication of ideas is well developed among Coyotes. The smell-telephone with the smelling posts is largely used, but they also communicate many ideas by example.

Their remarkable vocal powers are at least as important as any. The principal sounds they utter are described in another paragraph.

[7] Forest and Stream, April 10, 1897, p. 284.

MATING It is the opinion of all persons familiar with its habits that this animal is strictly monogamous. It is usually seen single or in pairs, mostly in pairs. Every scrap of recorded evidence that I can find, as well as all my own observations, go to prove that winter and summer it is the rule for 2 Coyotes to run, hunt, and live together. Even the 6 described by L. Wilmot were in 3 pairs. And since this union holds the year round, we may safely infer that the species pairs for life.

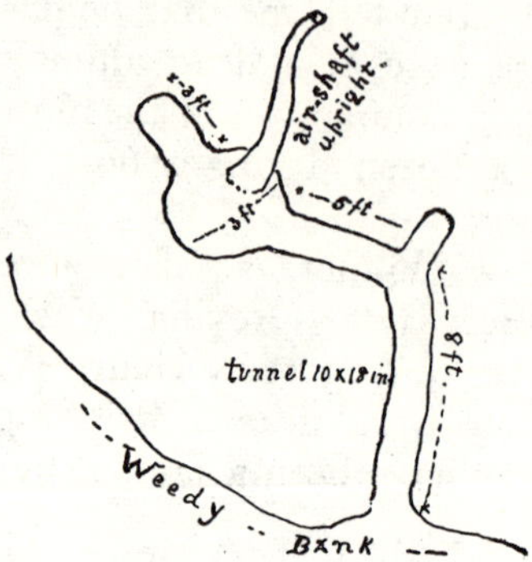

FIG. 200—Diagram of the Coyote Den opened by A. S. Barton, at Boissevain, Man.

As a natural corollary the male helps in raising the brood.

A. S. Barton says that the male aids the female with the young, at least till they are able to leave the den. In autumn, male and female are always found together, but he never saw the whole family together at this time. The young may have scattered, or those observed may have been a barren pair; such are frequent.

The actual mating season is about the middle or during the last two weeks of February.

DEN Their usual denning place is one dug by the Coyotes themselves, in some sunny bank, but they may use an abandoned Badger hole.

The entrance is about 10 by 20 inches and is commonly concealed in the bushes. The actual nest is sometimes lined with a little grass and fur, and sometimes is quite bare. Barton sends the accompanying plan and description of one which he examined near Boissevain, Man. (Fig. 200).

Apparently it had been dug by the present owners, and was much the same as the dozen or so others he had investigated. The air-hole, located after close search, was, as usual, an old Gopher hole, enlarged from below and directly over the nest; he supposes it was made to admit fresh air to the cubs.

I have not seen this ventilator, but may have overlooked it, as I had not heard of such a contrivance when last I examined a Coyote's den. It is well known that a family will have several dens, some of which are, as Barton says, "sleeping places for use during the heat of the day, which is one reason why so many attempts to dig out Coyote dens often fail of results."

Usually the young are born during the first half of April. YOUNG April 9 in the New York Zoölogical Park and April 20 in the Washington Zoo represent the extreme dates at hand.

They number from 3 to 10, but are usually 5 to 7.

They are blind and helpless, but covered with close, dark ash-coloured fur.

It is generally believed that the father is not permitted to enter the home for some days after their birth, but I have not been able to confirm this belief.

It is sure that he is never far away, and his devotion is vouched for by all who know him. Dr. W. T. Hornaday informs me that the father of the brood born April 9 took a keen interest in the young, and became very officious, even vicious, in their defence. Their eyes opened on eighth and ninth days, variously. When about three weeks old the mother would carry them out into the sun, or about the yard and back again. At five weeks they were old enough to walk out alone. They were not fed by regurgitation at any time, so far as known.

Keeper Carson, however, assures me that in the Philadelphia Zoo, where the Coyotes frequently breed, the mother disgorges food for them regularly, exactly as does the mother Gray-wolf. When some six weeks old, both parents begin to bring solid food to the little ones, and the entrance to the den becomes littered with feathers, fur, bones, and other remains of their prey.

The young are so keen to see and welcome father and mother back with the new catch, that they make little pathways from the den to all the near points that give a view. Here they will sit and watch, but are ever ready to skurry home on the slightest alarm.

A glimpse of their life at this time was secured by Professor John Macoun, of Ottawa, Ont., while exploring near Crane Lake, Sask., in 1884. On June 23, he came on a Coyote family —father, mother, and at least 3 young ones. At his approach all ran into the den.

As they get older, Barton says that the cubs scratch out little pockets leading from the main den. In digging after them, these are frequently covered over and escape notice, so that some of the little ones are never found.

I have a most interesting photograph by William McFadden, of Denver, showing 9 young Coyotes playing about the door of the house. This was taken in June. The young were about one-third grown; both parents were seen in attendance on them, and when they found that the home was discovered, they moved the young ones elsewhere. (Plate LXXI.)

This habit is quite general among Coyotes. Barton writes that on May 21, 1905, he found a den in a ravine a mile out of Boissevain, Man. The mother was running around and the pups squealing deep in the hole. But when he went back next day to dig them out, the litter had been moved evidently to a distance, for a careful search in the neighbourhood failed to locate them.

In July the young are half grown. They now begin to run with their parents and learn the arts of hunting. At this season the mother especially guards and trains them carefully. Her warning call of danger is a very distinctive cry—a prolonged, quavering yelp or squall, rising in pitch towards the end.

"I remember [says Barton] on one occasion I was hunting a young Coyote, when the mother coursed along a neighbouring height uttering this cry. I had two foxhounds in leash, they were after her, but a few minutes later came racing towards me in terror, closely pursued by the mother. They were so embarrassed by the leash and she was so active that she ran around and bit them as often as she chose."

PLATE LXXI.—COYOTE FAMILY—NINE PUPS—THIRTEEN MILES FROM DENVER.
From a photograph by W. R. McFadden.

In October the young are as big as the parents and the family is scattered. Food is still abundant, although the Ground-squirrels have retired to their winter quarters, and the Coyotes are sleek and fat, but from this time on the struggle for life grows hard and deadly.

"Wolves, and probably Coyotes [says Bailey[8]], do not breed till 2 years old, which accounts for the presence of roving bands during the breeding season."

The only migration that I know of in this species is the casual one in search of shelter or better hunting. In January, MIGRATION

FIG. 201—Tracks of Coyote. (Life size.)

1883, after a three days' blizzard at Carberry, Man., the Coyotes were seen moving all day from the north to the southwest. Eight individuals I saw, and the trails in the snow told of many others taking the same course. The wind was southwest. Barton says that in stormy weather there is a sort of local migration of the species from the Souris Plains to the sheltered region of Turtle Mountain.

The food of the species consists of every kind of fish, flesh, or fowl that it can master alive or discover dead. Ground- FOOD

[8] V. Bailey, Circular No. 63, Biol. Sur. U. S. Dep. Agr., April 29, 1908, p. 7.

squirrels, Mice, Rabbits, frogs, snakes, eggs, and fledgling birds are on its bill-of-fare, and the hen-yards as well as the sheep-folds are levied on in times of need.

HABITS

Mary Austin gives an admirable picture of a prowling Coyote in her "Land of Little Rain":[9] "Watch a Coyote come out of his lair," she says, "and cast about in his mind where he will go for his daily killing. You cannot very well tell what decides him, but very easily that he has decided. He trots or breaks into short gallops, with very perceptible pauses to look up and about at landmarks, alters his tack a little, looking forward and back to steer his proper course.

"I am persuaded that the Coyotes in my valley, which is narrow and beset with steep sharp hills, in long passages, steer by the pinnacles of the skyline, going with head cocked to one side to keep to the left or right of such and such a promontory.

"I have trailed a Coyote often, going across country, perhaps to where some slant-winged scavenger hanging in the air signalled prospect of a dinner, and found his track such as a man, a very intelligent man accustomed to hill country, and a little cautious, would make to the same point. Here a detour to avoid a stretch of too little cover, there a pause on the rim of a gully to pick the better way—and it is usually the best way—and making his point with the greatest economy of effort."

CUNNING

The Prairie-hare and others of the Plains beasts often find safety in superior fleetness when pursued by the Coyote. But the latter sometimes succeeds by cunning, when all its strength and speed might fail, as the following instances show:

John B. Goff, the hunter, tells me that while freighting between Rifle and Rawlins, Colo., some years ago, he saw 2 Coyotes chasing an Antelope. They worked a distance apart, keeping the Antelope running zigzag between them, so that it really did four times the running of either. It was nearly exhausted and ran up to his horses for protection. The

[9] 1904, pp. 30–31.

Coyotes then held off. He had no gun, but he threw a rope around the Antelope's neck and, being in need of meat, cut its throat, and threw it into the wagon.

In an article on "Coyote Partnership," Dr. George Bird Grinnell, referring to the Coyote plan of running an Antelope down by relay chasing, says:[10]

"Of course the Coyotes do not catch every Antelope they start. Sometimes the game runs such a course that it does not pass near any of the waiting Wolves, and only the one that starts it has any running to do. In such a case the pursuit is at once abandoned. Sometimes the Antelope is so stout and strong that it tires out all its pursuers. * * * Two or three years ago I camped one afternoon near Rock Creek, and as there was very little feed we turned the horses loose at night to pick among the sage brush and grease wood. Early in the morning before sunrise, while the man with me was getting breakfast, I started out to get the horses. They were nowhere to be seen, and I climbed to the top of the hill back of camp, from which, as it was the only high place anywhere about, I felt sure that I could see the missing animals. Just before I got to the top of the hill an old doe Antelope suddenly came in view, closely followed by a Coyote. Both of them seemed to be running as hard as they could, and both had their tongues hanging out as if they had come a long way. Suddenly, almost at the heels of the Antelope—much closer to her than the other Wolf—appeared a second Coyote, which now took up the running, while the one that had been chasing her stopped and sat down and watched. The Antelope ran quite a long distance, always bearing a little to the left, and now seeming to run more slowly than when I first saw her. As she kept running, it was evident that she would either run around the hill on which I stood or come back near it. At first I was so interested in watching her that I forgot to look at the Wolf that had stopped near me. When I did so he was no longer at the place where he had stopped, but was trotting over a little ridge that ran down from the hill, and watching the chase that was

[10] Forest and Stream, February 6, 1897.

now so far off. He could easily have run across the chord of the arc and headed the Antelope, but he knew too well what she would do to give himself that trouble. After a little it was evident that the Antelope would come back pretty near to the hill, but on the other side of it from where she had passed before, and the Wolf which I had first seen chasing her trotted out 200 or 300 yards on to the prairie and sat down. The Antelope was now coming back almost directly towards him, and I could see that there were two Wolves behind her, one close to her heels and the other a good way further back. The first Wolf now seemed quite excited. He no longer sat up but crouched close to the ground, every few moments raising his head very slowly to take a look at the doe, and then lowering it again so that he would be out of sight. Sometimes he crawled on his belly a few feet further from me, evidently trying to put himself directly in the path of the Antelope and this he seemed to have succeeded in doing. As she drew near him I could see that she was staggering, she was so tired, and the Wolf behind could at any moment have knocked her down if he had wanted to, but he seemed to be waiting for something. The Wolf that was following him was now running faster and catching up.

"When the Antelope reached the place where the first Wolf was lying hidden, he sprang up and in a jump or two caught her neck and threw her down. At the same moment the two Wolves from behind came up, and for a moment there was a scuffle in which yellow and white and gray and waving tails were all mixed up, and then the three Wolves were seen standing there tearing away at their breakfast."

In the October of 1893, while living in New Mexico near Clayton, I had an opportunity of watching a joint hunt of Prairie-dogs by 2 Coyotes, no doubt a pair. Early in the morning I was on a rugged hill overlooking a plain on which was a Prairie-dog town. One Coyote was in an arroyo or dry watercourse hidden from view. The other walked openly and calmly toward a Prairie-dog that was barking vigorously

PLATE LXXII.—THE RELAY CHASE.

To illustrate the Coyote's manner of combining drive and ambush with relay chasing, without which ruse they could not capture the swifter Antelope.

on its mound. The Coyote paid little heed but walked so as to pass within 20 yards. The Prairie-dog dodged down. Then Coyote No. 1 continued his leisurely walk, while Coyote No. 2 rushed forward and hid behind another mound. Very soon the Prairie-dog began to peep out, and seeing the Coyote at a safe distance he scrambled onto his high outlook to hurl defiant little barks after the foe. But the Coyote behind sprang and all but caught him before he scrambled into safety.

In this case the combination failed, but evidently it must oftentimes succeed.

On October 3, 1902, while driving near Meeker, Colo., I saw a cow defending her new-born calf from a Coyote. The calf was able to stand, and 2 or 3 steers lent some aid to the cow. The Coyote walked about openly and quietly or sat on his haunches some 20 yards away. The cow and steers went on feeding but kept an eye on the Coyote, and the mother always managed to be between the calf and his foe. Occasionally one of the defenders would throw up his head, shake his horns, snort, and even run a few steps at the Coyote, but there was a marked absence of action in the little drama. Doubtless it would continue so, unless the villain got a chance to rush in and inflict a deadly wound.

SHEEP-KILLERS

Coyotes rarely molest calves or pigs in Manitoba, but they are very troublesome among other live stock. In the summer they kill many turkeys that roam afield far from the protection of the house dog, and sheep are particularly subject to their inroads. They have, indeed, nearly put an end to wool-raising in the Province. They destroyed all Barton's sheep, invariably selecting the fattest and best. At first it was found sufficient to put bells on several of the flock. But the Coyotes have got so far accustomed to these that the bells now protect only the sheep that wear them; some shepherds aver that these sagacious little Wolves will get up at night and listen for the sheep bells, so as to know where an easy supper is awaiting them.

George H. Measham writes me from Shoal Lake in December, 1899: "Wolves are on the increase and becoming a regular pest. Many people have abandoned keeping sheep on their account. Although a bounty of $2 is paid for them, they manage to keep out of danger in a way only equalled by themselves."

STORAGE

Like many others of the family, the Coyote has the frugal habit of storing food for future use; whether it can thus effectually hide it from plunderers, or whether indeed it always remembers the spot afterwards I cannot determine, though I think it unlikely that an animal with its high mentality, its sense of locality and fine nose, could fail of going to the spot at will.

The following incident witnessed by A. S. Barton is a good illustration of the storage habit.

"I was mowing hay in my coulée, when I noticed, some distance away, a Coyote carrying something in his mouth. He trotted down the hill and with some difficulty through the long grass, but presently stopped and began to bury his booty in a mole-heap, covering it with his nose, as a dog does. On my approach he decamped and watched my proceedings from the nearest hill, and, curious to know what he had been burying, I unearthed his cache, and found, to my surprise, a fine turkey gobbler, still warm and uninjured, except that its neck was broken. I had no time and less inclination to advertise for an owner, but accepted 'the goods the gods gave' and carried my prize home. Our next Sunday dinner was much appreciated, and we cheerfully drank the health of the purveyor, and of my unknown neighbour also."

OMNIVOROUS

While a hunter by profession and by choice, there is nothing in the way of fish, flesh, or fowl, ancient or modern, that the Coyote disdains for food. In the South-western States it has gone farther, developing there the watermelon habit, and I was not surprised to find it a fruit-eater in the far Northwest.

Sample dung-pellets gathered on the Athabaska River, Alta., in late October, 1907, were sent to the Biological Survey at Washington for analysis. The following remains in them were identified by Edward A. Preble and W. L. McAtee:

Rosa (*acicularis?*), many seeds; *Ribes*, a great many berries; *Aralia nudicaula*, many berries and seeds; *Microtus drummondi*, some hair (apparently) and portion of skull; *Peromyscus*, incisor (apparently *Peromyscus*); Fish, portion of skull of some fish; Water-beetles (*Corixa*), remains of a good many; Grasshoppers (*Melanoplus borealis*), remains of about 20; Ant, head of one.

The winter is, of course, the season of peril for all creatures that do not store up a full supply of food, or hibernate. Can it stand the winter? is the crucial test of all Northern species. Probably the chief thing that carries the Coyote race through is the new supply of food brought in by their enemy—the winter—that is, winter-killed sheep and cattle. These are dragged forth from time to time, and at each carcass half-wild dogs contend in nightly feast with Coyotes, or both retire while a big Gray-wolf fills his capacious belly.

There are several disadvantages in this food supply: it MANGE affords a certain place for traps and poison to be laid; hundreds of Coyotes and not a few dogs are thus destroyed every year. The flesh of horses is credited also with giving mange to Wolves that over-indulge. Epidemics of mange have been known among the Coyotes. The popular view is that they come from over-feeding on dead horses. I do not know the evidence for this explanation or against it. The mange speedily ruins an animal's coat, and in a Manitoba winter, of course, this means death.

At Carberry, in 1892, I was generally assured by residents that Coyotes were quite common still, although the Foxes were growing scarce. As many as 70 or 80 skins were to be seen hanging in Carberry at the same time, and the price for the primest had then dropped from $1.50 to 75 cents. A local farmer, Thomas Kerr, said that one winter a Prairie-wolf without any hair on it hid itself under his straw stack, and, although driven away by his dog, it returned and ensconced itself under the granary, where he shot it. It had some disease which had robbed it entirely of its hair, excepting a little patch on the shoulders, and it was trembling with cold.

Another instance which shows how hard pressed the Coyote is at times by hunger was related to me by another Manitoba resident, Robert McCullough.

At Carberry one winter's morning he went with a boiler in his hand to get some feed from the driving shed. The door was a little open, and he saw as he approached an animal which dashed about in the gloom of the shed. Its actions showed it to be a wild creature. He ran to the door just in time to prevent its escape. Barring the passage with the boiler, he rushed to the stable and back with a fork. The creature, a Wolf, took refuge under a reaper and there McCullough speared it, but the fork only pierced the loose skin of its belly, and it turned on its enemy, who held the fork to the ground with all his strength, and was barely out of the reach of the Wolf's jaws, for the handle was short, but he dared not withdraw the fork to strike again, and he had nothing to finish the animal with, so it was a deadlock. After a struggle, however, the man got the end of the handle fixed under a beam and rushed off to get a club. On returning, the Wolf was gone, apparently for good. But the next morning it was found within a few yards of the same place, quite dead, for the fork had pierced its bowels. But why should it return to the shed?

MENTALITY

The Prairie-wolf is mentally a compound of Fox and Wolf. While gifted with a good deal of cunning that is shown in its avoidance of traps and its method of taking its prey, it is also a desperate fighter when at all evenly matched.

I have more than once seen a Coyote run across an open stretch of black ploughed land, then on a piece of dry yellow grass sink into concealment. It matched it perfectly in colour, but was probably actuated by the idea that it was cover.

The Coyote is less shy and cunning than the Gray-wolf. I find the following characteristic note in my New Mexico Journal:

"Clayton, December 14, 1893. This morning I found that a small Gray-wolf had run my drag till he came to the first trap; there he turned aside, passing three cheese baits. A second very large Gray-wolf struck the drag just before the second traps. He passed them unhurt, then came to a cheese bait, urinated on it, and had then left the drag altogether. A Coyote that was following him on the chance of pickings, came on the bait and was kept from it by the treatment the Wolf had given it, but went on a mile and a half, picked up a poisoned cheese bait, and then half a mile farther got caught in the next trap, where I found him stark and dead."

It takes a wonderfully good dog to kill a Wolf. Yet I knew a collie, 'Old Frank,' the property of my neighbour, John Thompson, of Carberry, Man., that had several times run down and killed Coyotes single-handed. I saw him actually perform this feat in the November of 1882. The Wolf faced him again and again, but he managed each time to escape serious injury from its jaws, and when the Wolf turned to fly he would snap at its rear. On skinning the Wolf I found that the dog's teeth had sunk deep into the Wolf's flesh each time, so that its hind-legs were disabled. This Wolf, however, died gamely fighting.

In my early days I caught a great many Wolves in traps—many scores, if not hundreds—and I found great diversity of behaviour among them at this trying time. Some were utterly cowed, and submitted to the death penalty in sullen silence, others struggled to escape, some yelled defiance, and not a few barked and growled savagely, trying to reach me, raging and defiant to the end.

I have often known a Coyote to tempt a dog to chase him, then, at a safe distance from the dog's human backers, turn on him and drive him back with noisy demonstrations that looked like a wild practical joke.

On March 18, 1883, I had an experience that shows somewhat of the mind of the Prairie-wolf.

About 9 A.M. I was leaving the barnyard with team and sleigh to get a load of wood. As I rounded the stable I came into full view of a dead calf that was lying on the open prairie 75 yards away. A Wolf was tearing at the calf; he saw me plainly but went on with his eating. Of course I had no gun; I knew that if I stopped now to get a gun the Wolf would run. So I kept straight on. I passed within 30 yards of him; he watched me, but kept on eating. After I was 300 yards away I turned back by another road, intending to go cautiously to the house and get the gun, but the moment I left the beaten road that watchful Wolf seemed to divine my purpose, and ran as though already the lead were flying after him.

A similar incident is narrated by Dr. W. T. Hornaday.[11]

"The delicacy of the Coyote's judgment in keeping always beyond fair gun-shots is truly wonderful. If he is not a mind-reader his actions belie him. Twice in Montana, each time for two weeks, have I tried my utmost to shoot a Coyote; but during those periods not one would offer more than a running shot at three hundred yards or more. Twice, however,—and immediately after the above,—when riding quite unarmed, have Coyotes sat down beside the trail, waited for me to approach within forty yards, then yawned in a bored manner, and slowly trotted off. It is my belief that those animals knew perfectly well my inability to shoot."

This is an incident of a kind that has led many to credit this animal (and others) with the power of telepathy. Before accepting such an explanation we must have many cognate instances recorded by trained observers and also have a fuller knowledge of the animal's sense-capacity in an ordinary way.

FIGHTING

In August, 1886, I saw 2 female Coyotes fight in a cage. They snapped chiefly at each other's necks. The victor, however, disabled and subdued the other by a serious bite in the

[11] Amer. Nat. Hist., 1904, p. 24.

PLATE LXXIII.—COYOTE IN SUMMER COAT.
Life study by E. T. Seton, September, 1886.

fore-leg, then amused herself by growling and urinating elaborately over the fallen foe; a treatment to which the vanquished submitted with every appearance of abject fear.

This disposition to spurn and insult a conquered enemy seems peculiar to the dogs. Nothing of the kind is observable in cats. When a cat's enemy flies, the victor is done with him, but flight of the foe is the strongest incentive in a dog to pursue for a time at least and cover himself with glory.

The speed of the Coyote is great, and has often been the subject of admiring comment, but I think it has been overrated. After collecting data of various kinds, such as actual known records of dogs and horses, also the comparative records of dogs and Hares, or horses and Foxes, Wolves and hounds, hounds and automobiles, I have attempted a scale of comparative speeds. This appears in a different form in the Antelope chapter, p. 233: SPEED

Blood Race-horse	covers	a mile	in about	1	minute	40	seconds
Pronghorned Antelope	"	"	"	1	"	50	"
First-class Greyhound	"	"	"	2	minutes	0	"
Jack-rabbit	"	"	"	2	"	10	"
Common Fox	"	"	"	2	"	20	"
Northern Coyote	"	"	"	2	"	30	"
Foxhound	"	"	"	2	"	40	"
American Gray-wolf	"	"	"	3	"	0	"

Many hunters would set the Kit-fox or Swift yet above the Greyhound, especially for a short race, but I have had no personal experience with the species in a chase. The little Prairie Cottontail, can, I believe, get away from the Swift in a hundred yards dash; they cannot keep it up for long, but their initial velocity is incredible and baffles the eye, not a leg, not a Rabbit is to be seen, nothing but a white streak across the prairie, till it promptly disappears in some burrow.

What actually counts in the race is, as usual, the *trifle more* speed that each animal can command.

For example, the Gray-wolf makes 650 yards to the minute and the Coyote about 700. But that 50 yards makes all the difference between living and dying. That 50 yards margin is probably the foothold on which the whole Coyote race has been built up.

These rates, it will be seen, trench on the especial realm of birds: small birds make only 25 to 30 miles an hour.

It is a well-known principle that the special development of an animal is its most variable part. Thus the peculiar bell in the throat of a Moose varies enormously; the bill of the Long-billed Curlew, the neck-feathers of the Ruff, the spots of the Ocelot, the white bands of the Skunk, the horns of the Elk, are so varied that rarely two are found just alike. Speed is one of the peculiarities of the Coyote as it is of the greyhound, and we must, therefore, look for great variations of rate. I have selected an average for my calculation, but there are occasional individuals, Coyotes of rare gifts, whose speed and endurance would put them very near the top of our scale.

An individual of this description lived for three years on the north slope of Turtle Mountain, near Boissevain, Man. He was known as the 'Greyhound Coyote.' A. S. Barton hunted him many times with first-class greyhounds, dogs which ordinarily had no difficulty in catching a Coyote, and, though the chase was several times over open prairies, he has always left the dogs behind in a straight three-mile run, and safely reached his retreat in the wooded ravines of Turtle Mountain, thanks, not to any stratagem, but to his speed. It is not known what his end was; he may, indeed, be living yet.

INCIDENTS

Barton relates a curious instance of a crippled Coyote living for the last two years near Boissevain. It is known as the 'Three-legged Terror.' One of its front legs is missing, probably it was lost in a trap, but in spite of this the creature can outrun an ordinary dog. Greyhounds or very fast dogs easily outstrip it, whereupon it finds a place to protect its rear and presents such a desperate front that it has hitherto escaped.

I suspect that this is a female, which might partly account for its immunity.

The winter is, of course, the chief enemy of the Northern Coyote. The shutting off of many food supplies, the severe weather, the exposure to view of the hunters, poisoned baits which in summer would be scorned but which are now swallowed in desperation, all unite to make havoc in the numbers, and those that are left by the end of February are the strongest and wariest. Next after winter, the worst enemy of the Coyote is man; next to this, dogs; next, disease and parasites; then probably the Gray-wolf, the eagle, and the horned owl. ENEMIES

It is improbable that these birds would attack a grown Coyote, or that the Gray-wolf could catch one, but the very young would fall an easy prey.

The Elk and Deer are to be reckoned in this list. They have well-founded hatred of all Wolves, and never fail to strike one when they can. One blow from the foot of an Elk or Deer, or even an Antelope, may disable a Coyote, and give the hoofed avenger a chance to finish his work.

I have never heard of a sane Coyote attacking man.

The following adventure with a mad Coyote was recorded by Malcolm Little, of Provo, Utah, and sent me by Mrs. S. Young Gates of the same city. MAD COYOTE

"The summers of 1891 and 1892 were extremely dry and hot in northern Mexico. A two years' drought had left the extensive valleys barren. The grass was crisp and bleached, the dust heavy and rose in clouds, and the Casas Grandes River was dry save for a few deep, stagnant pools. This being the only water within many miles, all the animals of the immediate vicinity came there to drink. As a result, on each side of the river, for a great distance, hardly a spear of grass could be found.

"The Coyotes were very numerous. They seemed to have been drawn from all the surrounding country to these watering places. So numerous were they, in fact, that in a few hours a man sitting in a tree near one of the pools shot 13 that came to drink. They roamed about in droves of from eight to twelve in search of food. They were gaunt and hungry-eyed, and their fur was long and shaggy. They appeared to have lost nearly all the sense of fear of man, for they would hardly move out of the way of a horseman. They came into the towns and several were killed in the yards of the dwellings. One, I remember hearing of, walked into the parlour at the front door, and while yet inside was killed with a garden hoe by the lady of the house. These many little encounters, however, were considered as only matters of pleasant gossip, and no one thought of any possible danger coming from them.

"In late June of 1892 three men camped for the night midway between the village of Ascension and the Boca Grande, near one of the watering places on the river. They were on the range in search of cattle, and, as is common with cowboys, after the horses were hobbled out and supper over, they spread their blankets on the ground. A wide bed was made and about nine o'clock they lay down to sleep, Derby Johnson occupying one side, and the Jacobson brothers the other two places.

"One of the Jacobsons was anxious about the horses and did not sleep soundly. In a few hours he was startled from a state of semi-wakefulness by a muffled sound, and seeing Derby sitting up in bed, and thinking something might be wrong with the animals, began to sit up also. As he did so he saw what appeared to be the tail of a Coyote moving to and fro. He sprang to his feet. The animal with its teeth fastened on the right jaw of his friend, just to the right of the chin, was clinging with the tenacity of madness; while Derby, apparently frozen by awful sensations, sat clasping the Coyote by the neck with both hands, one on each side. It was clear the only way the animal could be removed was by prying open its mouth. The brothers were strong men. Unconscious of the danger to themselves, or disregarding it, they took hold of the fastened jaws and broke them apart, the lower one being entirely wrenched from its place. They threw the Coyote to one side, but, with its lower jaw dangling downward and with menacing growls, it came towards them again. The young man,

Derby, now freed, took his knife from his pocket, cut the animal's throat, and then fainted.

"One of the brothers remained with him while the other went for the horses. About eleven o'clock that night they started towards home, a distance of fifteen miles. Derby could ride only a few hundred yards without having to stop to rest, and very frequently the journey was stopped by his fainting.

"They reached home early in the morning. Examination showed the Coyote had left a bad wound under his chin, where the teeth had penetrated, while in the upper part were only holes of the canines.

"In a few days the patient was around, apparently well, and doing his work.

"After the biting the Coyote was examined, and only dry cottonwood leaves were found in its stomach. From all appearances it had had the rabies.

"About a month after the occurrence the young man was riding with his companions, gathering cattle from out of a lake. The water was shallow, hardly knee-deep to the horses, and the cattle had gone there to feed on the rushes. All day this work continued, and while splashing through the lake he felt the first symptom of hydrophobia—a strong aversion for water. This feeling later was aggravated by some of his friends offering him their canteens from which to drink.

"He returned home immediately, and for a while was delirious. These spells continued intermittently. He grew worse for two or three days, constantly developing a still stronger dislike for all liquids, and he was able to swallow but very little. Towards the last his ravings became maniacal, and several men were required to hold him. A viscid secretion came from his mouth, the colour of his skin became purple, and his ravings were very loud—the latter, however, had nothing like barking about them, contrary to the common belief in cases of hydrophobia. During the last hour of his life he was quiet, and he died peacefully.

"After this happening a 'scalp hunt' was gotten up, and 109 Coyotes were killed."

VOICE

The voice of the Coyote is one of its most remarkable gifts. Barking is supposed to be limited to the dog and Coyote. This is not strictly true, for Wolves, Foxes, and Jackals bark at times, but it is true that the Coyote is the only wild animal that habitually barks.

We must assume, as general propositions, that nothing in nature is without adequate cause, and that it is always worth while to search that out. Most of the many calls of the Coyote are signals to its companions, but some of them seem to be the outcome of the pleasure it finds in making a noise. The most peculiar of its noises is the evening song, uttered soon after sunset, close to camp. This is a series of short barks, increasing in power and pitch till it changes into a long squall. One Coyote begins and immediately two or more join in, making so much noise that newcomers think there must be a hundred Wolves out there. It is kept up for perhaps a minute or two, then ceases till some new impulse seizes them. August 27, 1904, in W. F. White's menagerie at Winnipeg, I saw a Coyote pup, which, though little bigger than a house cat, and less than three months old, had a fully developed voice, and, much to the amusement of numerous bystanders, joined in the yapping chorus as lustily as his grown-up relatives.

Another note I have heard them utter towards dawn is a long, smooth sound, of truly musical quality. I have sometimes mistaken it for the fluty call of a loon to his mate.

I once knew a Coyote that would stay around the ranch till the small dog went valiantly after it. The Coyote would run till at a distance that made it safe from guns, then turn on the dog and drive him back ignominously to the shelter of the house. Of course, the dog soon learned that the enemy was not so 'easy' as he looked.

On each occasion when the Coyote turned, he uttered a series of gurgling, growling barks, that seemed to strike terror into the dog, and were to me an entirely new Coyote 'song.'

The sound the old one utters when the young are in danger is described by A. S. Barton as a loud, short, rough squall.

PLATE LXXIV.—COYOTE IN WINTER COAT.
In New York Zoological Park.
From a photograph by Elwin R. Sanborn.

PLATE LXXV.—COYOTE DEN.
In the sandhills, Carberry, Man.
From a photograph by E. T. Seton.

In their vespers he also notes an interesting habit. Two or three Coyotes will meet each night on a certain elevated place to sing. They have several of these recognized choir-lofts, but they never use the same on two nights in succession. Sometimes on Turtle Mountain, in dead calm moonlit nights, each Coyote gets up on his singing perch and pours out his loudest and finest notes. This is passed on from one point to another, till the whole mountain seems ringing with the weird music, and, from its very wildness and the vast stretch of the country that is concerned, the effect is truly impressive.

In captive animals these simultaneous outbursts are often observed. A favourite time is at noon, when the blowing of whistles seems to be the immediate cause, or possibly only the last touch that precipitates the event.

This species readily crosses with the dog. S. L. Bedson showed me a number of these hybrids at Stony Mountain, Man., in 1885. They were intermediate in character and continued to be interfertile with either stock, at least, for two or more generations. HYBRIDS

I never, however, heard of a cross between a Coyote and Gray-wolf, although the dog is as readily crossed with the latter as with the former relative.

Coyotes are kept down by poisoning, trapping, and hunting. KILLING

Poisoning, however, is now forbidden on account of the number of dogs destroyed.

Trapping is carried on as with Foxes, for the Coyote is quite as cunning as its red-haired cousin. Not less than a No. 2 trap will do, and it must be fast to a drag of 20 or 30 pounds weight, never staked solid. Trap, log, and all must be de-odorized by smoking and rubbing with blood. They must be carefully concealed in the soft ground and scraps of meat scattered about, but not on, the trap. A lump of meat buried *under* the trap is a sure attraction. The certainty with which they dig out buried meat, leads me to believe that the Coyotes are very lax in their ideas of property rights where a cache of eatables is concerned.

PELT The pelt of the Coyote should be cased. The more complete it is the better. It is prime from October to April, and, according to the Winnipeg market quotations of March 26, 1904, was worth $1.00 to $2.00. After consulting with many country storekeepers in the leading parts of Manitoba, I reckon that not less than 1,000 skins are shipped each year from the Province. It is a beautiful fur, but not durable enough for a rug.

XXXV.

The Canada Otter.

Lutra canadensis (Schreber).
(L. *lutra*, an Otter; *canadensis*, of Canada.)

Mustela lutra canadensis SCHREB., 1776, Saug. pl., CXXVI B.
Lutra canadensis SABINE, 1823, Franklin's Overland Journey, p. 653.

TYPE LOCALITY.—Eastern Canada.

FRENCH CANADIAN, *la Loutre du Canada.*
CREE, OJIB., & SAUT., *Ne-geek'.*
MONTAGNAIS (Gulf of St. Lawrence), *Un'chuch.*
CHIPEWYAN, *Nop'-e-ay.*
YANKTON SIOUX, *Pe-tang.*
OGALLALA SIOUX, *Ptan.*

The Weasel Family or *Mustelidæ* are carnivores of long form on short legs, with tail various, ears short, non-retractile claws, and teeth 32 to 38.

The genus *Lutra* (Brisson, 1762) comprises Weasels of large size, with short legs, long tails; adapted for life in the water, having dense oily fur, webbed feet, etc. The teeth are as follows:

$$\text{Inc.}\ \frac{3-3}{3-3};\ \text{can.}\ \frac{1-1}{1-1};\ \text{prem.}\ \frac{4-4}{3-3};\ \text{mol.}\ \frac{1-1}{2-2} = 36$$

In addition to these characters, the Canada Otter has the feet with more or less hairy soles.

The Muskrat is protected against the cold water by a fine robe of dense fur, the Whale by a thick layer of fat under the skin; the Otter is happy in the possession of both, and can enjoy the coldest of water in the coldest of weather.

SIZE Length, about 40 inches (1,000 mm.); tail, 12½ inches (317 mm.); hind-foot, 4 inches (102 mm.).

WEIGHT A female collected by Dr. Merriam, on Birch Creek, Ida.,[1] August 14, 1890, weighed 19 pounds (8½ kilograms); a young one with her weighed 10 pounds (4½ kilograms); both were fat.

COLOUR In general, the colour is dark rich glossy brown, becoming paler and grayer below; the brown of the head and muzzle changes on lips, cheeks, chin, and throat, rather abruptly, into a pale brownish-gray, almost a grayish-white. Some specimens are much paler.

In the American Museum is an albino, or nearly white, freak from Alaska.

The following races are recognized:

canadensis, the typical form.
lataxina F. Cuvier, smaller.
vaga Bangs, larger and redder than *canadensis*.
pacifica Rhoads, pale in colour.
sonora Rhoads, a large and yellowish race.

Besides which are two closely allied insular forms:

Lutra degener Bangs, very small; found in Newfoundland.
Lutra periclyzomæ Elliot, a large kind from Queen Charlotte Islands.

Life-history.

RANGE The range of this species includes nearly the whole continent. It is, or was, found in all parts of Manitoba, though its numbers are greatly reduced there to-day.

ENVIRONMENT Frequenting invariably the water, or the vicinity of water, the Otter finds its ideal surroundings in good-sized clear

[1] N. A. Fauna, No. 5, July, 1891, p. 82.

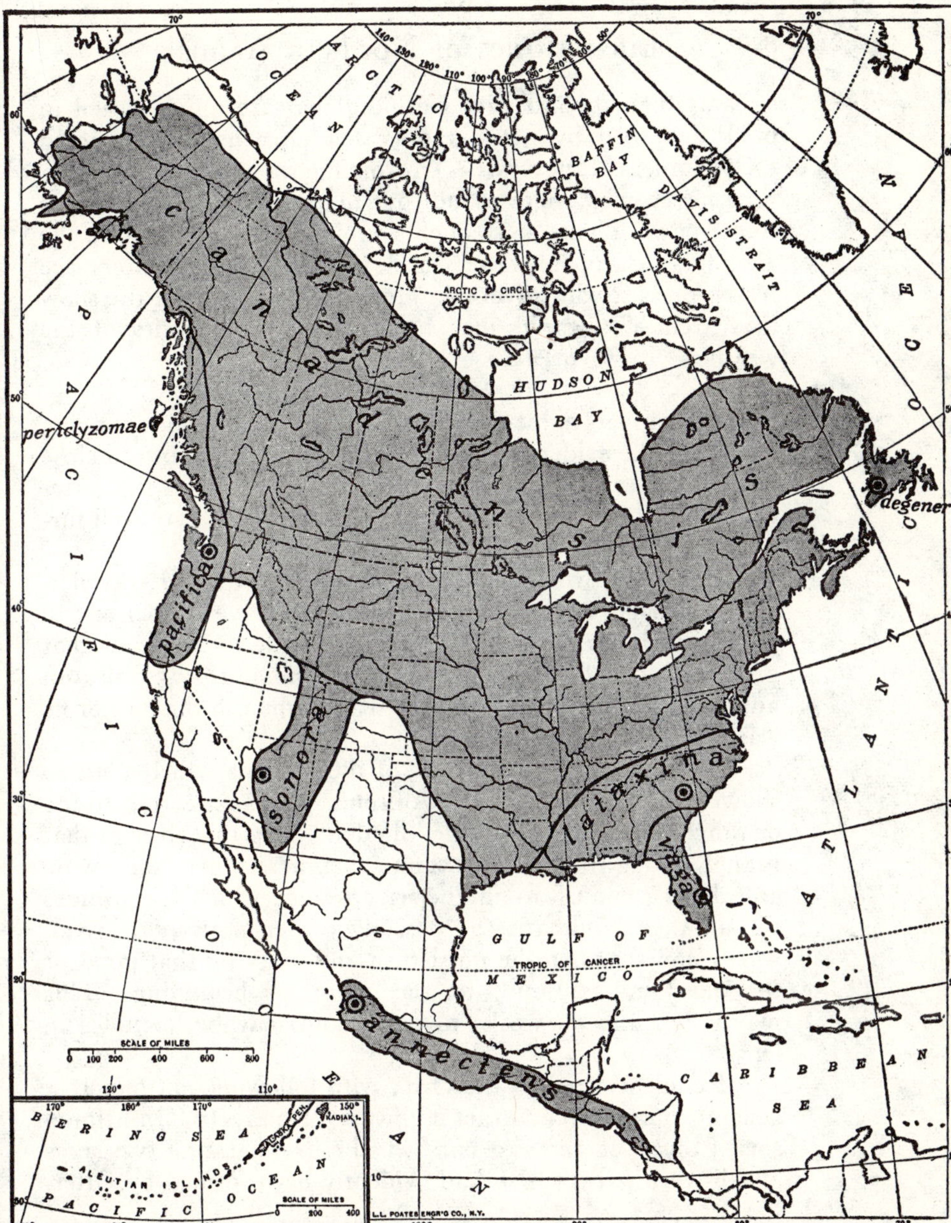

MAP 44—RANGE OF THE NORTH AMERICAN OTTERS.

(Excluding the Sea-otter)

This map is founded chiefly on records by E. Coues, C. Hart Merriam, D. G. Elliot, E. W. Nelson, R. MacFarlane, C. H. Townsend, E. A. Preble, J. Fannin, V. Bailey, A. P. Low, A. E. Verrill.

The lines are fairly well established, except in the Southwest. I find no records for the unmarked region. Anticoste should have been tinted.

The following are recognized:

Lutra canadensis (Schreber) with 5 races.

Lutra degener Bangs,

Lutra periclyzomae Elliot,

Lutra annectens Forsyth-Major, from Central America.

streams that abound with fish, and that are much varied in bank and bed by pools, rapids, log-jams, and overhanging rocky banks. It never *lives* far from the water, and in summer-time never goes far overland, but in winter its habit changes somewhat. Then "it frequents rapids and falls, to have the advantage of open water; and when its usual haunts are frozen over, it will travel to a great distance through the snow in search of a rapid that has resisted the severity of the weather."[2] (*Richardson.*)

HOME-RANGE All trappers agree that the Canada Otter is a wide-ranging animal. The evidence goes to show that its habits bear a close resemblance to those of its British congener, and as this latter has been closely studied by many naturalists, it may well furnish important side-light.

In Great Britain, it is well known to be a far traveller. T. W. Proger, of Cardiff, writes me: "I have known one to go 5 or 6 miles overland in a single night to a stream that promised good fishing. An Otter will range for 25 miles up and down a given river; and the scarcer the fish the farther he wanders."

Merriam gives[3] a number of illustrations which tend to show that each Otter in the Adirondacks has a certain route or range that is his own little kingdom. Up this river to that branch, along that to the swamp, then across over the divide and down some rill to another river, along which it continues till another landmark, or possibly owner-mark of a rival, warns him that here he must turn and cross by that pond or the old familiar rapid, to the point of beginning. This may take him two weeks to cover, and may be 50 miles in length.

Kennicott credits[4] the Otter with following not merely a general course, but an exact pathway. He says: "In Minnesota, I observed across a narrow isthmus separating two lakes a well-worn path, which had evidently been formed by Otter,

[2] F. B. A., 1829, I, p. 57.
[3] Mam. Adir., 1884, p. 88.
[4] Quad. Ill., 1859, p. 247.

and they seem generally to follow beaten paths when moving on land." ABUNDANCE

Possibly each individual has more than one of these routes.

In Manitoba, the Otter is becoming very scarce; the high value of its fur has told hard against its numbers. In the palmy days at the beginning of last century, Henry collected from 100 to 300 skins each year on Red River. The present output of the region is much smaller.

What, then, should be considered a fair population of Otter? In Essex, Eng. during the fall of 1904, I am told that 50 Otters were killed, though no one but the Otter hunters suspected their presence. As this 50 was far from exterminating them, there were probably at least 100 Otters in Essex, or one to each 16 square miles, and Great Britain, in like ratio, would show a total of 10,000 Otters. This is the reverse of a high rate of population. In Manitoba, they were very common at one time, as already noted, and I think it safe to say that in the primitive days there was an Otter for every 5 or 6 square miles, or at the rate of 3 pairs to the township. At present I doubt that there are 300 pairs left in the Province.

SOCIABILITY

Like the rest of the family, the Otter is neither sociable nor gregarious. The 3 or 4 that are often seen together in late summer are the mother and family. In their curious sport of sliding down hill, we possibly may find an exception to this rule, although some observers consider that the game is strictly an affair of the family or of mates.

SOUNDS, ETC.

The species makes a variety of noises. It utters a loud sniffing that sounds like clearing its nose of water, and it growls and snarls in menace. A female in the National Zoo at Washington, obtained in northern New York, often emitted a loud birdy chirp to express enquiry, desire, or hunger. Another female that I was sketching at the same time (April 28), made a low chatter or querulous grumble that seemed to express the same idea. The latter was from Florida.

A captive Otter kept by J. K. MacDonald, of Winnipeg, in 1886, at Bersimis on the Gulf of St. Lawrence, used to utter such a piercing whistle that my informant repeatedly heard across the river (a mile and a half away), as plainly, he said, as he could hear a man whistle if in the same room with him. He knew of no other animal sound so shrill, save the scream of the eagle or the loon.

MATING Most observers agree that the species pairs and that the mating season is towards the end of February.

The British Otters bred in captivity by A. H. Cocks gave an admirable opportunity for observation. The female showed that she was willing to consider a proposal of marriage by leaving little piles of well-mouthed straw here and there in the cage. These, as with the Marten, he believes, are to call the attention of the male. In a state of nature they would be left at convenient spots along the banks of the river or other main line of travel.

The male showed none of the savage ferocity of the male Marten. His behaviour, in general, was irreproachable.

DEN "It digs a burrow on the bank at the edge of the stream or lake, with the entrance under water, like that of a Muskrat. Its burrow is not so extensive, however, as that of the latter animal, and it never constructs a house of any kind. In a capacious chamber in its burrow, it forms a large soft nest of sticks, leaves, and grass; though sometimes it has been observed to take up its quarters in the bottom of a standing hollow tree or in a cavity in a fallen one."[5] (*Kennicott.*)

The British Otter is known to have a kitchen-midden, or garbage hole, near the den door, but this admirable sanitary arrangement has not yet been observed in our own species.

GESTATION The gestation of the British species was ascertained by A. H. Cocks to be 61 days. No doubt the period will be found about the same in *canadensis*.

[5] Quad. Ill., 1859, p. 247.

The young are born in mid-April, or sometimes as late as May 1. They usually number 1 to 3, but a female taken at Brokenhead and dissected by W. R. Hine contained 5 embryonic young. These were the size of a small Striped-gopher, and must have been near full time, as it was late in April. At birth and for some weeks afterwards their eyes are closed. Their colour is said to be very dark brown, almost black. Probably they are not weaned till four months old, and at nine months they are fully adult. But one brood is reared each year. YOUNG

Much discussion has taken place over the question whether or not animals train their young. It seems as though in very ancient forms retaining primitive habits, the young need little or no instruction from parents. Thus, an incubator duck will take to the water or snap at a fly when but a day old. On the other hand, those animals with highly specialized habits are slow to learn, and need some sort of stimulus. The young hawk or Weasel speedily learns to seize a bird, but the young osprey and Otter have departed further from the ancient way and are more in need of teaching. Whether this be conscious or unconscious on the part of the parents depends on our definition of these terms. TRAINING OF THE YOUNG

An interesting picture of their nursery life has been given by J. G. Millais, who studied Otter in Canada and in England, though, unfortunately, he omits the dates and places that would have added so much to the value of his remarks. His account, no doubt, refers chiefly to the British Otter, but it is almost certain to be found applicable in the main to our own species, when fuller observation shall have enlightened us on the early history of the young.

"As soon," he says,[6] "as they can see, the mother Otter takes her cubs to the water and teaches them to swim. At first they are said to be very reluctant to enter the water, and as a preliminary training, she often makes an exit hole upon the bank above her *holt*, where she allows them to play and run

[6] Mam. G. B. & I., 1905, II, pp. 19–20.

about for a few days before enticing them to mount on her back and embark on a voyage of discovery. For the first few days in the water she swims with them, but soon dives and returns to them again and again, until she has induced them to copy her movements.

"In their preliminary efforts at natation, young Otters are just as frightened as they can be, and keep querulously calling for their mother all the time she is out of sight. But at first she does not upset their baby natures, and only vanishes for a few minutes. As the young grow, these intervals become longer and longer, till she induces them to follow her in shallow water or in a still lake. I once had the good fortune to see an old female Otter playing with her three nearly full-grown young ones and evidently teaching them to dive.

"She teaches them to dive noiselessly, to circle in deep pools, and how to come up quietly behind sleeping fish or drive them into holes in the banks. Then they are taught to stir the mud with their pads, or turn over stones for hidden miller's-thumbs, and bury their heads in the mud after eels, or how to corner the darting salmon.

"That the swimming powers and the hunting of fish are acquired habits is shown by the fact that young Otters kept tame and allowed to run loose are almost full grown before they will take to the water; they grow up with Stoat-like habits, *i. e.*, hunting for their food on land."

A. H. Cocks says[7] of those (British) he bred in confinement that the young were blind until about 35 days old and entered the water of their own accord on the 58th day. He gives a suggestive account of the mother's efforts to make them eat two small fish, some four days later, "taking first one fish, then the other, then both together in her mouth, and moving them about close in front of the cubs to attract their attention, at the same time uttering a peculiar whine or growl."

SUMMER During the summer, the Otter family, mother and young, may be seen travelling and hunting together.

[7] Zoölogist, 1882, p. 203.

A glimpse of their life at this time was secured by Bert A. Dobson, of the Adirondacks.

One year, 1900, in early June, while fishing on Peavine Creek (Cranberry Lake), he saw a female Otter and two kits, one-third grown. She uttered a loud chirruping and, dashing down the creek, which was not deep enough for swimming, she led the young off, chirruping and clucking to them like a hen.

In Algoma, the young are seen with the mother in June, July, August, September, October, November, and December, but usually only one is near. Linklater, my chief informant in that region, has known the mother and 3 young to be killed at one shot, in September.

FATHER

It is always important to know the relation that the father bears to the family. Is he merely a selfish progenitor, concerned about his partner or partners only in the mating season? Or does he faithfully play the part of a helpmate and join with the mother in caring for the young.

It is impossible to decide for the Otter. Most observers think that the species pairs, that is to say, is not polygamous, but that the bond is broken with the waning of the honeymoon. Miles Spencer expressly says[8] that, about Hudson Bay, the female gets no assistance from the male while rearing the young. Other field naturalists think that the male does sometimes join in caring for his offspring. J. K. MacDonald, of Winnipeg, after many years among the Hudson's Bay Company's Posts, writes me:

"I do not think the male is ever far away from his hole or family. The fact that the female is seen with the young more often than the male, I think, simply carries out the natural law of progeny being more directly under the care of the female, while the male roams about, but never is far away. In the cases cited of females only being seen, there is nothing to disprove that the males, though not visible to the seer, may have been within a few yards of him."

[8] Low Expl. James Bay, Can. Geol. Surv., 1888, Pt. J, App. III, p. 77 J.

On September 27, 1905, George Crawford, the Mattawa guide, came to my camp at a place 40 miles north-east of Mattawa, and told of 4 Otter that he had watched the day before on one of the small lakes near by. They were 2 old ones and 2 young ones. He was quite sure that 2 were fully grown. Again, in October, 1904, Archie Miller saw 2 adults and 5 young together at Blue Lake, Quebec. This looks as though the father did sometimes accompany the family.

Nelson, in his "Natural History of Alaska,"[9] also says: "Toward the end of the winter they frequently make a burrow in a large snow-drift, and sometimes a party of 5 or 6 will be found occupying the place. Such parties usually consist of the 2 old ones and the last season's young."

Thus there is some evidence that the male Otter does not entirely neglect the duties of paternity, though there can be no doubt that the mother has the heavier burden and sometimes bears it alone. The young continue in her care until the winter; probably the bond is broken by the coming on of the new mating season in February.

HABITS

Like most of the family, this animal is chiefly nocturnal in habits, but can very often be seen hunting in broad daylight. It resembles its kindred also in its tireless activity. I do not think I ever saw an Otter lying still. The nearest to it are those that are busy feeding. When they rest I do not know.

SWIMMING

"It is eminently aquatic, more perfectly so than the Beaver, and its whole form is admirably adapted to a life in the water. The great weight of its bones, which would make it more clumsy on land, enables it to dive and swim under water with ease."[10] So says Kennicott.

In like strain also, the Chief Mittigwab, of Mattawa. He has several times tried to overtake the Otter while they were swimming in the open lake. Though paddling his best, about six miles an hour, he could not lessen the distance between. They swam like leeches, rolling over and under. He con-

[9] 1887, p. 250.

[10] Quad. Ill., 1859, p. 247.

siders that there is no creature so quick in the water as an Otter, except, perhaps, a loon. He tells me further that on September 17, he saw a young one and its mother capture a large speckled trout each, in Magnisipi River, Quebec, and, like all who have seen the Otter afloat, he gives enthusiastic expression to his admiration of its powers. It darts and turns, diving or floating, with speed that none of the river tribes can excel. Not only the glancing trout, but also the lightning swift salmon, are at its mercy. These it meets fairly and squarely in their chosen grounds, the clearest, coolest swirls and deeps, or the open flood, and, rejoicing in a worthy foe, it beats them at their own game, and glories, we have excellent reason for believing, not less in the noble catch than in the noble sport.

As well as quick, it is long-winded. A tame one belonging to "Antler" could remain under water for three or four minutes.[11] Merriam says[12] that in this respect its abilities almost equal those of a loon, and he has known an Otter to swim nearly a quarter of a mile without showing its head above the surface.

With such aquatic powers at command and with a keen sense of locality, it is easy to believe these hunters who maintain that the Otter can live for days under the ice of a lake, getting its breath at the cracks along shore.

RUNNING

On the land, though less at home, it is far from being helpless. All the records show that in snow time it travels great distances across country, and can go so fast that it takes a swift-footed man to overtake it.

ON SNOW

The marvellous power of this species to propel itself toboggan wise over the snow is something that I was deeply impressed with when a child, in Toronto, about 1873.

A citizen had a tame Otter that I was privileged to watch once or twice. I thought it the most beautiful creature I had ever seen, as it gambolled and dodged about the room in a spirit of good-natured frolic. It was entirely without the sullen ferocity of the Weasels, and when it was allowed to go

[11] Forest and Stream, December 11, 1879.

[12] Mam. Adir., 1884, p. 87.

FIG. 202—Otter poses, from life.

outdoors it rolled and tumbled in a snow-drift with evident delight. Sometimes it would run and slide on its breast with all its legs set backwards. Its progress at these times was singular, and continued for much longer than one might expect from the very slight push it began with. It seemed as though the muscles of the breast and belly were in some way helping it on, possibly the tail also aided with its sculling motion.

Richardson, the great authority on our northern animals, says,[13] concerning its winter travel overland in search of open water: "If seen and pursued by hunters on these journeys, it will throw itself forward on its belly and slide through the snow for several yards, leaving a deep furrow behind it. This movement is repeated with so much rapidity that even a swift runner on snow-shoes has much trouble in overtaking it. It also doubles on its track with much cunning and dives under the snow to elude its pursuers."

These very manœuvres were described to me by Linklater. He has often seen Otter sliding on the level snow, and once when he came on one in the woods it dived into the drifts, which were deep, and came up 40 or 50 feet away. By doing this again and again it dodged both the man and the dog, although the latter was a good one and the former on snow-shoes, till it reached a small lake; unfortunately for the Otter, this was frozen over, and on the ice the hunter killed it with a small club. It was a male Otter, and the time about the first of December.

"On the ice [says Merriam][14] they proceed by a series of what small boys called 'a run and a slide,' that is, they make several jumps and then slide ahead, flat on their bellies as far as their impetus and the smoothness of the ice permits, and then do the same thing over again, and so on." I may add that this seems to be their regular mode of progression, whether on land or water, ice or snow.

OTTER SLIDES

This method of travel brings us to a remarkable habit for which the Otter is celebrated. All hunters and naturalists in Eastern America record its singular amusement of coasting or

[13] F. B. A., 1829, I, pp. 57–8.

[14] Mam. Adir., 1884, p. 89.

tobogganing down a steep hill, preferably into the water, to remount at once and repeat the performance again and again, in company, perhaps, with its mate or young.

It is delightful proof of growth and uplift when we find an adult animal setting aside a portion of its time and effort for amusement, and especially for social amusement. A large number of the noblest animals thus relax from sordid life and pursue amusement with time and appliances after a fashion that finds its highest development in man. This is what the Otter is credited with doing, but there are naturalists who do not consider it proven. Therefore I give the evidence in full.

"The Otter is very fond of play, and one of their favourite pastimes is to get on a high ridge of snow, bend their fore-feet backward, and slide down the side of it, sometimes to the distance of twenty yards."[15] (*Hearne.*)

"Their favourite pastime is sliding, and for this purpose in winter the highest ridge of snow is selected, to the top of which the Otters scramble, where, lying on the belly with the fore-feet bent backwards, they give themselves an impulse with their hind-legs and swiftly glide head-foremost down the declivity, sometimes for the distance of twenty yards."[16] (*Godman.*)

This looks like a passage borrowed from Hearne, but he adds, "This sport they continue, apparently with the keenest enjoyment, until fatigue or hunger induces them to desist."

"In the summer this amusement is obtained by selecting a spot where the river bank is sloping, has a clayey soil, and the water at its base is of a comfortable depth. The Otters then remove from the surface, for the breadth of several feet, the sticks, roots, stones, and other obstructions, and render the surface as level as possible. They climb up the bank at a less precipitous spot and, starting from the top, slip with velocity over the inclining ground and plump into the water to a depth proportioned to their weight and rapidity of motion. After a few slides and plunges the surface of the clay becomes very smooth and slippery, and the rapid succession of the sliders shows how much these animals are delighted by the game, as

[15] Journey, 1792, p. 376.

[16] Am. Nat. Hist., 1826, Vol. I, pp. 225–6.

well as how capable they are of performing actions which have no other object than that of pleasure or diversion."

"This statement," say Audubon and Bachman,[17] "is confirmed by * * * recent writers who have given the history of this species, and is in accordance within our own personal observations.

"The Otters ascend a bank suitable for their diversion, and sometimes where it is very steep, so that they are obliged to make quite an effort to gain the top; they slide down in rapid succession where there are many at a sliding place. On one occasion we were resting ourselves on the bank of Canoe Creek, a small stream near Henderson, which empties into the Ohio, when a pair of Otters made their appearance and, not observing our proximity, began to enjoy their sliding pastime. They glided down the soap-like muddy surface of the slide with the rapidity of an arrow from a bow, and we counted each one making 22 slides before we disturbed their sportive occupation.

IN ALL LATITUDES

"This habit of the Otter of sliding down from elevated places to the borders of streams is not confined to cold countries, or to slides on the snow or ice, but is pursued in the Southern States, where the earth is seldom covered with snow, or the waters frozen over. Along the reserve dams of the rice-fields of Carolina and Georgia these slides are very common. From the fact that this occurs in most cases during the winter, about the period of the rutting season, we are inclined to the belief that this propensity may be traced to those instincts which lead the sexes to their periodical associations."

Kennicott follows with these remarks:[18] "It climbs to the top of some steep bank, made slippery by the mud and water from its own body, or, in winter, by snow and ice, and, lying down, with its fore-feet bent under, slides headlong to the bottom. Trappers inform me that they have often seen the Otter thus engaged for an hour or more, scrambling eagerly to the top again after each descent, and greatly enjoying the sport. By using their knowledge of this peculiarity, the hunters

[17] Q. N. A., 1849, Vol. II, p. 8.

[18] Quad. Ill., 1859, pp. 247–8.

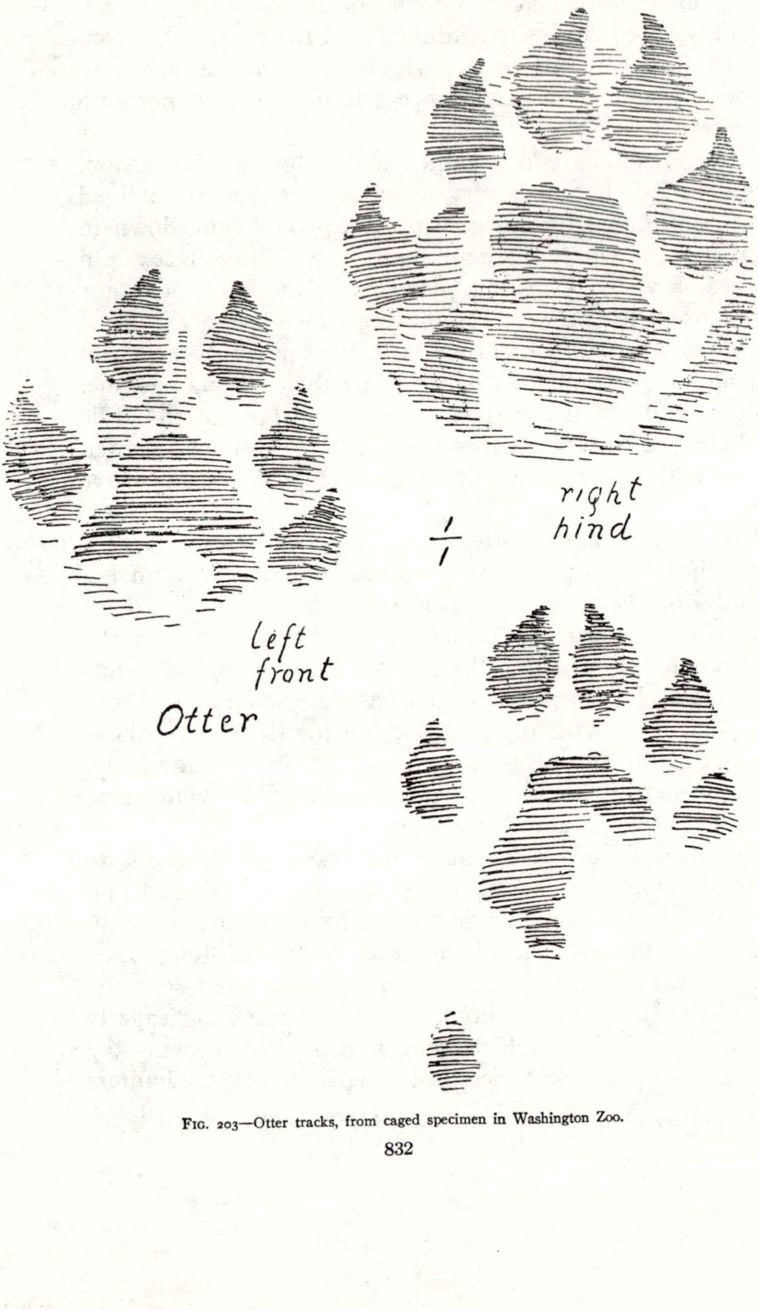

FIG. 203—Otter tracks, from caged specimen in Washington Zoo.

sometimes succeed in shooting or trapping the Otter at its sliding place, which may be easily recognized."

Similar testimony is supplied by my Adirondack guide, Bert A. Dobson.

One day in September, he was hunting Deer near Cranberry Lake, N. Y., and heard a loud splashing. On crawling near he saw 3 Otter (mother and 2 young) shooting down a slide into the water. They did it two or three times each before they smelt him and dashed into a hole.

Wherever I seek in Eastern America I find corroboration of this. Archie Miller, guide, from Mattawa, says that he has seen hundreds of these slides, and, no later than October, 1904, he watched a family of Otters at their slide at Blue Lake, 40 miles north-east of Mattawa. These were 2 old and 5 young ones. They slid down, rapidly climbing up again to slide down as before, for a full hour.

They are far from restricting this sport to snow time, however.

Kennicott says further: "This curious habit seems to be indulged in by the Otter at all times, when a suitable place can be found, though more in the love season [late February] than any other." SLIDE AT ALL SEASONS

Chief Mittigwab, after spending a lifetime in Otter country and seeing many Otter, says that they slide all the year round, for amusement; usually, but not always, at special places, most in the snow, during spring and fall. Archie Miller corroborates this in every detail.

At Swan Lake, Man., James M. Macoun saw slides that were in use during August, and J. G. Millais says of these 'chutes'[19] that he found numerous in Newfoundland, where he counted no fewer than 6 on one quiet brook flowing into St. John's Lake at the head of the Terra Nova River:

"All the slides I have examined in Newfoundland had been freshly used and were polished smooth, with the grass and moss all worn away, showing that they had been in constant use all the summer."

[19] Mam. G. B. & I., 1905, Vol. II, p. 21.

Thus each new scrap of testimony shows the sport farther removed from the influence of sex and season. And sport the exercise is—just as surely as is that of the small boy who drags his sled to the top of a hill for the pleasure of shooting down again, of having the delicious sensation of speed without effort. And whether the Otter slides from the top of a mere snow-drift into the adjoining hollow, or down a muddy bank into a stream, or, best of all, down a long icy hill, to plunge into deep cool water below, it is evidently done for sport, for the joy of feeling itself flying through space without labour and without violence, and with the very same exhilaration that such a thing would give to mankind. To this the creature fails not to add the crowning charms of good company and of friendly rivalry, for, so far as I can learn, no one has ever yet seen an Otter enjoying its slide alone.

It is the rule for *young* animals to play together, and occasionally the full-grown will indulge in a good-natured sham fight or a chase, for sport, but this is the only case I know of among American quadrupeds where the entire race, young and old, unite to keep up an institution that is not connected in any way with the instincts of feeding, fighting, or multiplying, but is simply maintained as an amusement.

Thus the case of the Otter in Eastern America has been fully set forth, but it is a remarkable fact that Sir John Richardson, who lived for years in the Great Lone Land, says not a word about the Otter's toboggan slides, though he describes its habits at length; Hearne's remarks I have quoted, but I can find no other hunter west of Lake Nipissing who has ever seen an Otter sliding for amusement, while I find many reliable naturalists of the North-west, notably Roderick MacFarlane, who have but little faith in it.

DRY WALLOWS

After remarking on numerous examples of Otter slides in the Adirondacks, Dr. Merriam writes[20] also "of their wallowing places, which are either level beds or slight depressions in

[20] Mam. Adir., 1884, p. 89.

which they play and roll. May's Lake, a small and secluded body of water abounding in trout, is fairly surrounded by them."

These I have never seen, nor do I find them commented on by other naturalists.

CLIMBING

The species is no exception to the rule that no animal can reach the highest development in more than one sphere. It is a king in aquatics, but on land is the least active of the Family except the Sea-otter, and I read with surprise J. G. Millais's statement that the Canada Otter commonly climbs trees.[21]

FOOD

This animal is pre-eminently a fish eater, yet at times varies its diet in many ways, feeding on frogs, crayfish, and shell-fish. "Crayfish, indeed," says Kennicott,[22] "sometimes form a considerable portion of its food, being taken in the water like fish. I have observed this to be the case when examining its excrement in Minnesota. * * * I have reason to suppose that it sometimes devours the Muskrats, in the house of which it is occasionally found in the Western prairie lakes and marshes. It would probably eat any flesh when impelled by hunger, but it has never been known to devour vegetables of any kind when in a state of nature."

Merriam corroborates the crayfish item, stating that:[23] "The numbers of crayfish (*Cambarus*) that the Otter destroys in the course of a summer is almost incredible. The Otter 'sign' that one finds so abundantly about our lakes and streams, on rocks and logs, often consist wholly of fragments of the chitinous exoskeleton of the Crustacean. At other times fish-bones are mingled with the broken crayfish shells. * * * When unable to procure these in sufficient quantity, it devours frogs, and is said to depopulate the poultry yard, and even prey upon lambs. * * * In confinement it will eat meat and is said to prefer it boiled." Richardson states[24] that: "In the spring of 1826, at Great Bear Lake, the Otters frequently

[21] Mam. G. B. & I., Vol. II, 1905, p. 18.
[22] Quad. Ill., 1859, p. 247.
[23] Mam. Adir., 1884, p. 88.
[24] F. B. A., 1829, p. 58.

robbed our nets, which were set under the ice, at a distance of a few yards from open water. They generally carried off the heads of the fish and left the bodies sticking in the nets."

FISH ITS FOOD

It is known to eat wild fowl, and doubtless, when pinched, will relish any kind of animal food. All of these, however, must be considered mere accidental variations, or emergency rations, on its real bill of fare, for nine-tenths at least of its food the year round is undoubtedly fish, the finest and freshest at that; not mud-cats, swinish carp, or logy suckers, but the sweetest and swiftest of all—the superb trout and the regal salmon.

The Otter is *the* fisherman *par excellence;* it is the successful fisherman, and therefore it is the hated of all other fishermen.

MENTALITY

Mentally this animal is at the head of its group. 'The most intelligent of the Family,' is the verdict of all who have studied all the Weasels, for the Otter is nothing but a big water Weasel. "From the nature of its habits and its sagacity," says Merriam,[25] "it is likely to remain after most of the other representatives of the Mustelidæ have been exterminated."

I wish it were so, and yet experience in Manitoba leads one to believe otherwise. The Skunk, the Mink, and the Ermine seem as abundant as ever, but the Otter is becoming a rarity.

PLAYFULNESS

At all ages Otters are playful animals. They chase each other in tireless games of tag, and sport and roll in the water like porpoises. If caught young they are readily reared and become the most fascinating of pets. It has been my luck to meet with two or three tame Otters, and in each case I was left with a profound admiration for the grace and playfulness of this exquisitely beautiful creature. On watching a gambolling Fox cub, a Fawn, an ocelot, a Marten, or even a well-furred pet Skunk, one is apt to be carried away and declare each in turn the most beautiful and graceful creature ever seen. But when all are gone from view, when nothing but the dim impression remains, it is the Otter that stands out pre-eminently

[25] *Loc. cit.*

as the most beautiful and engaging of all elegant pets. There seems no end to its fun, its energy, its drollery, its good-nature, and its postures of new and surprising grace. I never owned a pet Otter, but I never yet saw one without shamefully infringing article number ten of the Decalogue.

J. K. MacDonald writes me: "They make delightful pets. A tame Otter at York Factory, Hudson Bay, in 1871, used to lie about the Fort among the dogs. On first living among them several of them tried issue with him, only to find that he left souvenirs of the struggle on their legs, etc. The remaining dogs, benefiting by the experience of their friends, made no further attack on him, and so he moved around among them quite nonchalantly. He used to take long swims in the river, both to enjoy the pleasure of swimming and get some fresh fish of his own killing. He always returned and slept inside the Fort pickets. No doubt, at the mating season, he would have left if at large. This we were not permitted to prove, as a strange Indian arriving close to the Fort, saw what he thought was a wild Otter, and so he shot it."

"In growing old, however, they are apt to become ugly, and have been known to bite those who attempt to play with them." (*Merriam.*)

A FIGHTER

The Otter is a valiant beast. During its overland or oversnow journey in search of open water, if closely pressed by pursuit, it will turn on any assailant and defend itself with marvellous courage and power. Its gifts as a fighter are ably seconded by its protective armour of fur, hide, and fat, and it is very doubtful whether there ever lived a dog that could conquer an Otter in fair single fight.

"At all times and all occasions, furthermore, they manifest an insatiate and unaccountable desire to break the peace with any dog that chances to cross their path.

"If the misunderstanding occurs in the vicinity of the water, as it commonly does, there is a strong tendency for the participants to drift nearer and nearer the shore, for thitherwards the Otter artfully draws his antagonist. I have never witnessed one of these little altercations, but am told that a drowned dog is generally the result."[26] (*Merriam.*)

Otter hunters in Europe tell me that such is the power of its bite, that it is not a rare thing for an Otter to crush a hound's leg-bone in its jaws.

The most desperate achievement accredited to this animal appears in Nelson's "Alaska":[27]

"An Otter [he says] was one of the chief actors in a strange accident which occurred near the Yukon mouth during my residence in the north. A hunter went out to inspect his fish traps, and, failing to return in the course of a day or so, his friends began to look for him. He was found lying dead by the side of a small lake with his throat torn open and the tail of a dead Otter firmly grasped in both hands. One of the Otter's feet was fast in a steel fox-trap, and it was supposed that on his way home the hunter came across the Otter in the trap and, having no weapon with him and being a powerful young man, he tried to swing the Otter over his head and kill it by dashing it against the ground, but when in mid-air it turned suddenly and caught him by the throat, with the result as described."

CAPTURE

This fur-bearer is usually taken with a steel trap and fish-head bait, but traps that do not kill are cruel, and humanity would force all trappers back to the old deadfalls of a hundred years ago. My friend, A. W. Dimock, writes me thus from Punta Rassa, Fla.:

"Last year (1905) I held a fiercely struggling Otter crushed into the mud with a forked stick, while the cruel steel trap was taken from her lacerated leg and a cage placed over her. Two days later she ate her new-born pups, and in two days more, despite every attention, she died of grief and pain. Now I would make the use of a steel trap a penal offence and wearing pelt of a trapped wild creature a misdemeanour."

In some parts of the country Otter are killed by taking advantage of the sliding habit. Having found the slide, the hunter

[26] Mam. Adir., 1884, p. 90.
[27] Nat. Hist. Alaska, 1887, p. 250.

may lurk in ambush for the tobogganing party and shoot one, or perhaps two, before they can escape from range.

If its trail is discovered in the winter snow, when evidently the creature is frozen out of its safe refuge, the hunter follows, and within a mile or two he usually gets an opportunity to shoot the portaging voyageur.

The fur of the Otter is cased. It is one of the best, handsomest, and most durable in the market. No matter how others fluctuate, the price of Otter is always fairly high. The Winnipeg market quotations for March 26, 1904, were: Prime Otter, $6 to $12. FUR

At the London annual fur sales, held at Lampson's, March, 1906, 2,517 Otters were sold. The highest price reached was 210 shillings ($50.40) each, for 22 unusually good black, first-class Labrador skins. The run of first-class dark skins brought from 100 shillings ($24) to 150 shillings ($36), and with 60 shillings ($14.40) as the run of ordinary dark Otter, from which, according to size and colour, they graded down as low as 10 shillings ($2.40) and 5 shillings ($1.20).

During the eighty-five years, 1821 to 1905 inclusive, the Hudson's Bay Company collected 890,901 skins of this species, an average of 10,481 for each year. The lowest was 3,795 in 1829; the highest, 18,100, in 1830. The average for the ten years, 1895 to 1905, was 8,898.

Poland's lists show that during the seventy-one years, 1821 to 1891 inclusive, 444,372 skins were taken by the other American companies, an average of 6,258 each year. So that the average annual catch of Otter for fur is about 17,000.

XXXVI.

Common Weasel, Short-tailed Weasel or Ermine of the Woods; Bonaparte Weasel; Brown Weasel.

Putorius cicognanii (Bonaparte).

(L. *Putorius*, see ante; *cicognanii*, named in honor of Felice Cicognani, an Italian.)

Mustela cicognanii BONAPARTE, 1838, Iconogr. Faun. Ital. I, fasc. XXII, p. 4.

Putorius cicognanii RICHARDSON, 1839, Zoölogy of Beechey's Voyage, p. 10.

TYPE LOCALITY.—North-eastern North America.

FRENCH CANADIAN, *l'Hermine; le Roselet; la Belette de Bonaparte.*
CREE, SAUT., & OJIB., *Shing-gwus'.*
CHIPEWYAN, *Tel-ky'-lay.*
YANKTON SIOUX, *He-tong-ka-ska.*
OGALLALA SIOUX, *He-tu-kah'-san.*

This is the common Weasel about barnyards near the woods in Manitoba. At a short distance, in summer coat, it looks like a brown Squirrel with a white throat and paws, and a very small tail; it is mostly seen on the ground.

In addition to the generic characters (*Putorius*, see p. 872) it has the following:

SIZE Length, 12 inches (305 mm.); tail, 4 inches (102 mm.); hind-foot, 1½ inches (38 mm.).

The female may be one-fifth or even one-fourth smaller.

COLOUR General colour above, dark brown; tip third of tail black; no dark spot behind corners of mouth; under parts, including upper lip, chin, throat and front feet, white, sometimes tinged with yellow; hind-feet, pale brownish-white.

In winter it becomes the Ermine; it is now pure white with a yellowish tinge on rump, tail, and under parts; the black tail tip continues unchanged.

The change from brown to white is, of course, to enable the animal to live and hunt in the snow without being visible to all the creatures it would prey on. The change is effected by a moult, and, in Manitoba, occurs about the middle of October each year, without reference to the weather or the presence of snow, so that it is no uncommon thing to see a Weasel of dazzling whiteness running over the brown prairie or in the woods still carpeted with dead leaves. CHANGE OF COLOUR

The yellow tinge is believed to be partly, at least, external staining from the smell-glands with which the creature is so generously outfitted.

As late as mid-April, I have found them in full winter pelage. Soon after this they appear in the brown and white of the summer coat.

The following races are recognized:

cicognanii Bonaparte, the typical form.
richardsoni Bonaparte, much larger.
alascensis Merriam, like *richardsoni*, but white tips of feet more extensive and interorbital region very much broader.

The three species of Weasels found in Manitoba may be thus distinguished when in summer pelage (all are white in winter): THE THREE KINDS

1st. The Long-tailed Weasel (*P. longicauda*), the size of a small Mink, with tail 6 inches, or more, in length; lower parts, buff or pale yellow.

2nd. The Short-tailed Weasel (*P. cicognanii*), much smaller, about 12 inches long, with tail about 4 inches long, and lower parts white.

3rd. The Least Weasel (*P. rixosus*), still smaller, almost as small as a Mouse, about 6 inches, with tail 1 inch long, without any black tip; lower parts, white.

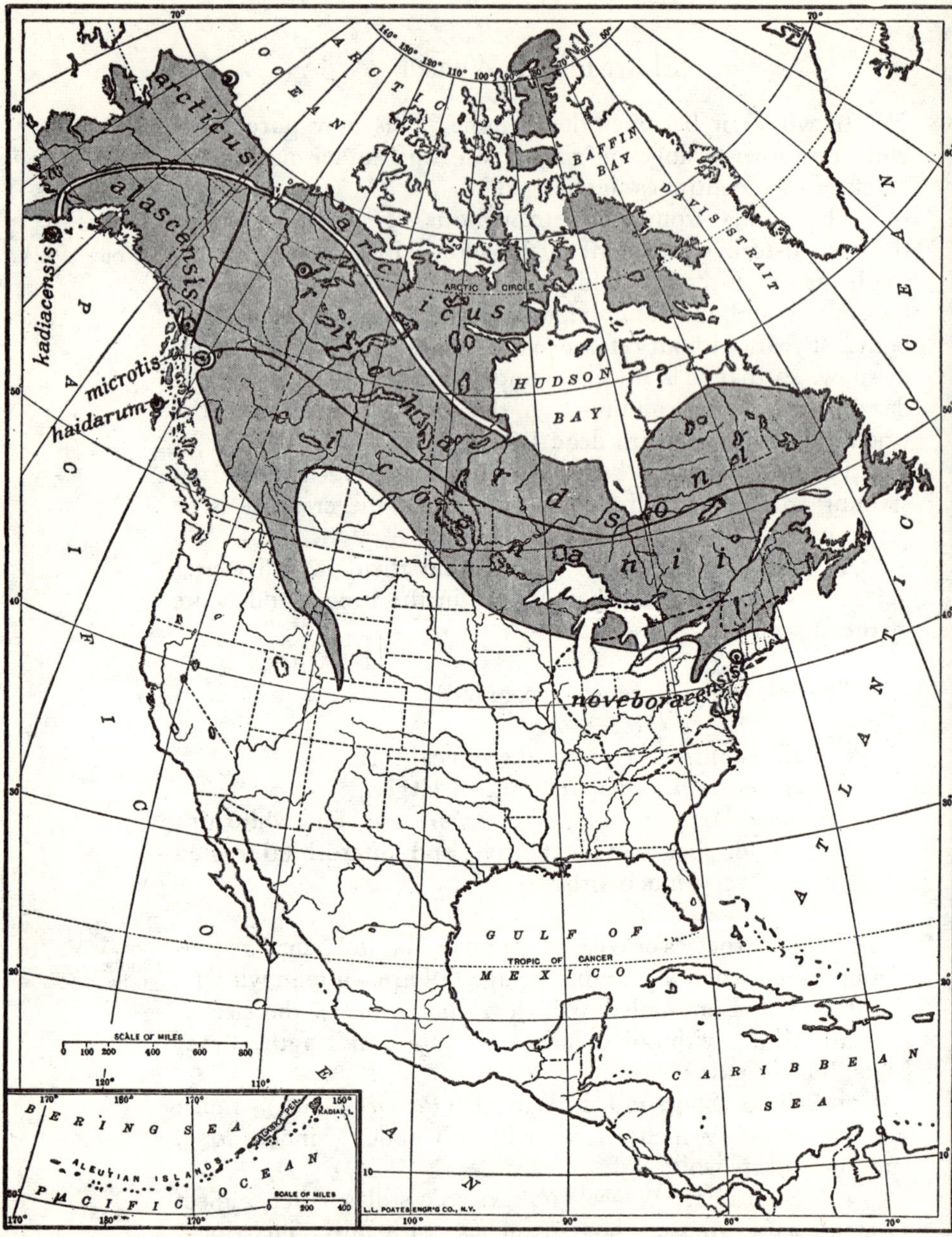

MAP 45—RANGE OF THE WEASELS FOUND IN CANADA, EXCLUSIVE OF THE LONG-TAILED AND LEAST WEASEL GROUPS.

This map is diagrammatic and provisional. The range of *arcticus* must be modified in all borders—and that of *cicognanii* is little known in the south and south-west.

It is founded chiefly on Merriam's Synopsis with records by J. Richardson, R. Kennicott, E. W. Nelson, R. MacFarlane, C. B. Bagster, O. Bangs, G. S. Miller, S. N. Rhoads, A. P. Low, J. P. Howley, E. A. Preble, C. C. Adams, D. G. Elliot.

The following are entered:

Putorius cicognanii (Bonaparte) with 3 races,
Putorius microtis Allen,
Putorius arcticus Merriam with 2 races,
Putorius noveboracensis Emmons with 3 races,
Putorius haidarum Preble.

There is another from which it should be carefully distinguished, that is the New York Weasel, or Blacktailed Weasel (*P. noveboracensis* Emmons), familiar to many in Ontario. In contrast, these are the points:

cicognanii has tail $\frac{1}{4}$ of total length; terminal $\frac{1}{3}$ of tail, black; under parts, pure white in summer.

noveboracensis has tail $\frac{1}{3}$ of total length; terminal $\frac{1}{2}$ of tail, black; under parts, in summer, often yellow, and a brown spot back of the mouth—sometimes this island becomes a peninsula. This is about $\frac{1}{4}$ larger than the preceding. For range, see dotted line on map No. 45.

LIFE-HISTORY.

RANGE

The range of this Weasel extends in the great coniferous forest from the Atlantic to the Pacific, thus taking in all of Manitoba except the true prairie region.

ENVIRONMENT

This is a forest animal, found chiefly on the ground, but capable of climbing, when it must, with quickness little inferior to that of the Squirrel. Though a wood-dweller, I have seen it a mile from cover on the open prairie, where it seemed very much at home in the holes of the Striped Ground-squirrel (*C. tridecemlineatus*).

HOME-RANGE

The habit of all Weasels, as far as known, is to quarter themselves on a good cover or game range, killing everything they can catch, until driven out by a stronger one or till their havoc has spoiled the hunting; then they travel on in search of new grounds. They will go a mile or two in a night, and apparently without clear intention. If they find a barnyard, or a promising place of any kind, they remain and slaughter as before. If they be killed, the place may continue unweaselled for months, but another wandering devil is likely to appear and repeat the destruction of the first.

There is, no doubt, a limit to the wandering of the Weasel, otherwise each species would be continental in range, but I do not know where the limit is drawn. The argument of analogy would lead us to believe that in the case of the Brown Weasel, a ten-mile round would probably represent the wanderings of the individual.

POPULATION

There are doubtless several Weasels to each square mile in the wooded country. I found this species common about

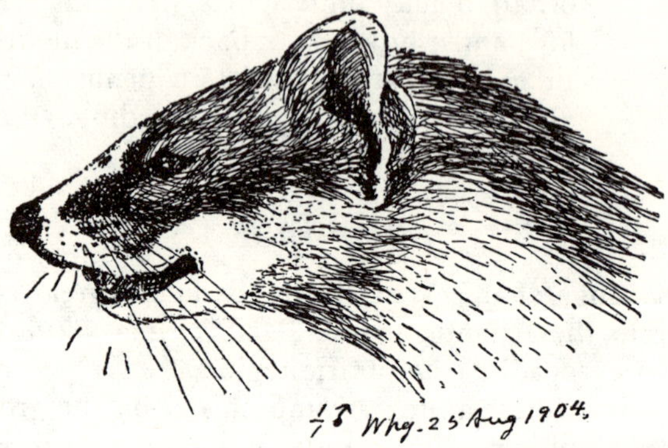

FIG. 204—Head of Short-tailed Weasel ♂.
Sketched in the flesh at Winnipeg, Aug. 25, 1904 (life size). This is somewhat abnormal in colouring. The light area on throat and on cheek under ear (as dotted) is sulphur-yellow, the rest of the cheek, the chin, lips, and *central* part of snout, pure white.

Carberry, Dauphin, Winnipeg, and Ingolf. Kennicott says it appears to be abundant along Red River.

In the winter of 1900–1, it was extraordinarily numerous in Manitoba.

George H. Measham writes me from Woonona, Shoal Lake, January 17, 1901: "Charles Tweddell has trapped over 40 of them just around the shore where the boats were. It is strange how these animals seem suddenly to get numerous and as suddenly disappear."

R. MacFarlane records[1] that in 1903 the Hudson's Bay Company exported 33,883 Ermine skins; probably half were of this species. But this destruction makes no obvious difference in their numbers.

[1] Mam. N. W. Ter., Proc. U. S. N. M., Vol. XXVIII, 1905, p. 713.

SOCIABILITY

The Weasels of this group are neither sociable nor gregarious; they are, indeed, an unlovely lot, no matter how we look at them.

STOAT

Nevertheless, for guidance of future observers, it may be well to remember that the British Stoat, their near kinsman, is known to unite in numbers on occasion for the common good. Thus E. T. Booth records[2] a case of 20 to 30 uniting to attack a terrier dog. Other cases are recorded of their attacking men. And Millais refers[3] to the well-known fact that a pack of 6 or 8 will unite to hunt.

INTERCOMMUNICATION

There is little doubt that the insufferable smell of the 'stinking' Weasels was developed originally as a method of intercommunication, as a means of getting at their friends; though not a few, notably the Skunk, and in a less degree the present one, have carried it so far that they now find it an effectual means of getting at their enemies.

The sharply demarked white and brown with the black tail-tip are believed to be the uniform or directive marks of this species. They are shown by no other mammal of the size in North America. So far as I know, they are not civilized enough to use the smell-telephone (see Wolf).

VOICE

All of the Weasels growl, snarl, hiss or puff, or utter a reiterated sharp sound, which is like a bark in the large species and a shrill screech in the small ones, and record is made herein later of a female, presumably of the present kind, calling her young by a "sort of grumbling coo."

British Stoats, according to Millais,[4] "will stop and bark for a long time at some object that puzzles them." When playing, Stoats give out a chuckling, happy sound, uttered in a high and a low key. When angry, they make a loud chattering noise, and, when hunting in packs and in full cry, are said to 'give tongue.'

MATING

Little is known of the mating of this animal. There is some reason for believing it takes place about the third week

[2] Field, October 6, 1883.

[3] Mam. G. B. & I., 1905, Vol. II, p. 122–3.

[4] *Ibid.*

of March and that the species pairs; but the male does not trouble his head about wife or family after the love season is over—though there may be some exceptions to this, as the following cases show:

Charles G. D. Roberts tells me that one day when he was a boy, living in Westmoreland County, N. B., he was sitting in a field near a stone heap, by the edge of Tantramar Marsh, when two large Weasels (Ermine) came out of the stone pile and ran round him in a manner so threatening that

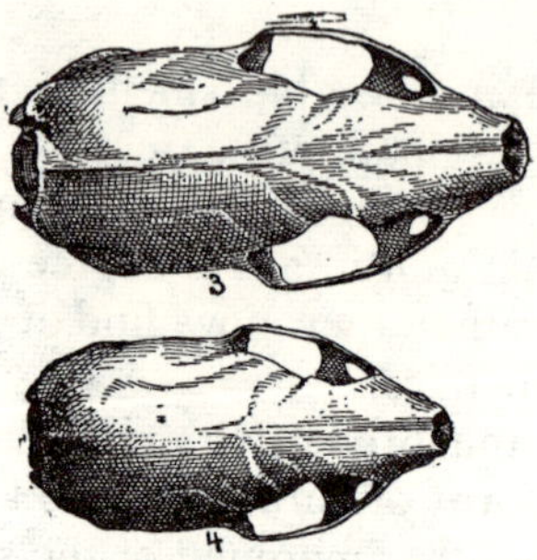

FIG. 205—Skulls of Short-tailed Weasel.
3. Adult male.
4. Adult female.

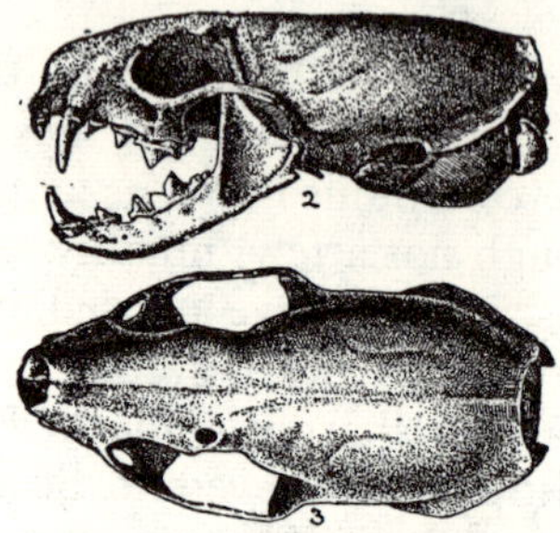

FIG. 206—Skull of Short-tailed Weasel adult ♂, from Elk River, Minn.

From C. H. Merriam's Synopsis, N. A. Fauna, No. 11, 1896. Plates II, and p. 11. Cuts supplied by the Biological Survey, U. S. Dept. of Agriculture.

he was afraid of them. He remained very still and at length they disappeared into the stone pile. He took them for a pair, as they were obviously associated, and he thinks the nest was in the pile. As *cicognanii* is the only species of the size known from New Brunswick, the identification is good.

E. A. Samuels records[5] having seen *a pair of Weasels* pursuing a Chipmunk about the middle of June, 1901, near Fordham, N. Y. The present was probably the species he saw, and the fact that two were united at that season for a common purpose is a shred of evidence that the species pairs.

John Burroughs tells me that one summer, fifty-five years ago, when he was a boy in the Catskills, he saw 2 old Weasels and 3 young ones together run across an open lane.

[5] Forest and Stream, July 27, 1901.

The following note, which I made on a pair of Black-footed Ferrets at the New York Zoological Park, may be adduced as collateral evidence, for mating habits do not vary much in the same family: The male Ferret is very aggressive. He utters a loud, harsh, barking a dozen times in rapid succession, also a loud hissing. Nothing enrages him more than any interference with the female. As this is true the year round, it points to permanent mating.

From these facts, then, we must assume that this Weasel pairs and that the male, sometimes at least, takes an interest in the young.

The only detailed evidence I can find on the dens of this Weasel is as follows: DENS

John Burroughs, in November, 1893, saw a Brown Weasel carrying Mice into its burrow, as narrated in the paragraph on storage. He dug after it for several hours one day. Next day he returned with better tools and tried again, moving over a ton of rooty earth and exposing many more galleries, but finding no larder. He found, however, several little "expansions and at last one of his banqueting halls, a cavity about the size of one's hat, arched over by a network of fine tree roots. The occupant evidently lodged or rested here also. There was a warm, dry nest made of leaves and fur of Mice and Moles. I took out two or three handfuls. In finding this chamber I had followed one of the tunnels around till it brought me within a foot of the original entrance. A few inches to one side of this cavity there was what I took to be a back alley where the Weasel threw his waste; there were large masses of wet, decaying fur here, and fur pellets such as are regurgitated by hawks and owls. In the nest there was the tail of a Flying-squirrel, showing that the Weasel sometimes had this game for supper or dinner."[6] After this the Weasel's labyrinth seemed to grow more complicated as well as expand to include the neighbouring country, and the digger had to give it up without finding the store of Mice.

[6] Squirrels and Other Fur-bearers, 1900, pp. 77–8.

A burrow so extensive, Burroughs thinks, must have been the work of many seasons, and therefore a permanent home of this Weasel.

SANITATION One of the most interesting features in it was the midden-heap. When we find an animal far enough advanced to purify its nest by appointing and keeping apart a place for garbage and filth, we must honour it for having advanced the first degree in sanitation, and even for having taken a step towards civilization.

GESTATION The period of gestation in our Ermine is unknown, but A. H. Cocks found it to be about 40 days in the British congener.

YOUNG The young number from 4 to 8, are usually 5 or 6, and doubtless, like Stoats, they are blind for some nine days after birth.

In her ordinary life the old Weasel walks to and fro, far and near, throughout the earth seeking whom she may destroy; but in spring, as we see, she responds to the home love, and for a time is chained to the nest with the young. She hides them with care, nurses them with tender and truly musteline assiduity, and guards them with a marvellous courage, until in late summer when they are about six or eight weeks old; then they are strong enough to follow her about, and she resumes her travels at the head of her half-dozen budding cutthroats, and leaves a trail of destruction still wider than before.

On June 28, at my home in Connecticut, John Crawford saw in a dry stone wall by my house an old Weasel (*noveboracensis*) and 5 young ones about half grown. She ran across an open space to the woodshed, where she called the young ones in a "sort of grumbling coo." Each time she did so they popped out their heads.

Crawford tried to hit them with a stick. The mother ran forward a number of times, calling the young to come, and at last dashed back into the wall where they were. As we saw no more of them, the old one doubtless went on with her band of pirates.

In the case observed by John Burroughs, the mother showed her ready devotion, for he fired at the young ones, wounding one of them so it could not run, but the mother seized it in her mouth and bore it away to safety.

As already noted, the father, in some cases at least, is active in the care of the young.

So far as known, there is but one brood each year.

HABITS

At Carberry, I have often seen this energetic little creature seeking for Mice in the deep, soft snow. Its actions are much like those of an Otter pursuing salmon. Sometimes it galloped along a log or over an icy part of the drift; then plunged out of sight in a soft place, to reappear many yards away, bounding here and there, over and under, restless and tireless as the waves of the sea—forever changing his place, pose and direction, an embodiment of lithe grace and endless assiduity. At such times, if he disappears in some crevice, hole, or maze, he is easily persuaded to come forth again, if you remain still and squeak like a Mouse.

The smell of blood must be as far reaching as it is attractive to these sanguinary little creatures. I have frequently hung new-killed Rabbits and partridges temporarily in trees, and, after an absence in some cases of a few minutes only, have found an Ermine mauling the game, though there was no sign of such a visitor when the cache was made.

The Weasels have the unloveliest disposition of all our wild animals. Outside of their strength and courage, we find in them little to admire. Most other animals have a well-marked home-region and friends, but the ordinary life of a Weasel is that of a wandering demon of carnage. Dr. Coues has tersely summed up Weasel, body and soul, in a few characteristic lines:[7]

"A glance at the physiognomy of the Weasel would suffice to betray their character. The teeth are almost of the highest

[7] Fur-bearing Animals, 1877, p. 129.

known raptorial character; the jaws are worked by enormous masses of muscles covering all the sides of the skull. The forehead is low, and the nose is sharp; the eyes are small, penetrating, cunning, and glitter with an angry green light. There is something peculiar, moreover, in the way that this fierce face surmounts a body extraordinarily wiry, lithe, and muscular. It ends in a remarkably long and slender neck, in such a way that it may be held at right angles with the axis of the latter. When the creature is glancing around, with the neck stretched up and the flat triangular head bent forward, swaying from one side to the other, we catch the likeness in a moment—it is the image of a serpent."

The thugs of India claim to be devotees of the Goddess of Destruction; and profess, therefore, that it is their duty to kill as many human beings as possible. The Weasel is the Thug of the Wild World. While other animals may kill to excess for the gratification of appetite, the Weasels alone seem to revel in slaughter for its own sake, to find unholy joy in the horrors of dying squeak, final quiver, and wholesale destruction. Gifted with tremendous strength and activity; at home in the tree top, under the snow, on the earth, under ground, or in the water; keen of wits, tireless of wind and limb, insatiably cruel and madly courageous, they are all too well equipped for their chosen Herodian task.

FOOD

The Weasel preys on every kind of bird and beast that it can master, and this means everything from turkey and Rabbit down to tomtit and Shrew. On the list of its prey we find recorded all kinds of domestic poultry, all wild birds that it can catch, Rats, Mice, Squirrels, Chipmunks, etc. It is the most villainous of murderers when it finds an open way to the chicken house.

Bachman tells[8] of 40 well-grown fowls having been "killed in one night by a single Ermine. Satiated with the blood of probably a single fowl, the rest, like the flock slaughtered by the Wolf in the sheepfold, were destroyed in obedience to a law of nature, an instinctive propensity to kill. We have

[8] Quad. N. A., 1849, Vol. II, p. 58.

PLATE LXXVI.—THE DEMON OF MURDER.

Drawn by E. T. Seton, to illustrate Audubon and Bachman's incident of the Brown Weasel that wantonly massacred the Chipmunk family.

traced the footsteps of this blood-sucking little animal on the snow, pursuing the trail of the American Rabbit, and although it could not overtake its prey by superior speed, yet the timid Hare soon took refuge in the hollow of a tree, or in a hole dug by the Marmot or Skunk. Thither it was pursued by the Ermine and destroyed, the skin and other remains at the mouth of the burrow bearing evidence of the fact."

Kennicott records[9] that "in a single night and the early part of the following evening one of these killed nearly 50 chickens, several of which were adults, and many half grown."

"Wherever the Ermine has taken up its residence," says Bachman,[10] "the Mice in its vicinity for half a mile around have been found rapidly to diminish in number. Their active little enemy is able to force its thin vermiform body into the burrows, it follows them to the end of their galleries, and destroys whole families. We have on several occasions, after a light snow, followed the trail of this Weasel through the fields and meadows, and witnessed the immense destruction which it occasioned in a single night. It enters every hole under stumps, logs, stone heaps, and fences, and evidences of its bloody deeds are seen in the mutilated remains of the Mice scattered on the snow. The little Chipping or Ground-squirrel, *Tamias lysteri*, takes up its residence in the vicinity of the grain fields, and is known to carry off in its cheek pouches vast quantities of wheat and buckwheat to serve as winter stores. The Ermine instinctively discovers these snug retreats, and in the space of a few minutes destroys a whole family of these beautiful little *Tamiæ;* without even resting awhile until it has consumed its now abundant food, its appetite craving for more blood as if impelled by an irresistible destiny, it proceeds in search of other objects on which it may glut its insatiable vampire-like thirst. The Norway Rat and the Common House-mouse take possession of our barns, wheat stacks and granaries, and destroy vast quantities of grain. In some instances the farmer is reluctantly compelled to pay even more than a tithe in contributions towards the support of these pests. Let,

[9] Quad. Ill., 1859, p. 244.

[10] Quad. N. A., 1849, Vol. II, pp. 59–60.

however, an Ermine find its way into these barns and granaries and there take up its winter residence, and the havoc which is made among the Rats and Mice will soon be observable. The Ermine pursues them to their farthest retreats, and in a few weeks the premises are entirely free from their depredations. We once placed a half domesticated Ermine in an out-house infested with Rats, shutting up the holes to prevent their escape. The little animal soon commenced his work of destruction. The squeaking of the Rats was heard throughout the day. In the evening it came out licking its mouth, and seemed like a hound after a long chase, much fatigued."

At Ingolf, Ont., September 16, 1904, I saw a specimen of this Weasel that had been shot by the station agent. He told me that there were five or six Rabbits each night about the station. But one day the Weasel took up its abode near by and the Rabbits disappeared. The Weasel came into the station one night and, by help of the dog and a stick, the man injured it, but it escaped through a hole in the mosquito bar. Next day it was back and he killed it with a shot-gun. The pertinacity of the animal in returning was very characteristic.

STORAGE HABITS

The storage habit is not what we look for in a creature so reckless and wasteful as a Weasel, and yet it seems fairly well-developed in this species. Bachman, after the above experiment with the Ermine as a ratter, says:[11] "A board of the floor was raised to enable us to ascertain the result of our experiment, and an immense number of Rats were observed, which, although they had been killed in different parts of the building, had been dragged together, forming a compact heap." And again he says[12] that he has known the Ermine to kill and cache in the snow a Cottontail Rabbit, pressing the snow tightly down over it.

John Burroughs gives[13] another curious case of Weasel storage, in connection with the den already described. He saw the creature (*cicognanii?*) carrying a Mouse into a hole

[11] *Ibid.*, p. 60.

[12] *Ibid.*, p. 58.

[13] Squirrels and Other Furbearers, 1900, pp. 72–5.

about every fifteen minutes till 4 were stored within; the next day the same thing went on until 4 more were carried in before his eyes, and doubtless others he had not seen, so he set to work to dig out and examine this larder, but the farther he went the more branches the tunnel had, and after many hours' digging he gave it up.

Kennicott credits[14] the kindred species (*noveboracensis*) with collecting in a particular spot the Rats and Mice it has slaughtered, until a hundred or more of the victims are in the pile.

As the Weasel usually craves hot, fresh blood, and a living prey, I am puzzled to understand its occasional wabblings toward the frugal habits and virtuous ways of much better citizens.

COURAGE

Much as we may hate it for its sanguinary disposition, we are bound to respect the Weasel for its courage. It will ordinarily face any animal up to twenty or thirty times its size; a mother Weasel will face and fight an elephant; she will fly from nothing that may threaten her young.

That this animal will sometimes attack man is shown in a case related by Burroughs:[15] The Weasel turned savagely on a man that had interfered with its feasting on a newly killed Rat. It dodged his blows of stick and stone in a way "singularly uncanny and startling. It was like some infuriated imp of Satan, dancing before him and watching the chance to seize him by the throat or to dash into his eyes."

SPEED

Whatever a Weasel does, is done quickly—whether it be to seize the bounding Squirrel, clinch on the rash terrier's nose, elude the rifle ball at the flash, or save its young—it is known to act like lightning and with nearly uniform success.

Measured by miles per hour, I doubt not its speed on the ground would be low, but in dodging it is quick to bafflement of the eye and the gun. In the trees it is perfectly at home,

[14] Quad. Ill., 1858, p. 106.

[15] Squirrels and Other Fur-bearers, 1900, p. 84.

much more so than a Chipmunk, and almost as much so as a Red-squirrel.

Bachman writes[16] thus of one: "To avoid a dog that was in close pursuit, it mounted a tree and laid itself flat on a limb about 20 feet from the ground, from which it was finally shot."

SWIMMING

In the water it is a good swimmer. The following was observed by J. W. Curran, of Montreal, while camped at Lake Couchiching, Ont., in July, 1899:[17]

"About 50 yards away from us a Chipmunk jumped off a tree overhanging the water and plunged boldly in, followed at a distance of not more than 3 feet by a Weasel. It was a great jumping contest, and our hearts were with the little fellow in front. However, we remained neutral. For 25 yards things looked black for the Chipmunk. The Weasel pulled up slightly, probably a foot—and we prepared to go out and give a hand. It was a fast race, too, the pair easily beating the best swimming I ever saw a dog do. The Weasel, I think, showed more of his body and seemed to exert himself more. After the first spurt the Chipmunk managed to hold the lead, and at the end of one hundred yards or less the Weasel, completely blown, suddenly threw up the sponge and wheeled around for the shore, his successful competitor keeping right on for another island a quarter of a mile away.

"I think a Chipmunk and probably a Black-squirrel can beat a Weasel swimming, and also that Weasels do not depend entirely on smell when after a meal."

ENEMIES

If Weasels were to be greatly multiplied they would quickly destroy every small bird, beast, and reptile in the country. Fortunately, they are nowhere abundant. Although prolific, and comparatively safe from the attacks of bird and beast of prey, they never become numerous. The reason lies, partly, I believe, in their own ferocity. More Weasels are killed by

[16] Quad. N. A., 1849, Vol. II, p. 58.
[17] Forest and Stream, June 2, 1900.

Weasels than by any other foe. Two Weasels cannot live in the same thicket; one of them must fall or flee. It is probable, too, that an unusual increase in Weasels results in such wholesale extermination of their prey that local famine awaits these bloodsuckers. In the fights for life that follow, the slightest difference in weight counts, and it appears that females are often overpowered and destroyed by their unchivalrous lords, so that a ruthless check is put on the further multiplication of their race, and their number once more brought to its proper low adjustment.

I never saw one of these Weasel fights, but I have heard of them, and have seen a duel between Martens in which the female was killed. Sex probably counts for nothing among these Weasels, except in the breeding season.

One of the most curious cases of a Weasel meeting his doom is this recorded by T. McIlwraith.[18] He does not give the species of Weasel, but, from the place, it was most likely the present one: "Twenty years ago, I knew a youth who shot one of these birds [Bald Eagle] as it flew over him while he lay concealed among the rushes on the shore of Hamilton Bay watching for ducks. On taking it up he found an unusual appendage dangling from the neck, which proved, on examination, to be the bleached skull of a Weasel. The teeth had the 'death grip' of the skin of the bird's throat, and the feathers near this place were much confused and broken.

"The eagle had probably caught the Weasel on the ground and, rising with his prize, a struggle had ensued in the air, during which the Weasel had caught the bird by the throat and hung there till he was squeezed and clawed to pieces."

There is a curious and interesting side to Weasel nature well-known in England, and doubtless to be discovered in our own species, as soon as it has been observed as fully as its British cousin. The Stoat often practices a piece of perfectly Satanic dissimulation as a ruse to approach some intended prey that is in an open place. ANTICS

[18] Birds of Ontario, 1894, pp. 209–10.

Millais[19] gives a number of instances of Stoats openly approaching the intended quarry by rolling on the ground and gambolling in various antic ways, so as to arouse the curiosity and lull the suspicions of the victim as it draws ever nearer.

A most interesting sort of game played by two males and a female Stoat is described by C. B. Moffat,[20] who witnessed it in June, 1890, near Ballyhyland, Enniscorthy, Ireland. All were apparently full grown. They were chasing each other like dogs or kittens, knocking each other over, and turning somersaults. "A curious crowing sort of note, '*Curoo, curoo, curoo*,' uttered very quickly, was frequently uttered, and invariably when they ran at full speed. Great part of the game consisted in all three animals careering across the road again and again, frequently crossing each other, when they sometimes sprang high in the air and cannoned against one another, all evidently in the height of fun. Then there was a ceremony, which I could not quite understand, of pressing their noses on the bare ground and running along for a foot or so, making a slight grating noise. I do not know how; they all did this."

There is much evidence of adult British Stoats having games of this sort in summer and in winter, apparently at times when the sex instincts are dormant.

VALUE TO MAN

The grewsome chapter of carnage and woe that appears under the section on habits and food might lead one to list the Weasel among things to be destroyed at all times, and by any means, as a thing without redeeming qualities. But Kennicott, after years of close acquaintance, says:[21] "I have frequently found the half-eaten remains of Meadow-mice in their burrows or under corn-stalks, which had doubtless been destroyed by this Weasel [*P. noveboracensis*], or perhaps the smaller one (*P. cicognanii*). It is a surprising thing that an animal so large as this should be able to force its way into the burrows of Meadow-mice, and yet it appears to do so without difficulty.

[19] Mam. G. B. & I., 1905, Vol. II, pp. 116–17.
[20] Zoölogist, 1890, p. 381.
[21] Quad. Ill., 1858, p. 105.

"Stacks and barnfuls of grain are often overrun with Rats and Mice; but let a Weasel take up his residence there and soon the pests will disappear. A Weasel will occasionally remain for some time in a barn feeding on these vermin without disturbing the fowls.

"Indeed I am inclined to think that notwithstanding their occasionally predatory inroads, they should not be killed when living permanently about meadows on cultivated fields at a distance from the poultry."

The Weasel, then, like so many of our carnivores, will eat any living thing it can master, but probably counts on Mice as its steady diet the year round. FOOD

This is one of the species that supply the famous Ermine fur of commerce, but it has not the enormous value that one sees ascribed to it in reckless print. Not several dollars, but a few cents, are the usual equivalent of a skin. The value is so low that few trappers think them worth skinning. FUR

I am told by D. A. Boscowitz, the fur-dealer of Victoria, B. C., that at the London fur sales, in Lampson's, March, 1906, 80,000 and odd Ermine were sold. The highest price was 7 shillings and sixpence ($1.80) a skin for prime white Siberian without yellow tint. Prime American and Canadian skins brought only 4 shillings (96 cents). Other grades ranged from that down to sixpence (12 cents) for third-class. ERMINE

XXXVII.

The Least Weasel, or Mouse-hunter.

Putorius rixosus Bangs.

(*Putorius*, see ante; *rixosus*, Latin for 'quarrelsome' or 'aggressive,' though in this respect the species does not seem to be any worse than its betters.)

Putorius pusillus BAIRD, 1857, N. Am. Mamm., pp. 159–161 (in part).

Putorius rixosus BANGS, 1896, Proc. Biol. Soc. Wash., pp. 21–22.

TYPE LOCALITY.—Osler, Saskatchewan, Canada.

FRENCH CANADIAN, *la petite Belette*.

CREE, OJIB., & SAUT., *Shing-gwus-ance* or Little Weasel, diminutive of *Shing-gwus*, the Weasel. This name is applied to all of the small species indiscriminately.

CHIPEWYAN, *Tel-ky'-lay-az-zy* (little Weasel).

YANKTON SIOUX, *Ke-tong-ka-ska*. (Given to all small Weasels.)

This species has been partly known to naturalists since 1857, when Baird described a specimen, believing it to be the same as its larger cousin, the Least Weasel, of Great Britain. In 1893, W. C. Colt, at Osler, Sask., captured a summer specimen which he sent to Outram Bangs, of Boston. In 1896, this naturalist, convinced that it represented an undescribed species, gave it the name *rixosus*. It proves to be very different from its European cousin in size as well as in colour and cranial characters; it is considerably smaller and is, indeed, the smallest known beast of prey in the world.

PLATE LXXVII.—THE LEAST WEASEL.
Drawn from a specimen taken at Morden, Manitoba.

It is readily recognized by its very small size and short tail without black tip. For generic characters of *Putorius*, see page 872.

Head and body, about 6 inches long (150 mm.); tail, $1\frac{1}{4}$ inches (31 mm.); hind-foot, about 13-16 inch (20 mm.). SIZE

In summer coat, upper parts and tail even umber brown; under parts, pure white without any yellow tinge, and sharply demarked from the brown; or, in detail, lips, cheeks, chin, throat, fore-neck, chest, belly, inside of each leg, and the toes, white. COLOUR

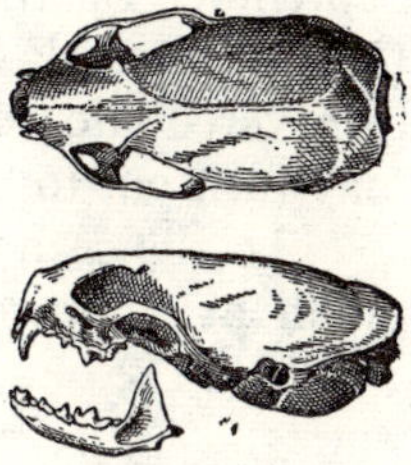

FIG. 207—Skull of *P. rixosus*, the type.
From Merriam's N. A. Fauna 11, Biol. Surv., Dept. Agr., U. S.

A female which I got at Old Fort Reliance, Great Slave Lake, September 15, 1907 (No. 1090, Seton Collection), was: length, $6\frac{7}{8}$ inches (175 mm.); tail, $1\frac{3}{16}$ inches (30 mm.); hind-foot, $\frac{13}{16}$ inch (21 mm.). The body, after skinning, was $\frac{3}{4}$ inch (19 mm.) through the deepest part and but $\frac{1}{2}$ inch (13 mm.) through the chest.

Three races are recognized:

rixosus Bangs, the typical form.
eskimo Stone, a larger, redder race with shorter tail.
allegheniensis Rhoads, larger, darker, and more thinly furred than *rixosus*.

Life-history.

The species is found in Arctic and boreal America from the Bering Sea, at least to Hudson Bay, probably to the Atlantic; and from the Arctic Ocean to the southern limits of the Canadian Fauna. RANGE

Professor S. F. Baird had a specimen from Pembina, Minn.[1] Outram Bangs records[2] it from Osler, Sask., Alaska,

[1] Pacific R. R. Rep., Vol. VIII, 1857, p. 160.
[2] Rev. Weasels, E. N. A., Proc. Biol. Soc. Wash., February, 1896, p. 22.

Moose Factory, and Fort Albany. According to Kennicott,[3] it is found in Indiana and northern Illinois, and, in fact, from New York westward. G. F. Dippie showed me a specimen at Calgary where he says it is common. J. M. Macoun found it at Jasper House. E. A. Preble took specimens in 1901 at Fort Smith and Fort Resolution.[4] E. A. Preble and I secured one at Old Fort Reliance, Great Slave Lake, September 15, 1907; and W. H. Osgood got one at Tyonek, Cook's Inlet, Alaska.[5] In Pennsylvania, S. N. Rhoads recently discovered and described[6] a small Weasel (*allegheniensis*) that may turn out to be a race of *rixosus*.

At the American Museum, New York, is a specimen taken at Johnstown, Ohio, by N. C. Buxton, January 25, 1907.

IN MANITOBA

Thus all Manitoba is well within its range. A winter specimen was sent me from Woonona, Man., by W. G. Tweddell. Two brown-coated specimens in my collection were taken near Morden by D. Nicholson in November, 1903. He has seen several in the region of Pembina Mountain. I have about a dozen records from Winnipeg. Dr. Gordon Bell shot one near Delta Lake, Man., September 15, 1902, and J. S. Charleson says it is quite common in fall about Macdonald, Man. Although the Least Weasel has a wide extension in countries that have long been studied by naturalists, it is so elusive and hard to observe that until recently it has escaped our acquaintance, and as yet we have practically no knowledge of its habits. The specimen from which I made the drawing was taken at Morden, Man., by D. Nicholson, about November, 1903, and is still in full summer coat. He had several brought him at different times, but said that they "went bad" so quickly that most were lost. When a small animal turns putrid in two or three hours after death, it usually means that *its food is insects*.

[3] Quad. Ill., 1859, p. 245.
[4] N. A. Fauna, No. 27, 1908, p. 234.
[5] N. A. Fauna, No. 21, September, 1901, pp. 69–70.
[6] Mam. Penna., 1903, pp. 173–6.

MAP 46—RANGE OF THE LEAST WEASEL.

Putorius rixosus Bangs.

This map is founded on records by R. Kennicott, O. Bangs, Wilfred H. Osgood, S. N. Rhoads, E. W. Nelson, E. A. Preble, J. M. Macoun, G. S. Miller, and E. T. Seton.

The outline shows the theoretical range; the spots are actual records.

FOOD This proneness to spoil therefore gives us the first ray of light we have on its mode of life. I have also known it to feed on Mice, and doubtless it adds small birds to its list.

E. Wilson tells me that, at Winnipeg, Arthur Hutchings caught a Least Weasel in a mouse-trap about April 15, 1907. It was no larger than a Field-mouse and pure white. The trap had broken its leg. He nursed it till it was well, and then set it free. It now lives about his woodshed and is remarkably tame.

My Fort Reliance specimen had come on one of our mouse-traps in which was a dead Mouse, had eaten the head, and then dragged trap and Mouse some 20 feet, where itself was killed by another mouse-trap. As its stomach was quite empty, I think there may have been two Weasels there.

The following incident, witnessed by my friend George L. Fordyce, of Youngstown, Ohio, furnishes additional light on the food habits of this pigmy:

"While out in the field this morning (December 26) walking along the bank of a ravine at the edge of our golf course, I saw a Field-mouse run out of the bushes into the rough grass that is just outside of the fair green of the course. In another instant, what I thought at first to be a white Mouse came out at the same place. The Mouse ran into a wheel track and disappeared under the grass, coming out about 6 feet from where it went in. The white animal followed through the same course, and when it came out, I saw that it was a small Weasel, very little larger than the Mouse, and that it was following the trail of the Mouse by scent.

"For a time the Mouse ran in circles and zigzagged about, often coming around within four or five feet of the Weasel, but the latter seemed so intent on the trail that it did not notice the Mouse to one side. After a time, the latter started toward the open golf course, and when the Weasel reached the point where the trail was straight, it sighted the prey, made a sudden dash forward, and, although 25 feet behind, overtook the Mouse while it was going three or four feet.

"For a few seconds they seemed to fight, until the Weasel got the Mouse by the throat, and started for the bushes, dragging the body. When it came within about three feet of me, I moved a little to see what it would do. It dropped its victim and ran into the ravine. The Mouse had a drop of bright red blood in the centre of its white throat. I waited near by for fifteen or twenty minutes, thinking the Weasel might come back, but it did not show up again; even an hour later the Mouse had not been disturbed."[7]

W. H. Osgood, while collecting in Cook's Inlet, Alaska, secured an example, of which he says:[8]

"One adult female was taken in a swampy place near Tyonek, September 19. It was caught in a small mouse-trap in a *Microtus* runway, and doubtless would have escaped had it not thrashed into a pool of water and drowned."

INDIAN SUPERSTITION

"The natives," he adds, "regard the capture of one of these rare animals as a piece of great good fortune. One old Indian who frequently visited our cabin told us that his brother who had caught one, when a small boy, had in consequence become a 'big chief'; and he assured me that since I had caught one I must surely be destined to become a man of great wealth and power."

This brief account contains everything that is on record about this wide-spread but furtive species.

Direct observation fails us here, and for further light we must look to the next best thing—the analogy of its kindred. The nearest well-known relative of our tiny carnivore is the Little Weasel or Mouse-hunter (*P. nivalis*) of Great Britain. This differs chiefly in being slightly larger; and there is no reason to doubt that in habits, as in anatomy, they are very similar.

BRITISH WEASEL

According to Thomas Bell,[9] the Little Weasel in Great Britain preys chiefly on Mice, for which reason he regards it as

[7] Personal letter, December 26, 1907.
[8] N. A. Fauna, No. 21, September, 1901, pp. 69–70.
[9] British Quadrupeds, 1874, pp. 183 *et seq.*

an animal to be encouraged about barns; but it also devours insects, small birds, and, on rare occasions, chickens. It hunts by scent, climbs, and swims with ease.

Gestation in the Weasels is about 42 days. The young are usually 5 in a litter, but vary from 4 to 6. The nest is in a hole in a bank. It is lined with grass and herbage. The mother will defend them fearlessly and desperately against all assailants.

Most Weasels are solitary hunters, but the mother will go hunting with her whole brood in late summer and early autumn. These family gatherings are doubtless the 'weasel packs' one occasionally hears of.

HUNTING IN PACKS

J. G. Millais points out[10] that in parts of England the Common Weasel (*P. nivalis* Linn.) is called 'fairy,' and states that: "The habit of Weasels to travel and hunt in company at night, most likely explains a superstition which still lingers in the West of England, to the effect that Hares are hunted at night by packs of little fairy hounds, locally called 'Dandy Dogs,' and these some of the country people will assure you they have seen and watched with awe."

VALUE

This Weasel is never known to attack well-grown poultry, or mammals larger than a rat, so that it must be considered a friend of the farmer, an animal, therefore, that is worthy of full protection.

These observations will help us to an understanding of our own still more diminutive species, and any reader who happens to have first-hand facts about this animal or its habits, can render good service to the cause of Natural History by putting his knowledge on record.

[10] Mam. G. B. & I., 1905, Vol. II, p. 135.

XXXVIII.

Large Weasel, Large Ermine, Long-tailed Weasel or Yellow-bellied Weasel.

Putorius longicauda (Bonaparte).
(L. *longicauda*, from *longus*, long; *cauda*, tail.)

Mustela longicauda Bon., 1838, Charlesworth's Mag. Nat. Hist., II, p. 38.
Putorius longicauda Rich., 1839, Zoöl. Beechey's Voy., p. 10.
Type Locality.—Carlton House, Sask.

French Canadian, *la Belette à longue queue.*
Cree, Saut., & Ojib., *Shing-gwus'.*
Yankton Sioux, *He-tong-ka-shah.*
Ogallala Sioux, *He-tu-kah'-san.*

This large Weasel is readily recognized by its very long tail and, in summer coat, by its rich buffy-yellow under parts, very different from the white or pale sulphur tint often seen on the under parts of other Weasels.

Seen afar, in winter, it might be mistaken for a white Squirrel.

In addition to generic characters (see p. 872), it has the following:

Length, about 18 inches (457 mm.); tail, 6 inches (152 mm.); hind-foot, 2 inches (51 mm.). Female about one-seventh smaller. SIZE

All above, pale warm yellowish-brown (much like that of *cicognanii*), darkest on crown and back, lightest on legs; all below, rich warm buffy-yellow; tip of tail for one-quarter of length, black; chin, cheeks, and upper lip, white. In winter, pure white, except the tail-tip, which continues black. Female similar. COLOUR

The following races are recognized:

longicauda Bonaparte, the typical form.
spadix Bangs, darker and more richly coloured.
oribasus Bangs, darker and duskier than *spadix*.

Besides these are the following very nearly related forms, at present ranking as species:

Putorius arizonensis Mearns, similar to *longicauda* in colour and marking, but much smaller.
Putorius saturatus Merriam. Like *arizonensis*, but larger and darker, with distinct spots behind the corners of the mouth.
Putorius alleni Merriam. Like *arizonensis*, but upper parts more suffused with yellow and audital bullæ flatter.

LIFE-HISTORY.

RANGE The range of the species is the Great Plains from Kansas northward to the Saskatchewan. In Manitoba, I found it abundant throughout the prairie region, seeking by preference such cover as the edges of broken land, thickets, or river banks. I have seen it out on the open prairie, but not more than a mile or so from timber. I am told, however, that it follows the Richardson Ground-squirrel wherever it goes, preying on it and living in its burrows. As might have been expected, the ranges of these two animals coincide in the north.

I never saw or heard of a specimen taken in the forest country. All those examined by me in Manitoba were from Winnipeg, Carberry, and Morden. It is probably found throughout our Alleghanian or Transition Region.

HOME-RANGE I have no evidence on the home-range of the individual, but imagine, from the nature and abundance of its food, that in summer, at least, it need not travel so far as do most of its kin.

POPULATION A pair of Long-tailed Weasels to every square mile of prairie would, I think, represent the utmost number of this species. This is, however, a mere guess, founded on the number of Weasel tracks in the snow. Settlement seems to have done nothing towards thinning their ranks. They are, I think, as numerous now as ever.

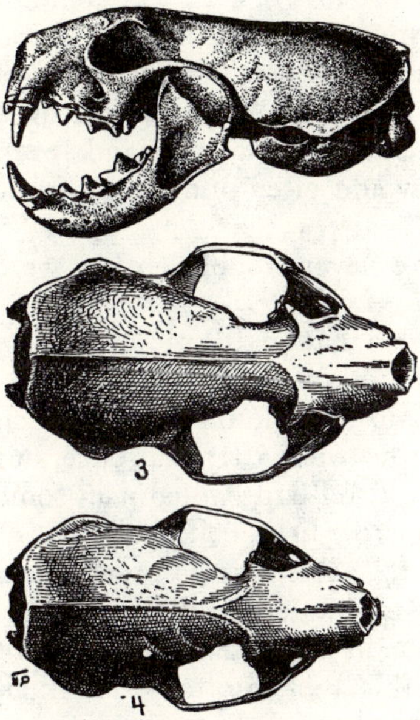

FIG. 208—Skulls of Long-tailed Weasel.
Uppermost, side view of adult ♂ skull.
Middle, top view of adult ♂ skull from Carlton House, Sask.
Lowest, top view of adult ♀ skull from Carlton House, Sask.
(Cuts from Merriam's Synopsis of Weasels. N. A. Fauna, No. 11, 1896, p. 20, and Plate III. Supplied by Biological Survey, U. S. Dept. of Agriculture.)

SOCIABILITY In sociability, means of intercommunication, mating, breeding, etc., the Long-tailed Weasel appears much like the smaller Brown-weasel, but there are very few facts at hand for help in comprehending its ways of life.

HABITS Some interesting observations, which probably refer to the present species, are contributed by Dr. G. B. Grinnell:

"In certain portions of the West the Common Weasel, or Ermine, frequently takes up its abode in the villages of the Ground-squirrels, which are such a pest, and preys on the young and perhaps the adult Squirrels. If, for any reason, the Squirrels desert their villages and move onward—as they frequently do through lack of food—the Weasels are likely to migrate with them.

"This year the Ground-squirrels have been a pest on the Blackfoot Reservation, in western Montana, and have devoured many of the gardens, root and branch. There seems to be no efficient way of destroying them, though, by means of the trap and a small rifle, my friend, J. B. Monroe, had managed to kill in his small garden about 300 up to July 1.

"In this village there are several Weasels, and Major R. A. Allen, who devoted much time to shooting Squirrels, frequently saw them. One seemed to have little fear of him. Sometimes he would see the creature run into its hole, and, going there, would hold a dead Squirrel down in the hole, and the Weasel would come up and seize it with its teeth. At length, the little animal became so tame that it would come to him and reach for the Squirrel held above, and would often jump into the air trying to catch it.

"Of course, the Weasels were never troubled—they were useful in destroying Squirrels."[1]

Professor John Macoun tells me that on July 29, 1906, at a place 20 miles south-west of Saskatoon, Sask., he saw a Richardson Ground-squirrel plunge into its hole on the approach of a Long-tailed Weasel. The latter went after it at once, but soon came out. Evidently the Squirrel had baffled it in some way; perhaps by plugging the burrow behind itself.

FOOD

In addition to Ground-squirrels, this species preys much on the Snowshoe-hare or Bush-rabbit, especially in winter. The Weasel actually runs it down in open chase, in spite of the Hare's superior speed. Of this I have often seen track record in the snow. Once only did I see the pursuit.

In the winter of 1886, while hunting in the poplar woods north-east of Carberry, I saw a Hare running through the brush, pursued at some distance by an Ermine or Long-tailed Weasel. The chase circled about the place where I was camped. I stopped to watch it. The Hare was at the point of giving up when, all at once, it ran towards me and took refuge under the sleigh, near my feet. The Weasel ran around at a distance but, before I could get hold of my gun, he decided to seek his dinner somewhere else.

While Ground-squirrels in summer and Hares in winter may be staples of its diet, I doubt not the Long-tailed Weasel is ready to prey on any living creature it can catch, from Mouse

[1] Forest and Stream, September 14, 1901, p. 205.

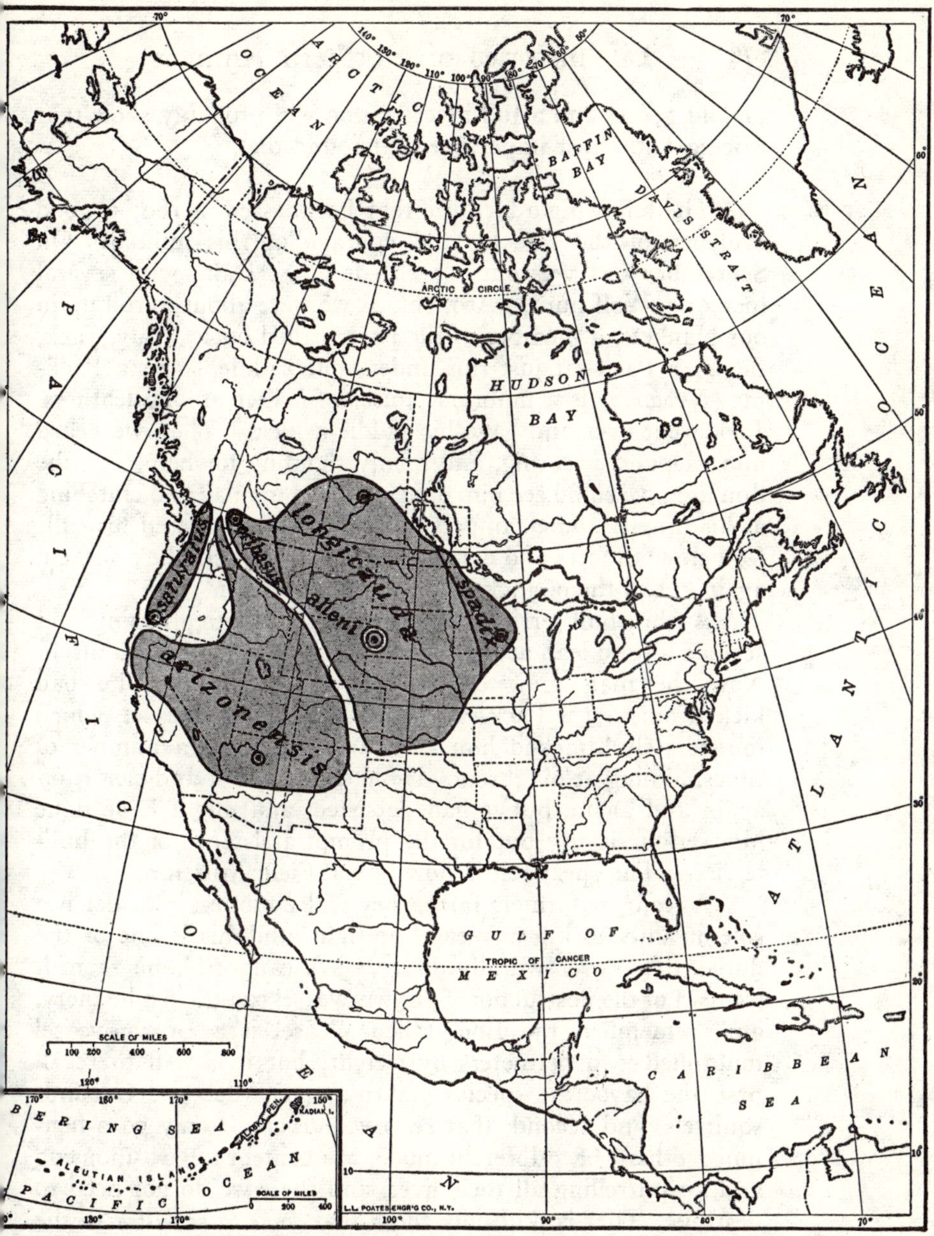

MAP 47—RANGE OF THE LONG-TAILED WEASEL AND ITS NEAR KIN.

Putorius longicauda Bonaparte.

Founded chiefly on C. Hart Merriam's Synopsis, papers by O. Bangs, with original records by E. T. Seton. The cross above Lake Nipissing or an extralimital record of *Spadix*. See Miller, Mam. Ont. Proc. Bost. Soc. Nat. Hist., 1897. Vol. 28, No. 1, p. 44.

to wild goose, and a full investigation will probably show that Mice are its most important year-round diet.

COURAGE

The following adventure that I witnessed in 1897 is good evidence of the ferocity and courage of this animal: On September 5, I was out near Medora, N. Dak., with several men on a Wolf hunt. At night, as we were about to roll up in our blankets, a member of the party called out: "Say, Jack, there's a Pack-rat just run under your saddle." As a Pack-rat (*Neotoma*) is a notorious mischief-maker among leathers, Jack went over and gave his saddle a kick. Then we heard him gasping, swearing, and finally shouting for help. In the dim light we could see him dancing like a maniac and clutching at his throat. The campers all sat up and answered his calls for help with jeers and derision. "Look at Jack; he's got 'em again. Kill them, Jack; the air's full of them," etc.

A white bull-terrier with us now rushed forth growling, and seemed also to leap at the man's throat, then to shake himself. Now the man grew calm, and we learned that he had kicked out, not a Pack-rat, but a Long-tailed Weasel, which immediately attacked him. It ran up his legs a number of times, aiming at his throat. He had clutched it and cast it off again and again, but it had persisted, and might have done him serious injury but for the prompt assistance of the bull-terrier. The specimen is now in the Field Museum.

As long as farmers farm, they will doubtless consider it a solemn duty to kill a Weasel at sight, and this is one of the duties they never wilfully dodge. We cannot blame them if we read of the destruction a Brown-weasel can do in a hennery, and, remember, the Long-tailed Weasel is a Brown-weasel multiplied two diameters by weight; but it is well to recall first, the havoc the species makes among Mice and Ground-squirrels, and second, that *certain individuals* only go a-henning; others, of a different mood, are content to go a-mousing and a-squirrelling all their lives, and these we do not need to destroy. That is, kill only those that *come to be killed* in the barnyard.

The Long-tailed Weasel does not allow us to forget that his name is *Putorius*, and *Putorius* is related to *Mephitis*. Merriam says:[2] "I met one high up in Salmon River Mountains, September 5; he was in pursuit of a Richardson Squirrel in a damp, moss-covered place in a dark spruce forest, and stood bolt upright when he saw me. I wounded him with my auxiliary, and he immediately emitted his powerful stench and disappeared in a hole at the root of a spruce."

In the mountains of Wyoming (1898), I watched a Long-tail, hunting in the snow around me, and in spite of heavy frost made the accompanying sketch to illustrate his pose and tracks. His manner of diving under the snow and of coming up at unexpected and remote points was remarkable and suggested an eel in the mud.

Though the fur is fine and of exquisite yellow and white in winter, it is of too low value to be of commercial importance. (See small Ermine, p. 857.) **FUR**

[2] N. A. Fauna, No. 5, 1891, p. 83. (May have been *arizonensis*.)

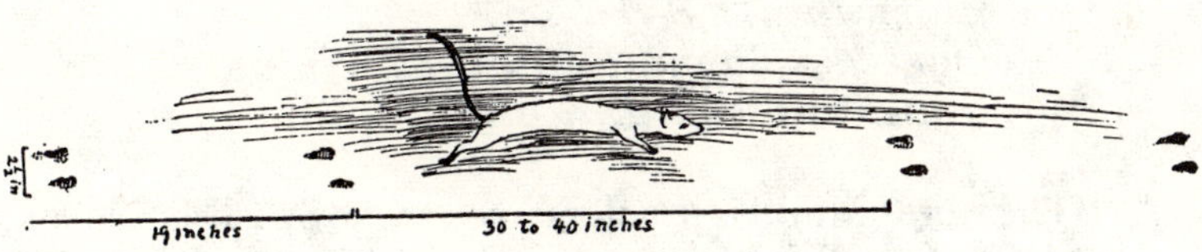

XXXIX.

The Mink, Minx or Vison.

Putorius vison (Schreber).

(L. *Putorius*, a 'stinker,' applied, for good reason, to all the Weasels; *vison*, meaning?)

Mustela vison SCHREBER, 1778, Saugthiere, III, p. 463.
Putorius vison GAPPER, 1830, Zoöl. Journ., V, p. 202.
TYPE LOCALITY.—Eastern Canada.

FRENCH CANADIAN, *le Foutereau.*
CREE, *Sang-gwiss'.* According to Richardson, *Shakwæshew* or *Atjackashew.*
OJIBWAY, *Shang-gwes'-se.*
SAUT., *Sang'-way-soo.*
CHIPEWYAN, *Tel-chu'-say.*
OGALLALA SIOUX, *Lo-chin'-cha.*
YANKTON SIOUX, *Doke-sesch.*

The word 'Mink' is usually traced to the Swedish 'Maenk, though it is not clear how or when the word crossed the ocean.

The genus *Putorius* (Cuv., 1817) comprises Weasel-like animals with long, slender bodies, short legs with five toes on each foot, more or less bushy tails, short ears, and teeth as follows:

$$\text{Inc. } \frac{3-3}{3-3}; \text{ can. } \frac{1-1}{1-1}; \text{ prem. } \frac{3-3}{3-3}; \text{ mol. } \frac{1-1}{2-2} = 34$$

SIZE The Mink unites these characters with the following:

Length, about 24 inches (610 mm.); tail, 7 inches (178 mm.); hind-foot, $2\frac{1}{2}$ inches (63 mm.). The females are smaller.

PLATE LXXVIII.—THE MINK.

An ordinary male weighs about 2 pounds, but I have seen adults that were only 1½ to 1¾ pounds. The largest I ever weighed was taken at Winnipeg, November 1, 1907; it turned the scale at 2 pounds 6 ounces. The females are considerably less than the males, weighing, according to Resseque,[1] about 1 pound 10 ounces. WEIGHT

FIG. 209—Right paws of young Mink ♂. (Life size.)
Desbarats, Ont., Aug. 17, 1904.

In general the Mink is nearly uniform umber-brown, darker and glossier on the back, and deepening on the tail nearly to black; the chin is more or less white, and there may be some white spots anywhere on throat, breast or belly, but these are very irregular; some specimens are totally without white. In the American species the white does not reach the upper lip. In the Siberian species the upper lip is normally white. This animal does *not* turn white in winter. COLOUR

The impression it gives as it dodges in the woods along the water is of a long, thin rat, with brown fur and hairy tail.

The following races are recognized:

vison Schreber, the typical form.
lacustris Preble, a larger race.
vulgivagus Bangs, paler, with heavier dentition.
energumenos Bangs, very large and very dark.
ingens Osgood, very large and somewhat dark.
lutreocephalus Harlan, larger than the type with shorter and paler fur.

LIFE-HISTORY.

The range of the Mink includes all of Manitoba; it is found even along the sloughs of the prairie region, although it is RANGE

[1] Coues, Fur-bearing Anim., 1877, p. 183.

less abundant there than in the woods to the northward and eastward.

ENVIRONMENT The peculiar environment for which it is adapted is the border-land between water and woods, between Otter and Weasel. Although it can live in the water and catch fish, like the Otter, it can also hunt on the land like the Weasel, following its prey into cover of rushes and woods, pursuing it over logs, into burrows, and occasionally even climbing up some sloping

FIG. 210—Young Mink ♂. (Life size.)
Desbarats, Ont., Aug. 17, 1904.

tree to get a better view of the situation. Nevertheless, it must be admitted that on the land it is as inferior to the Long-tailed Weasel as in the water it is to the Otter, and many of its meals are of a kind that either of its cousins would despise, being neither fish nor flesh.

HOME-RANGE The home-range or locality of each individual is probably very large for so small an animal. Its habit seems to be that of all Weasels; it hunts a given area till the game grows scarce by destruction or flight, then it moves on a mile or two, along stream or overland, in search of new hunting grounds. In this way it will change many times during a season, but always, I believe, keeping within a well-defined area that it knows and considers its range. Judging from the time a Mink

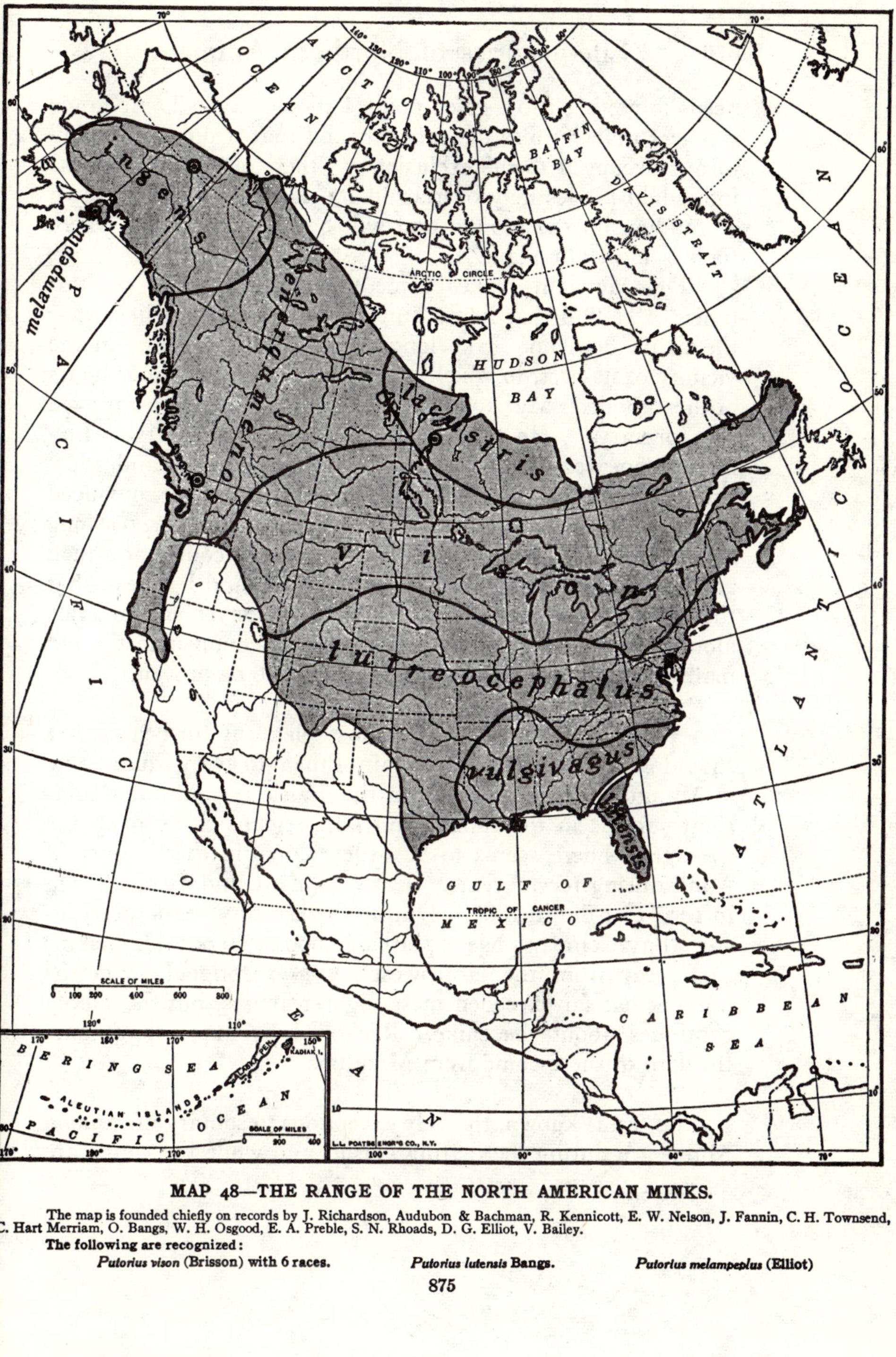

MAP 48—THE RANGE OF THE NORTH AMERICAN MINKS.

The map is founded chiefly on records by J. Richardson, Audubon & Bachman, R. Kennicott, E. W. Nelson, J. Fannin, C. H. Townsend, C. Hart Merriam, O. Bangs, W. H. Osgood, E. A. Preble, S. N. Rhoads, D. G. Elliot, V. Bailey.

The following are recognized:

Putorius vison (Brisson) with 6 races. *Putorius lutensis* Bangs. *Putorius melampeplus* (Elliot) needs to get back to a given point, I should say its home-range was less than 5 miles in diameter, and that it did not by any means occupy it to the exclusion of others of the kind; these individual ranges may overlap like a number of rings thrown at random on the ground and will be most numerous where food is most abundant.

Dr. Merriam makes some remarks that bear on this topic:[2] "I find that many hunters and trappers believe that the Mink does not make long journeys, but remains in the vicinity of its nest, to which it returns every twenty-four hours or thereabouts. My experience, in certain cases, at least proved the contrary." He then gives an account of a large Mink that reappeared at intervals of two or three weeks and adds: "This and other more or less similar experiences have convinced me that the Mink frequently, if not commonly, makes long excursions like the Otter, following one watercourse and then another, and returning over the same route, and I believe that they have a number of nests scattered at convenient intervals along these circuits. This habit may be confined to the old males, but whether it is so or not remains to be proven."

ABUNDANCE

The Mink is one of our most plentiful fur-bearers. I should guess that there is one pair of them to every square mile in Manitoba; less, no doubt in the prairie region, but a sufficient surplus in the timber and lake regions to keep up the average. There seems to be little change in the number of Mink during recent years. I saw as many and as much sign in 1904 as I did in 1882. During the last fifty years the Hudson's Bay Company has exported 40,000 to 90,000 Mink skins each year from the Northwest. On exceptional years the number has far exceeded these highest figures, but the supply continues about the same. Reckoned by area, about one-thirtieth of these come from Manitoba.

SOCIABILITY

So far as known, the only exceptions to solitary life among Minks are during the mating season, and while the young are

[2] Mam. Adir., 1884, pp. 65–6.

with the mother. As these are strictly family groups, they are not real exceptions, and the Mink must be considered an unsociable animal.

The rudiments of the mud-pie telephone, as described in the Muskrat chapter, are found among Mink, but this appears to be their sole impersonal mode of intercommunication, and a very poor one at that.

VOICE

The only sounds I have heard the species utter are a growl, a deep savage snarl, a louder snarl of defiance that is almost a scream, and finally a shrill screech when it is in a trap.

Kennicott credits[3] it with uttering a remarkable shrill, twittering squeak, not unlike that of a bunting, but this only when hurt or excited. To these we must add the loud sniffing, which, while it is merely an attempt to smell clearly, also conveys to another Mink the idea that here there is something which is probably worth while approaching and smelling.

MATING

Most naturalists believe that the Mink is polygamous or possibly polyandrous. Thus Kennicott says:[4] "The Mink is not at all gregarious and does not even live in pairs. During the love season, which occurs in February or March, according to the climate, the female is accompanied by one or more males." And in the Resseque Minkery one male commonly served six females.[5] However, the fact recorded by many observers that during the mating season the males fight desperately to a finish, is directly opposed to any theory of polyandry.

We are told, further, that the females in the Minkery "come in heat with great regularity, all being ready for the male within ten days, and the period of excitement lasts about four days."

The following from the pen of Bachman illustrates their habits at this season:[6] "The latter end of February or the beginning of March in the latitude of Albany, N. Y., is the rutting season of the Mink. At this period the ground is usually still covered with snow, but the male is, notwithstand-

[3] Quad. Ill., 1858, p. 103.
[4] *Ibid.*, p. 102.
[5] Coues, Fur bearing Anim., 1877, p. 182.
[6] Quad. N. A., 1849, Vol. I, p. 258.

ing, very restless, and his tracks may everywhere be traced, along ponds, among the slabs around saw-mills, and along nearly every stream of water. He seems to keep on foot all day as well as through the whole night. Having for several days in succession observed a number of Minks on the ice hurrying up and down a mill-pond, where we had not observed any during the whole winter, we took a position near a place

FIG. 211—Mink about one-fifth of life size.
From pen drawing made by E. T. Seton for the Department of Agriculture in 1888. Published in the Report of the Ornithologist and Mammologist 1889, p. 488.

which we had seen them pass, in order to procure some of them. We shot 6 in the course of the morning, and ascertained that they were all large and old males. As we did not find a single female in a week, whilst we obtained a great number of males, we came to the conclusion that the females during this period remain in their burrows."

The question is, however, far from being settled. Many observers have seen Minks in pairs together, sharing each other's lives and fortunes to some extent.

Thus Charles G. D. Roberts tells me that one year about midsummer as he was carrying his canoe around a log jam on the Nashwaak River, N. B., he saw two full-grown Mink travelling together, and evidently associated. They left the log jam just as he had done, to travel down the bank; as they came near he stood still to watch them. One ran by him and over his foot; the other, a large one that he took to be the male, ran past, some ten feet aside. Beyond him they resumed their journey and took to the water again about forty yards below.

Finally, we are told that in the Cancandea, N.Y., Minkery:[7] "About the middle of March the females are separated from the males until the young are reared. The necessity for this arises from the fact that the males seem inclined to brood the young almost as much as the dam, when both are permitted to remain together." This is strong evidence that in a state of nature the male Mink is a model father, which necessitates that he be also a faithful mate.

In Manitoba, pairing takes place in March and may extend even into mid-April.

There is hardly any low situation near the water where a Mink will not make its den. Long burrows in banks, holes under logs, stumps or roots, and hollow trees are favourite places, but crevices of rocks, drains, and nooks under stone piles and bridges are frequently selected.

If the burrow is one dug by the animal itself, it is about 4 inches in diameter and, if in ordinarily easy digging, it may continue along for 10 or 12 feet at a depth of 2 or 3 feet.

On the prairie Kennicott found the Minks "living in burrows, often 6 or 8 rods in length, on high ground, from which long galleries extend to the edge of a slough or pond. These galleries, however, are not formed by the Minks, but by Muskrats, which dig them in order to place their nests beyond the reach of high water and yet have subterranean communication with the stream."[8] Similarly, he credits it with occupying the burrows of the Badger and Skunk when in a suitable locality, and finally with frequently digging dens in old ant-hills, presumably because these are dry elevations near the water.

NEST

At the end of the long, crooked, 4-inch tunnel is the nest, or nursery den, of the family. This is usually described as a

[7] Coues, Furbearing Anim., 1877, p. 184.

[8] Quad. Ill., 1858, p. 102.

chamber about a foot in diameter, warm and dry, and well lined with fine grass, feathers, and any other soft material that is handy.

Sometimes the nest is in a Muskrat home, whose rightful owners have been dispossessed, probably eaten, but the lining and finish are said to be the same as in the underground den. So far as known, it is made and guarded solely by the mother.

The male, however, has a den of his own, probably a number of them, scattered over his home-range, as already noted. Indeed, it is tolerably certain that at all seasons each Mink has one or two refuge dens at convenient parts of its home-region, where he or she can count on a comfortable nest no matter what weather may be in possession of the outside world.

GESTATION, ETC. According to all authorities, the period of gestation is exactly 42 days.

In Manitoba, the young are born about the last week in April or first of May. They are usually 5 or 6 in number, but may vary from 3 to 10. On arrival they are about the size and shape of a little finger, pale in colour, blind, naked, and helpless. Their eyes open when they are five weeks old, and now the little creatures begin to look like Mink, for they are covered with a close fine coat of fur.

On June 28, 1883, I found a young Mink lying under some brush on a sort of trail between two ponds among the Sandhills, near Carberry. I was led to it by its plaintive squeaking. It was well formed, but its eyes were not open, and I saw nothing of the mother or of any other young Mink. Why it was there is a puzzle, as this did not seem to be a nest. Possibly the mother was moving her brood to new quarters and left this in a temporary resting place. I carried it home. Its eyes opened about July 1. Reckoning backward, this one must have been born about May 29 and engendered in mid-April.

About this time their eyes are opened and the mother begins to supply them with solid food.

The following, by Dr. T. S. Roberts, of Minneapolis, is a graphic picture of the mother's life and labours for her young at this season:[9]

"While engaged in geological work on the Cedar River, near Osage, Ia., my attention was attracted by the peculiar actions of a Mink (*Putorius vison*). By careful manœuvring we were enabled to approach to within a short distance of where it was engaged, and there watch its behaviour unobserved. It was an old mother Mink engaged in fishing for her young. On the ripples in the centre of the stream, where the water was not more than two feet in depth, was a flat drift boulder rising a few inches above the surface. On this rock the mother Mink would take her position and here watch for small fish to approach, when she would dive into the water, be gone for a moment, and then reappear on the opposite side of the rock, usually with a fish in her mouth, which she would deposit in the centre of the stone and its struggles instantly stopped by a quick, sharp bite at the back of the head, which caused immediate death. This process was repeated without intermission, except to stop for an instant to shake the water from her furry coat, until 7 fish, varying from 4 to 7 inches in length, were deposited on the rock. Then, without stopping to rest, taking one fish in her mouth, she plunged into the water and swam ashore, climbed up the steep bank, and ran hastily to her young, in a burrow under an old stump on the bank of the stream, fifty yards away. In a moment she was seen returning, plunged into the stream and swam to the rock, took a second fish in her mouth, entered the river once more, and returned to her young as at first. This was repeated until all the fish had been carried away. A few moments after having removed the last fish, she returned and began her work once more. This time, however, her labours were without result, so, shifting her position to another rock in the stream, a short distance away, she continued her fishing. But, although more than a quarter of an hour was spent in energetic effort, her labours were without avail,

[9] Mam. Minn., 1892, pp. 127–8.

and she was this time compelled to return to her young empty-handed.

"From the bank of the stream, where egress from the water was made, to the burrow fifty yards distant, a well-beaten path had been formed by the mother Mink in her daily excursions in quest of food for her young."

As late as June 28 they are still in the nest with the mother in the country about Methy Lake, as evidenced by a note in Richardson's last "Journey":[10] "A female Mink (*Vison lutreola*) was killed as it was crossing a bay of the lake. It had 8 swollen teats and its udder contained milk; so that probably its death insured that of a young progeny also."

YOUNG Soon after their eyes are open the young Mink seem to go forth into the world under the guardianship of the mother. W. R. Hine tells me that he has often seen the old one with her brood in June. At such times she shows fight facing one, and uttering a snarling screech; the young meanwhile escape to the water. By now they have doubtless abandoned the nest.

One of my Kippewa guides, Archie Miller, relates that in a great forest fire near Temiscaming Lake, in July, 1901, he saw an old Mink with her 4 young swimming up a creek towards a lake. The woods were blazing on both sides and they were travelling up stream to escape, coming up for a breath, then diving and swimming under. They were hard pressed, nearly worn out indeed, and could not swim more than 25 feet without coming up. They were the size of a Red-squirrel; only *one* parent was seen.

Another guide, Edouard C. Crête, contributes some interesting observations on the home life. He says that one year he was staying at a hay camp, 11 miles west of Deux Rivières, Ont., from July 12 to 22. Every day, for a week at least, after the 15th, an old Mink came with 5 young ones to feed on the rubbish thrown out. They were there twice a day, morning and evening, regularly until he went away.

[10] Arc. Search Exp., 1851, Vol. I, p. 109.

Yet another interesting glimpse of the family life is furnished me by A. Barton Hepburn, of New York. When he was a boy living on the home farm at Colton, N. Y., he was going with his father one day late in June across an alder brook by the road bridge when they saw in the bushes to one side an old Mink with 5 young ones that were about one-quarter grown. They were following her, but when they came to the road, they held back and would not quit the cover to cross the road. She made several efforts to coax and lead them on, but they were timid. At length she seemed to lose patience; she seized them, one at a time, by the neck and so carried them across to the opposite thicket, where they continued their journey. He saw nothing of the father Mink and does not remember whether or not the mother made any sounds.

The little ones continue with the mother until the middle of August; they have now learned something of the ways of life, the family breaks up, and henceforth all are seen wandering alone. They are now about half grown in point of weight. As usual, the females mature sooner. We learn from Resseque[11] that they attain to their full stature in ten months, and reproduce when one year old, "while the males are not full grown until they are a year and a half old. It is noted that in every litter one or the other sex predominates in numbers, there being rarely half of them males and the other half females."

There is but one brood each season.

Fish are perhaps the Mink's choice food, and it delights in taking them by open pursuit in the clear water. Although it is inferior to the Otter in this craft, Audubon and Bachman record that they have seen one catch a trout a foot long.[12] A quadruped that can catch a trout that size can catch anything that swims in the smaller streams. Those that live along the prairie sloughs feed chiefly on frogs, tadpole, Mice, and Muskrats. The latter it follows under water into their FOOD

[11] Coues, Fur-bearing Anim., 1877, p. 182.

[12] Quad. N. A., 1849, Vol. I, p. 255.

burrows or their homes, killing and devouring them in spite of a most desperate resistance. The Muskrat is a noted fighter and always dies game; still it dies when it meets the Mink.

Dr. E. Coues thus[13] condenses M. A. Howell, Jr.'s account of a Muskrat-Mink adventure: "Whilst Snipe hunting on a marshy island below the Kickapoo Rapids of the Illinois River, the writer noticed an object which appeared like a ball some six or eight inches in diameter rolling towards the water, and soon ascertained that it was a Mink and a Muskrat clinched together, and so completely covered with mud as not to have been at first recognized. At his approach the Mink released its hold and made its escape, but the Muskrat was already dying of severe wounds in the head and neck, from which the blood was flowing profusely. The Muskrat had evidently been captured and overcome in fair fight by broad daylight, and the Mink would have devoured its victim had not the hunter interfered."

E. W. Deming, the New York animal-painter, informs me that on Green River, of Illinois, the Muskrat is a regular food of the Mink. He once found the remains of 10 Muskrats in a Mink den. This is a typical and extremely carnivorous record, but nothing in the way of flesh, fish, or fowl comes amiss. It is delighted with the chance to rob the sportsman of a string of fish or a wounded duck, even seizing the latter before the gunner's eyes, and I have followed its track through the snow in Ontario to read the grewsome story of its running down and devouring a Gray-rabbit. From all accounts it often amuses itself with preying on house-rats where they abound. All kinds of birds and eggs are most acceptable food when it can find them. Not rarely it quarters itself on the hen house, killing each night for food, and especially relishing the blood and brains of its victims.

It will prey on snakes and clams when nothing better turns up, and I have several times followed its tracks at Lake Winnipegosis to learn that frogs, crayfish, and carrion were staple articles of its diet.

[13] Fur-bearing Anim., 1877, p. 178.

HABITS AND CHARACTER

The Weasel is a sanguinary little incarnation of fury and valour, with but little cunning; it is low in intelligence and incapable of friendship with man or any one else. The Otter, though a Weasel in pedigree, seems to have responded to the elevating and gentling influences attendant on the fisher-life. It is the least destructive, the most docile and intelligent of the Family. The Mink is half-way between in habits and character, as it is in food and haunts. After sojourning in the reeds along the river for a time catching fish and killing Muskrats in Otter-fashion, or running down Rabbits and Mice Weasel-fashion, it may set out across country to find better hunting and happen, in its travels, to discover the real Happy Hunting Ground in the form of some farmer's barnyard. Very naturally, it settles down in this ideal spot—didn't it set out to find this very thing? —this highly populated wilderness of buildings and sheltered nooks is perfect and here "every prospect pleases—only man is vile." The Mink's attitude toward this game preserve is quite different from that of the lesser Weasels. *They* are mad to kill—kill—kill; they will, if possible, kill everything there in one night, then leave the ruined place to seek some new field of carnage. Not so the Mink. It has but little of the killer spirit. It kills because it must eat, and, having found the well-stocked henneries, it says to itself, "Here now will I settle down, eat, drink, and make merry, for these are mine own preserves by right of discovery, and I will defend them against all invaders." On the list of invaders it puts the farmer and his family, and his dogs and his cats, and all those that put their trust in him. From safe hiding under the barn or in the log-pile it sallies forth at night to kill and eat; sometimes one fowl each night for many nights in succession; sometimes it yields to the blood-lust (not unknown among mankind), and kills half a dozen of the defenceless prey, feasting only on those choicest parts of all, the blood and brains, just as the Buffalo killer would shoot down half a dozen Buffalo because it was so easily possible, and then take nothing but the tongues.

Usually the Mink is killed before leaving the barnyard precincts, but it often happens that a number of narrow escapes from shot-guns or dogs decide it to move on. In the hours of the night it goes forth, bounding with high-arched back. Its speed is not great, but, like all Weasels, it is possessed of endless strength and doggedness, and though a man can outrun it on the open and outwalk it travelling, its steady bounding may take it miles away before morning. As it journeys it is ever on the alert for guidance from its nose. There are a thousand accidents to turn its steps one way or another; the cluck of a grouse, the rustle of a Mouse in the grass, an easier path, a promising odour in the wind, the wind itself, may each and all give trend to its tireless bounding and bring the hunter at last to some marsh-land of promise, or mayhap another barnyard, wherein it may settle down again to comfort of a kind, taking, undoubtedly, its life in its teeth while doing so, a condition that it has not the wit to think about, and if it had it would simply dismiss the thought, viewing this merely as a normal condition of all existence. The Mink certainly never spent a moment of its life without being under the shadow of impending death, and as certainly it never lost a wink of sleep through thinking about it.

If surprised during its hunting or suddenly brought face to face with man, it often rises up on its hind-quarters to get a better view; in this position it looks extraordinarily long. I once met one out on the prairie. It rose up to scan me from every one of its twenty-four inches of stature, and stood so till I came within ten feet and removed the top of its head with a thimbleful of sparrow-shot.

This species is active and hunts chiefly by night, but is often seen in the daytime, especially in the mating season and in the fall.

CHARACTER

A tame Otter makes one of the most engaging of pets. No normal man ever made a pet of a Weasel, but tame Minks have often proved most docile and interesting little creatures, capable of forming strong attachment, although ever ready to bite when provoked.

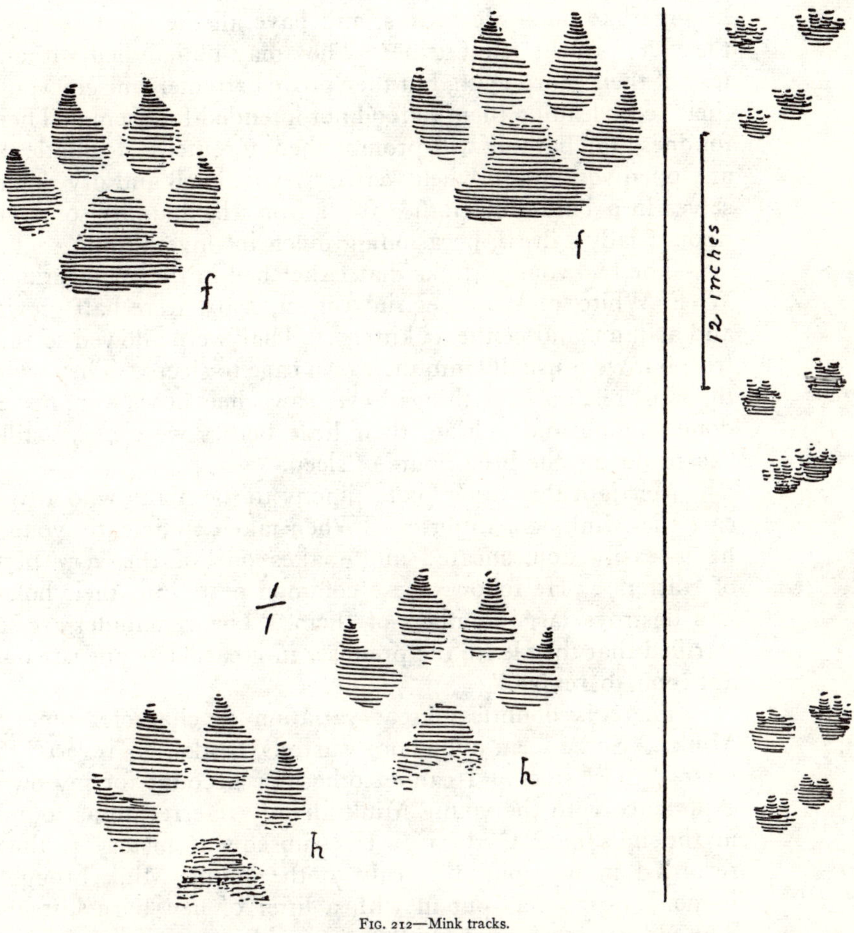

FIG. 212—Mink tracks.

Winnipegosis, Man., Sept. 2, 1904.

The four large ones are exactly life size, and were made by a very large Mink. Only four toes show in each, though it has five toes on each foot.

The small series on the right show the ordinary speeding gait of the Mink; about 12 inches is covered at each bound.

Resseque, of the Minkery, says:[14] "If taken in hand when their eyes are first open, they are readily tamed; they should not subsequently be allowed to remain with the mother or in each other's society. By continual petting and handling they become like domestic ratters, and have all the playfulness of the young of the feline tribe. They may be handled without fear of their sharp teeth, but they prove extremely mischievous, their scent leading them to food not intended for them. Their fondness for bathing will prompt them to enter a teakettle or any open vessel, and when wetted they will roll and dry themselves in a basket of clothes fresh from the laundry, or even upon a lady's dress, occasioning much inconvenience."

The two young Minks that I sketched in the menagerie of W. F. White, of Winnipeg, in August, 1905, were half grown and as tame and gentle as kittens. They were allowed to run free, but went usually into their own cage to sleep. They were the most restless little things I ever saw, their noses were never done sniffing and poking, their little bodies were never still, except during the brief hours of sleep.

Merriam thus adds his testimony to the many who advocate the Mink as a ratter:[15] "When taken sufficiently young he is easily domesticated and makes one of the very best of 'ratters.' He follows these common pests into their holes and destroys large numbers of them. The remainder are so terrified that they leave the premises in great haste and are not apt soon to return."

There is doubtless great variation of character among Minks. Some seem to be born wastrels, freaks, or 'reverts' to their Weasel stock. I cannot otherwise account for my own experience with the young Mink already referred to as found in the hills near Carberry. His subsequent history is thus recorded in my journal: July 7, the young Mink brought home June 28 was put in with a litter of new-born kittens. The old cat received it kindly and suckled it as if it were her own. In a few days its eyes were opened. It was now strong and its horrid nature began to show itself. It often milked the

[14] Coues, Fur-bearing Anim., 1877, p 183. [15] Mam. Adir., 1884, p. 67.

cat till it was satisfied, then turned on its gentle foster-mother, trying to tear her. Several times the cat cried out but, seeing no marks, I thought it must be a trifling hurt; still she would not injure the little brute, but continued to nurse it tenderly. More than once I had to save one of the kittens from its jaws. To-day I found the hens in an uproar, and, on going to the place, found an old clucker rushing about with the diminutive Mink fastened to her neck. I choked it off and put it back with the kittens. This afternoon I chanced to lift the cat up and

FIG. 213—Mink poses. (From life).
Winnipeg, Aug. 24, 1904.

was horrified to find the Mink had bitten off four of the foster-mother's teats, lacerated her whole breast, and eaten a hole nearly into her bowels, and yet mother love made the poor creature keep on without harming the devil she was rearing.

All of this ends to-day.

The Mink has few enemies that are dangerous, for it is a dangerous fighter, but the great horned owl must be reckoned with, as is shown by the following from A. N. Cheney.[16] A friend of his was at the opening of his tent one nightfall, at Triton Club Lake. He had been "fishing, and his trout were in front of his tent, when a Mink came from somewhere and seizing a trout in its mouth made off with it, but before the Mink was out of sight a big owl swooped down, grabbed the Mink and made off with it into the air, and the last he heard of the Mink it was screaming pitifully up towards the zenith." ENEMIES

[16] Forest and Stream, May 7, 1898, p. 371.

When cornered and at bay he is the embodiment of savage ferocity. Dr. Coues has given us[17] a masterly picture of a Mink in a trap. "One who has not taken a Mink in a steel trap can scarcely form an idea of the terrible expression the animal's face assumes as the captor approaches. It has always struck me as the most nearly diabolical of anything in animal physiognomy. A sullen stare from the crouched motionless form gives way to a new look of surprise and fear accompanied with the most violent contortion of the body, with renewed champings of the iron, till breathless, with heaving flanks, and open mouth dribbling saliva, the animal settles again and watches with a look of concentrated hatred, mingled with impotent rage and frightful despair."

This is the picture of a man who had seen it. He had eyes, but surely he lacked both ears and nose, else he had recorded the piercing screech of fear and fury, and the all-pervading, far-reaching, skunk-emulating musky stench with which the trapped Mink never fails to saturate the air, the place, the trees, the breeze, and so proclaim afar and afterwards that hereabouts a Mink was held in direst straits.

BATTLING When Mink meets Mink, the battle is worthy of such desperate and valiant warriors, and they meet much and often at certain seasons. So it is surprising how few have witnessed the deadly engagement. The following description is that of a man who beheld it with his own eyes.[18]

"It was my good fortune to witness a fight to death between a couple of male Minks a fortnight ago, one of the most fearless as well as silent struggles I ever saw. In one of the largest stone quarries in western Ohio, a small clear brook grows larger, deeper, and wider by the help of the clear sparkling water of many a spring, issuing from the crevices of the limestone and flowing through little hollows and around pieces of stone that have been thrown out as worthless. In the course of years the accumulation of rubbish or 'dump,' as it is termed, has made a perfect retreat for Minks, Weasels, and Rabbits.

[17] Furbearing Anim., 1877, p. 176.

[18] Williamsburgh, Ind., Forest and Stream, Nov. 24, 1892, p. 444.

Having a considerable amount of dirt in it, the quarry affords a paradise for such animals that are in the habit of burrowing close to small streams. One afternoon as I was on my way to another block to give some instructions to my men, I accidentally glanced towards a small sandy spot running out into the stream, forming a peninsula or cape by the junction of a little brooklet and the main or larger brook. The spot was bare, with the exception of a few Spanish needles, and was overhung with small willows. Standing within six or eight feet of me were two large male Minks, eyeing each other intently, their small eyes looking like coals of fire, and with mouth slightly open, their backs arched, necks stretched forwards, their hind-legs close to the ground, they looked very much like miniature tigers waiting an opportunity to spring towards each other.

"The first motion was a quivering of the bodies, followed by increasing rapidity in breathing. Then slowly each advanced towards the other until 18 inches apart, when each seemed to be gathering all the strength possible to make a final leap. After standing motionless for a full minute, one of them turned his head a little to the right, and instantly the other did the same. Probably ten seconds elapsed while in this attitude when, quicker than the eye could follow, they sprang towards each other, and, catching by their throats just back of the lower jaw, held to each other with a grip that meant death to the one that should let go his hold. With a determination to conquer or die, they wound each other round and round until exhausted from loss of blood, when, spreading their fore-feet to brace themselves, they stood perfectly still until one of them staggered, then fell, pulling the other with him, laying them both down in the sand, growing weaker and weaker until they ceased to breathe. When they were dead I picked one of them up, and the jaws of each were so set that it was with difficulty I pulled them apart. The battle was probably of a minute's duration, in which time the animals never noticed my presence, or, if they did, they made no sign of their knowledge of it. I presume the object was for one to catch the other so close to his lower jaw that he could get no hold upon his enemy's throat, when a very short time would decide the battle."

STORAGE HABITS

There is a phase of the storage habit that is well developed in the Mink, and it must be distinguished from the carrying home of provender for the young. The first illustration at hand is given by Merriam,[19] who found by the autumn nest of a solitary old Mink "the remains of a Muskrat, a Red-squirrel, and a downy woodpecker." I should like much to know the three very different chapters of hunting represented by these three captures.

SPEED

I have several times run after Mink on the open prairie and found it quite easy to overtake them. From this I should estimate their best speed on land at 7 or 8 miles an hour. Their bounds in ordinary travelling are from 10 to 15 inches clear. Though easily overtaken, catching them is quite a different matter, for they dodge with marvellous adroitness, and they are quick, too, at reading the little nature finger-posts that tell which way to run for a Badger hole or other haven in the earth.

On October 3, 1884, while following a Mink through the snow in the Sandhills south of Carberry, I came to where it had tobogganed itself down a long hill, for a distance of 18 feet, after the manner of an Otter.

On the water I should estimate its best travelling speed at 1 to 1½ miles an hour. This is not reckoning the dive or under-water spring that it can and must make to catch fish. It is much swifter than the Muskrat, but apparently cannot dive so far. I once saw a young male Mink hunted down among some floating logs. He might have escaped had he dived and swum fifty feet to the cover along shore, but he did not, apparently because he could not swim so far under water.

STRENGTH

Its strength is illustrated in the following incident:

While Duck shooting at Swan Lake, Man., October 18, 1901, H. W. O. Boger, of Brandon, shot a mallard which fell

[19] Mam. Adir, 1884, p. 66.

on the newly forming ice. An hour or two afterwards he saw a Mink come from the shore on the ice, which was then half an inch thick; it seized the mallard by the neck and dragged it away on the ice. Boger was seventy-five yards away; he shouted, but the Mink gave no heed; it dragged the duck to the rushes and disappeared with it.

Coues mentions[20] a similar case wherein the Mink dragged a mallard half a mile to get it to its hole. As a full-grown Mink weighs but 2 pounds, and a mallard over 3, it is as though an ordinary man had dragged a 200-pound man for half a mile, and did it with little difficulty.

Charles Hallock records[21] that he has known a Mink to come and steal his trout as fast as he caught them, until it had gone off with an aggregate weight of 12 pounds.

This species is not a climber in the sense that a Marten is, but it can and does occasionally go aloft. Dr. T. W. Gilbert, of Carberry, brought me a Mink that he shot out of a poplar tree at a height of 15 feet. In Minkeries it is found that though they cannot climb on a smooth surface, they easily go up a rough tree trunk or fence. CLIMBING

All of the Weasels have anal glands which give off a very strong and more or less offensive smell when the animal is excited. The Skunk, of course, is the grand master in this department, but the armament of the Mink is not to be despised. It cannot squirt its liquid musk out to a distance as can the Skunk, but it can and does pour forth a loathsome plenty when the proper occasion has, in its opinion, arrived, and this is whenever it considers itself in peril of its life, or is suffering grievous bodily harm, or enraged against a rival, or struggling in a trap. Merriam considers it a more unbearable stench than that of the Skunk, and adds:[22] "It is the most execrable smell with which my nostrils have as yet been offended, and is more powerful and offensive in some individuals than SCENT-GLANDS

[20] Fur-bearing Anim., 1877, p. 179. [21] *Ibid.*, p. 180.
[22] Mam. Adir., 1884, p. 67.

in others—the difference probably depending upon season and age. In one specimen the fetor was so intolerably rank and loathsome that I was unable to skin it at one sitting, and, I am free to confess, it is one of the few substances of animal, vegetable, or mineral origin that have, on land or sea, rendered me aware of the existence of the abominable sensation called *nausea.*"

MIGRATION

During October and November I have sometimes thought I saw signs of migration among the Mink, but it was quite erratic and may have been nothing more than the general rush for good places in which to settle for the winter, before the frost imposes on them a marked change in life. At this time I have often seen them out on the open prairie or in the daytime far from cover. At one or two farmhouses near Carberry, where I learned that Mink had arrived after the first snow, the tracks came from the north-west, but this may not have meant anything.

TRAPPING

Though more wary than Skunk or Weasel, the Mink is easily trapped. Some of the old ones that have had painful experiences, become cunning, but most of them are unsuspicious of danger in any inanimate form, and will enter the most obvious of traps, especially if they be baited with the head or blood and brains of some large bird, delicacies that have as strong an appeal to the Mink as catnip for a cat, or honey for a Bear.

In the fall of 1886, I put out a steel trap for a Mink that used to travel up the old DeWinton Slough, back of Carberry. The trap was set, by luck, just the day before he passed that way, but, unfortunately, a ruffed grouse chanced to run through the thicket and get into the trap, so the Mink, coming on the scene, discovered a feast ready prepared for him. On returning next day I found the remains of the grouse with other details of the affair, so reset the trap in the same place. Lutreola was lurking near; next night I caught him by the front foot, but he gnawed the foot off and escaped. I was prevented returning to the trap for several days. Then I found the same Mink caught in it by the hind-leg. He had gnawed off the leg, but *beyond the trap*, and was still held by the stump. He was dead, and buried too, for he had raked together all the leaves, grass, earth, and sticks within the length of the chain. Trap and all were hidden; only his head and his remaining front arm were out of his self-made funeral mound.

The old-fashioned deadfall is the trap that should be used, as it does not injure the fur and it kills the animal instantly, so that there is no unnecessary suffering. The box-trap is effectual and humane if visited regularly. It should have, at the back, a window covered with ¾-inch mesh wire netting. It has the advantage of protecting its catch from passing marauders. The steel trap, if used, should be visited often. The less the creatures suffer the better the fur. In the unhappy case cited above the pelt was worthless.

The fur of the Mink is a staple of the trade; it is a close, strong, beautiful fur, of great durability. Its market value varies much with the caprices of fashion, but it has always fetched a price that makes it worth catching. The high-water mark for Mink was about twenty-five years ago, when prime dark skins brought from $4 to $10. Dr. Merriam records[23] having then sold one of unusual size for $14. FUR

Mink to-day is not high but rising. At the London annual fur sales, held at Lampson's, in March, 1906, 126,161 Mink skins were sold. The highest price paid was 40 shillings ($9.60) each, for 102 extra large prime dark skins; 20 shillings to 30 shillings ($4.80 to $7.20) was a more usual price for first-class skins; while third- and fourth-class skins went as low as 3 shillings (72 cents). But this is a fur of standard attractions, and other furs are getting scarcer; therefore we can count on a steady rise in Mink.

This is a cased pelt. It is prime from the first of November to the first of April. It is unlawful to trap or destroy it in Manitoba at any other time.

[23] *Ibid.*, p. 66.

During the eighty-five years, 1821 to 1905 inclusive, the Hudson's Bay Company collected 3,503,660 skins of this species, an average of 41,219 for each year. The lowest was 4,549 in 1822; the highest, 90,080, in 1876. The average for the ten years, 1895 to 1905, was 57,729.

Poland's lists show that during the seventy-one years, 1821 to 1891 inclusive, 7,993,719 skins were taken by the other American companies, an average of 112,587 each year. So that the average annual catch of Mink for fur is about 154,000.

In the year 1889 the total catch was about 400,000, or to be exact, 395,470 were marketed, with York Factory returns not included, as they did not arrive.

The Mink returns show a steady general increase which seems to prove an increase of the Mink population since white men have possessed the country.

BREEDING FOR FUR

The high price that the fur commanded some twenty five years ago led several persons to try breeding Mink for the market. They multiply readily in captivity and are easily managed, so that the project seemed assured of success, when suddenly the fashion changed. Mink 'went out,' the price dropped below the possibility of profit, and ended the scheme. There is, however, every reason to believe that this was a temporary drop. On all hands we are confronted by these facts: the wild fur supply cannot be materially increased, the demand is getting greater, the prices are steadily rising. The fluctuations caused by fashion do not affect the main issue: a high-class fur will always fetch a high price. The breeder can greatly improve his stock by selection and so make all his product high-class. Many kinds of fur are breedable; one, at least, is sure to be in fashion.

For the guidance, therefore, of those who wish to embark in such an enterprise, I give a brief account of the successful methods of the Minkery.

Mink-farming.

Mink are easily raised in captivity, and when their fur brings a good price, as at present (1908), mink-farming is a fairly profitable field for small capital. At least a dozen Minkeries have been operated since the days of Resseque, who carried on the first of which we have detailed accounts. This was in the early 70's, at Verona, Oneida County, in northern New York State.[24]

The essential principles of a successful Minkery are intelligent individual care of each animal, perfect cleanliness, and moderate and varied feeding.

A convenient Minkery for 10 breeding females and 2 males might be thus planned:

In the end are 12 cages, in 2 tiers. Each cage is 5 feet wide, 3 feet high, and 10 feet deep; made of close galvanized wire of ¾-inch mesh, with solid wood floor. In each is a wooden nesting box, 18 inches long, 9 wide, by 6 high. This has a 3-inch doorway at one end of the long side, with a sliding door, which may be worked from outside to shut the animals in the nest, if desired. The door of the cage should be so large that this box can be lifted in and out. A long trough in front of each cage should contain running water.

The two larger pens are open courts into which are doorways opening from each cage. These should be surrounded by a tight 6 or 7 foot board fence, which is sunk in the ground, and there rests on a 2-foot underhang of stone, cement, zinc, or galvanized mesh wire. At the top of the fence, and across the front of the top cages, should be a 2-foot overhang of smooth boards or tin, and at the corners it should be 3 feet wide.

Mink are good climbers and diggers, but this arrangement will keep them in. Of course, a stone and cement floor, with a mesh wire roof, is surer, but more costly.

[24] Described in Coues's Fur-bearing Anim., 1877, p. 181; Forest and Stream, October 22, 1874, and Fanciers' Journal and Poultry Exchange, October 15. 1874.

The two runs are separated by a ¾-inch mesh wire with tin overhangs, and a tank of running water at the boundary affords bathing for both runs. These are shown here, about 15 by 20 feet, but would be all the better if ten times as large,

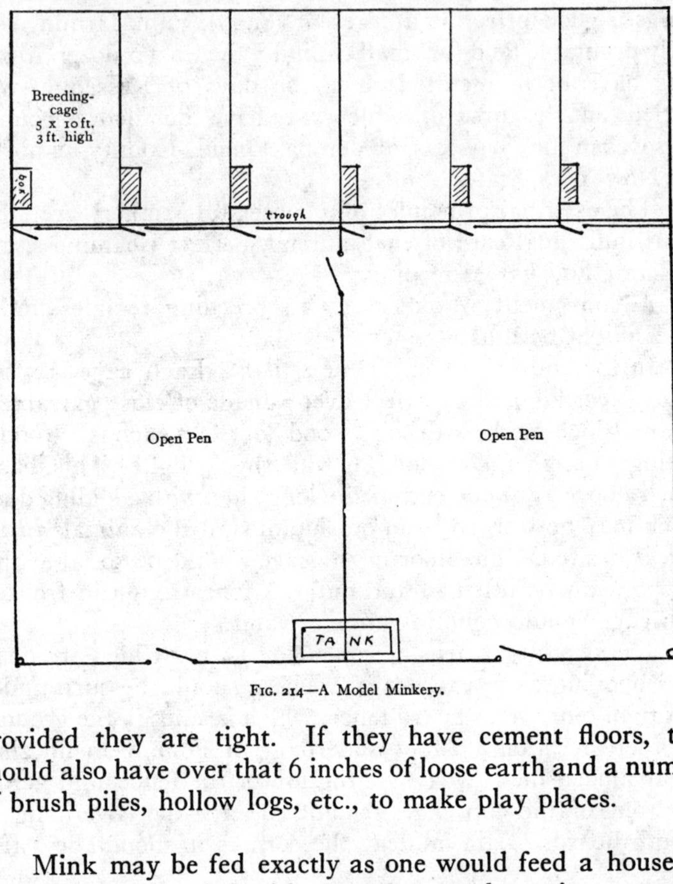

FIG. 214—A Model Minkery.

provided they are tight. If they have cement floors, they should also have over that 6 inches of loose earth and a number of brush piles, hollow logs, etc., to make play places.

FEEDING Mink may be fed exactly as one would feed a house-cat —table scraps, varied with meat two or three times a week. Fish is very much to their liking, and may be given nearly every day if other things are used. Bread and milk, johnny-cake, etc., should be added for variety; even raw liver may be given sparingly at intervals, but cooked food, as a rule, is safer.

Two light meals, morning and night, or one substantial meal, late in the day, is sufficient; and at all times an abundance of clean water.

STOCKING

Mink may be bought either from advertising dealers or through advertisements in the country newspapers. They bring from $1 to $10 each, according to size, age, and beauty. One male to half a dozen females is a usual allowance.

The best come from Labrador, the poorest from the Mississippi and the South-west.

GENERAL MANAGEMENT

It has proved far the best to keep the old ones apart. One Mink, one cage, should be the rule for them; except, of course, in February—the breeding season—during which month the male may be allowed to run with the half-dozen females, all in one large pen. They should be watched, however, and quarrellers removed at once. Females that are known to have been served may be removed, at least for a few days; after which their behaviour, on again meeting the male, will show whether they have need for further attention.

As a rule, all females come in heat within two weeks. By March 7 the season is over and the animals should be returned each to its own cage. Non-breeders that are sure not to quarrel may be left together in one cage.

After six weeks' gestation the young are born; they number 2 to 6. They are blind and helpless for some weeks. When about six weeks old, usually near June 1, they begin to come out of the nest and take an active interest in life, following the mother and eating her food.

They are now easily tamed, gentle, and playful.

By August, they are weaned and all may be taken from the mother and turned loose in the main run. The more they are *gentled* by handling the better.

Their fur is ready for marketing at Christmas, is good any time from November to March; but Mink fur is greatly improved by keeping the animal till two years old, and, of course, by castrating the male.

Keep the best always for breeding. Do not use exceptionally fierce individuals for breeders. A fast day once in two weeks is a good thing for fat animals.

Some breeders clip off the nail and first joint of the two middle toes on each front foot; this prevents the Mink climbing or digging. It is, however, a disfigurement as well as a cruelty.

The cages must be kept clean and the earth in the runs overturned or refreshed once a week.

These animals are exceedingly clean and cause no smell, except, perhaps, when mortally hurt. They are easy to keep, need but little space and food, and breed regularly.

PROFITS When the fur is down to Skunk prices—that is, $1 to $3 a skin—there is no money in breeding Mink. At present the market is up, prime dark skins being $4 to $8; exceptionally fine dark skins have gone as high as $15 and $18. At these prices there is reasonably good money in the business; more especially if the Minkery is on a small scale and carried on in conjunction with a hen-yard or a squabbery, whose by-product will almost feed the fur-bearer.

XL.

The Marten, Saskatchewan Marten or American Sable.

Mustela americana abieticola Preble.

(*Mustela*, Latin for Weasel, applied by Linnæus to the European Marten, *Mustela martes; americana*, American; *abieticola*, from *abies*, a fir-tree, and *colere*, to inhabit, that is, 'fir-inhabiting.')

Mustela americana TURTON, 1806, Linn. Syst. Nat., I, p. 60.
TYPE LOCALITY.—Eastern North America.

Mustela americana abieticola PREBLE, 1902, N. A. Fauna, No. 22, p. 68.
TYPE LOCALITY.—Cumberland House, Sask.

FRENCH CANADIAN, *la Marte: la Fouine.*
CREE & MONTAGNAIS, *Wah-pe-stan'.*
OJIB. & SAUT., *Wah-be-jay'-she.*
CHIPEWYAN, *Tha.*
YANKTON SIOUX, *Mah-ha-pah-skay-chah.*

The genus *Mustela* (Linnæus, 1758) is much like *Putorius*, but consists of larger animals, with bushier tails that are half as long as the body; large ears, and arboreal habits; they do *not* turn white in winter, that is, their winter coat is not very different in colour from that of summer. The dentition differs from that of *Putorius* chiefly in having 4 more premolars, and is as follows:

$$\text{Inc. } \frac{3-3}{3-3}; \text{ can. } \frac{1-1}{1-1}; \text{ prem. } \frac{4-4}{4-4}; \text{ mol. } \frac{1-1}{2-2} = 38$$

SIZE In addition to the generic characters, the Marten has: Length, about 25 inches (635 mm.); tail, 8 inches (203 mm.); hind-foot, $3\frac{3}{4}$ inches (95 mm.). (For skull, see Plate LXXXI, facing p. 968.)

COLOUR In general it is of a rich dark yellowish-brown, shaded into blackish on the tail and legs, and into gray on the head, with ear linings of dull whitish, and a large irregular patch of pale buff or orange on the throat and breast. But the general colour ranges in any given locality from golden-yellow to blackish-umber, and the patch on the throat from orange to white. The claws are white.

When seen in the trees it gives the impression of a large dark Squirrel, with white ears and a yellow throat; when on the ground it looks like a dark-coloured young Fox.

The following races are recognized:

americana Turton, the typical form.
abieticola Preble, which differs from *americana* chiefly in being much larger and in having heavier teeth.
abietinoides Gray, a dark-brown race.
brumalis Bangs, a large dark race.
actuosa Osgood, a large gray race.
kenaiensis Elliot, a small pale race without throat patch.

LIFE-HISTORY.

RANGE In one form or another the species ranges from ocean to ocean wherever there are heavy pine or fir forests, in the Canadian and Pacific Faunas, and northward.

IN MANITOBA Its range in Manitoba broadly coincides with the unbroken coniferous forests.

ENVIRONMENT This is one of the few species that really prefer the glooms of firs. It will not rest in broken woods, it shuns the borderlands, and the merest beginning of a clearing about the settler's home is enough to drive it away; consequently, it has always been one of the first to retreat before civilization. It is found in all the north-western half of the Province and on the Porcu-

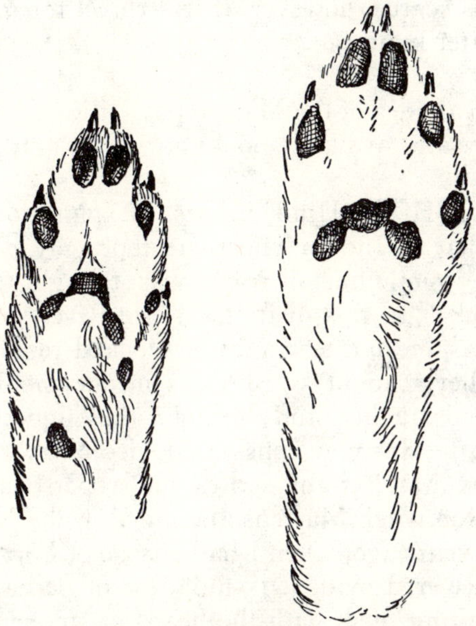

FIG. 215—Right feet of Rocky Mountain Marten ♀ (*M. c. origenes*). (Life size.)
Drawn by E. T. S., at Marvine, Colo., Sept. 19, 1901.

pine, Duck, and Riding Mountains. J. S. Charleson got a specimen from the Carberry spruce woods and D. Nicholson heard of one or two on the Pembina Mountain in 1879, but not since, nor was it ever common there, for Alexander Henry records[1] from Park River in that same region, November 21, 1800, "They take a chance Marten, but the latter are very scarce." This same traveller complains[2] that when he got to Icelandic River, on the west shore of Lake Winnipeg, on August 13, 1808, "The miserable country is destitute of large animals. Martens only are numerous and there is no good fishery."

[1] Journal, 1897, p. 155.

[2] *Ibid.*, p. 452.

HOME-RANGE I have no evidence on the home-range of the individual Marten beyond the opinion of trappers that it will range a mile or more around its head-quarters. George Linklater thinks that two square miles, in Algoma, will cover the usual beat. When food is scarce, however, it will travel for many miles in search of better hunting.

ABUNDANCE To form an idea of the Marten population I have consulted all authorities and available books as well as trappers and fur-traders.

Professor H. Y. Hind, in his "Exploration of Labrador,"[3] says that an Indian who there took only 22 Martens all winter on a 30-mile line of traps, was thought to have had very bad luck. In this district, Moisie River, Martens were scarce. The same hunter, in a well-stocked region, the Manicouagan Valley, caught 57 Martens in one month. As they were of good quality and brought the highest price then current ($5), he was now considered successful.[4]

Roderick MacFarlane writes me a personal experience during a season when Martens abounded in the North:

"Many years ago, when I had charge of Fort Good Hope, Mackenzie River, I made a 15-mile line of perhaps 100 dead-fall traps, baiting them with the heads of smoked fresh-water herring. A few days later I went to the end of the track, and on my way out I found quite a number of whiskey-jacks taken, and one or two Squirrels, while 8 Martens were secured dead. Most of the baits had, however, been removed by Mice and birds. On my way out I placed new baits in all of the traps, and 4 more Martens were found in them on my way back to the post. They were freshly taken. On my next visit a *cussed* Wolverine had preceded us, the few Martens secured were stolen by him, and the entire line demolished. The brute kept around most of the winter, so that my total catch of the season was only about 60 skins. Had he kept away, I might have secured between 100 and 200 skins.

[3] 1863, Vol. I, p. 195.

[4] *Ibid.*, pp. 47–8.

MAP 49—RANGE OF THE AMERICAN MARTENS.

This map is founded chiefly on records by J. Richardson, J. Bachman, S. F. Baird, R. MacFarlane, E. W. Nelson, R. Bell, H. C. Yarrow . Low, C. Hart Merriam, O. Bangs, E. A. Preble, J. Macoun, W. H. Osgood, C. B. Bagster, D. C. Elliot, J. Fannin, J. D. Figgins, E. T. Seton . Rhoads, A. E. Verrill, and E. R. Warren.

The following are recognized:

Mustela americana Turton, with its 6 races,
Mustela atrata Bangs. The Newfoundland species,
Mustela caurina Merriam, in 2 races, *caurina* and *origenes*,
Mustela nesophila Osgood, in Queen Charlotte Islands.

"I have also known a *Saulteaux* prisoner from Lac la Pluie, banished to Mackenzie River for some crime in the fifties of the last century, capture 39 Martens out of 40 deadfall traps made by him on a round of a dozen miles, just as the snow fell. That was what he got on his first visit, and the fortieth trap showed that a Marten was trapped but succeeded in getting away.

"Two hundred Marten skins was the hunt (average) of the best Indian hunters at Fort Liard for several years when these animals were abundant in the early fifties. For output 1853, the late William F. Lane (an Irishman) traded 12,000 skins of the Marten at Liard, the best trade ever known there. The next year, under Robert Campbell, it yielded over 10,000, and some 2,000 less the next. During the decades of the forties and fifties, the Mackenzie River Marten trade was by far the best obtained before or since. In years of scarcity the trade has not averaged a fifth of those of plenty."

From these and many parallel facts I conclude that 200 Martens, taken in one winter on a 25-mile line of traps, would be a large haul; more, indeed, than the best Marten country could stand. A few years at this rate and the region would be trapped out. The area involved would be a strip 2 or 3 miles wide. In other words, if 4 Martens were killed in two or three successive years on each square mile of a region that was thickly populated by the species, it would probably exterminate them in that locality. From which we may argue that 6 Martens to the square mile would be a high rate of population. I doubt if the number in Manitoba's pine woods to-day is a twentieth of this.

The species is, indeed, becoming scarcer every year in all the southern parts of its range concerning which I have information.

FLUCTUATION

One of the interesting unsolved problems of animal life, and especially of Marten life, is the periodic rise and fall of the population. The Marten continue to increase for seven or eight years, until they seem ten times as numerous as at the beginning of the period; then they decline quickly for one or two years until again near zero.

This fluctuation must be due either to migration, epidemic, starvation, or destruction by trappers, or a combination of these.

J. K. MacDonald, of Winnipeg, after 35 years' experience as a chief trader for the Hudson's Bay Company, maintains that migration is the cause. He writes: "I think there is sufficient proof that they do migrate. A question annually put to the Indians returned from the woods in summer or fall was, 'What signs of Marten have you seen?' and where but few of these animals may have been seen in the previous winter, I would be told, 'They are travelling north, south, east, or west,' as the case might be, and so definite was their knowledge that these Indians would go that winter to head off the wanderers, and they never failed to come in contact with them. These movements of large bodies of the Marten go on in summer and till severe weather sets in, beginning again in March and continuing, as far as the males are concerned, till such time as the snow is not fit to travel on; and then on again during the summer. It is accepted as a matter beyond cavil by all Northerners—that is, Hudson Bay hunters—that the Hare, Lynx, and Marten do migrate, and the fluctuation in their numbers is not considered to be caused by epidemics—save in the case of the Hare.

"The Rabbit is always numerous where Lynx and Marten are plentiful, and it is looked on as a *sine qua non* by hunters and traders that it is following up the Rabbit or Hare that causes these migrations—that the migration is, in fact, quest for food."

Bernard Ross comes to a different conclusion. Writing of the periodical disappearance, he says:[5] "It occurs in decades, or thereabouts, with wonderful regularity, and it is quite unknown what becomes of them. They are not found

[5] Can. Nat., 1861, VI, p. 28.

dead. The failure extends throughout Hudson Bay Territory at the same time and there is no tract or region to which they can migrate, where we have not posts or into which our hunters have not penetrated."

This seems to prove that they are not migratory, and the aggregate fur returns of the whole country afford conclusive evidence that though there may be some local migrations, the fluctuations are general. That the whole Marten population increases and decreases with fair regularity in periods of eight to ten years. MacFarlane thinks that there is some migration but that other things enter into the problem. He believes that the abundance of the Martens is a direct result of abundance of Rabbits, and when the Rabbits fail, many Martens die, others migrate. His remarks[6] are as follows:

"The scarcity and abundance of Marten and Lynx depend upon the scarcity and abundance of the Rabbit or Hare. Many Indians assert that Marten and Lynx (of which, by the way, not a few die off, especially when Hares are scarce) migrate, as well as most of the Rabbits which are not snared, etc., by the natives, or carried off by disease, and as they are not uniformly abundant all over the five territories (apart from the fact that they suddenly appear in localities where they had previously for a season or so been conspicuous by their absence) there seems to be good ground for the supposition that they migrate. There are other circumstances, also, such as an unfavourable season for breeding, a scarcity of the required food, and the destruction by fire of the extensive areas of forest, which would, of course, more or less affect the abundance of these and other species of animals in certain localities."

Moreover, he does not consider trapping the cause of the disappearance. He writes:

"The theory of exhausting any wild tract of country by overtrapping will not apply to the territories of the Hudson's Bay Company. When Marten are abundant in good years,

[6] In recent letter.

a favourably circumstanced hunter will use the same line of traps as successfully season after season, and results only decline as the Marten disappear, and increase again as they come back. In this connection a very important factor is the

FIG. 216—Marten.
Drawn by E. T. Seton for the Biological Survey, U. S. Dept. Agr. N. A. Fauna, 16, 1899, p. 106.

periodic greater or lesser fertility of the female. The Indians have made the same remark in respect to Beaver, Musquash, Fox, Mink, and Lynx."

The problem is far from solved. Doubtless it is complex, but the facts as known may be thus summarized: There are many irregular local migrations and variations of the Marten population which, however, are not large enough to change materially the regular periodic rise and fall of its aggregate numbers.

The Marten increases with the increase of the Rabbit, but begins to disappear while yet the Rabbit is abundant. I suspect that MacFarlane is near the truth when he makes it turn on the *greater or less fertility of the female*, which may be a result of the two causes, *starvation* and *overfeeding*. In famine years the females do not breed, and in years of overfeeding they do not breed; so the *increase is ended*. There is no evidence of epidemic to account for the disappearance of those already existing. Probably it is the result of many causes. The increase has been stopped and, owing to the growing scarcity of Rabbits and Mice, the Fisher, Foxes, and Lynxes—also greatly multiplied—prey now largely on the Marten. Cannibalism and starvation set in, and if any die by disease, they are never seen by man because devoured by beasts. Meanwhile hunger makes them ready to enter any baited trap, the trapper makes great catches, the surplus is soon worked off, and the Marten are reduced again to near zero.

SOCIABILITY

This interesting creature appears to be the least sociable of this unsociable family. Otters will meet to enjoy their slide in a merry party, Skunks will gather for warmth, the smaller Weasels will help each other in distress or in hunting, but, so far as I can learn, no man ever yet saw two adult Martens meeting with feelings other than those of deadly hate; the one essential supreme exception to this is doubtless found in the moment of sexual congress.

INTERCOMMUNICATION

Not having many ideas that it wishes to communicate to its kind, the Marten has few methods of communicating them. The scent-glands, so important to its kindred, are greatly reduced in this species.

SCENT-GLANDS

Nevertheless, the British Marten that I saw in A. H. Cocks's menagerie, made use of the musk gland at least every five minutes as they galloped about, pressing the parts on some projection of the cage. The sketches (Fig. 217) illustrate two in the act; both were males. It is easy to see how this may be a means of letting others of the kind know that a Marten has been here recently.

VOICE

Its voice is heard in a growl, a hiss, and a snarl which sometimes becomes a shrill screech. Cocks's Martens bleat like a lamb when hungry. Its grand directive label is the blazing patch of yellow, framed in dark-brown, which ornaments its chin and throat. By this ye may know the Marten.

The mating of Martens is shrouded in mystery. There is **MATING** no positive evidence that they are monogamous, polygamous, polyandrous, or promiscuous. Apparently no one has ever

FIG. 217—British Martens rubbing their musk on projections on the cage. (From life.)

yet seen two adult Martens living together freely and amicably. Analogy seems to prove that the species pairs, but the slight evidence at hand would add that they part again in a few days; and the less they see of the cutthroat sire, the better for mother and young. This is all so unnatural and so largely founded on cage-observation, that we must hope for a pleasanter picture when better study of wild individuals supplies more reliable facts.

A curious item bearing on the relation of the sexes is supplied me by Madison Grant. In the winter of 1901 and 1902, two Bitter-root guides set about catching Marten on the head-quarters of the Clearwater, with a view to starting a fur-farm. They said that they captured two dozen during the winter, but all were males, so in the spring they killed what they had and marketed the fur. They believed that the females were lying up and the males were feeding them.

Miles Spencer says[7] that according to the Indians near Fort George, H. B., the Marten there mate about the first of March. I learn from the guides in Algoma, 600 miles farther south, the Marten pair from the 15th to the 20th of February. Captain R. Craine, of Wayagamug, Mich., tells me that in the winter of 1894 he captured a female Marten on the upper Fraser in British Columbia and kept her three months before she escaped. Of this individual he made many interesting observations. She came in heat early in March. At this time the anal parts were inflamed and swollen; she had a curious way of rubbing them on the floor. Sometimes she would back slowly up the wall of the cage, tail first, until she touched the top, and from time to time during the night she would utter a prolonged screeching, keeping it up until some one shouted at her.

This is all the direct testimony I can find on their mating habits.

But the collateral evidence of the British Marten bred in captivity by A. H. Cocks (of Henley-on-Thames, England) is next best thing. "Litters of this species," he says, "have been bred in my collection. * * * All attempts at breeding were extremely hazardous: the allowing of a pair to run together was apt to result in the death of the female, in consequence of one or more of the long canines of the male penetrating her brain, the damage being inflicted so instantaneously that there was no possibility of a timely separation."[8] The union he believes takes place at night. I give a condensation of his account.

"At last (first week in January), this year we noticed little mouthfuls of short straw deposited here and there in the cage of the female Marten, a sign of her being in season. * * *

[7] Low, Expl. James Bay, Can. Geol. Surv., 1888, Pt. J, App. III, p. 77 J.
[8] Proc. Zoöl. Soc., December 4, 1900.

Accordingly a male was admitted from the adjoining cage on January 5 * * * and the pair finally separated on the 18th. * * * Young were born early on April 22. * * * The probable period of gestation of this species is, therefore, a few hours over 103 days, the extreme of possibility ranging from 96 to 106 days."

A. H. Cocks very appositely remarks here that this duration is a surprise, as the Pole-cat goes 40, the Otter 61, and the Ferret about 42 days.

All observers agree that the nest is made by the female alone. The favourite place is a hollow tree, but sometimes a burrow in the ground is selected; it is carefully lined with grass and moss. NEST

The young number from 1 to 5; are usually 3 or 4, and are born in late April. Spencer says,[9] on the evidence of his Indians, that at first "they are the size of a new-born kitten, brown and black in colour [not white], helpless, with closed eyes. The female suckles the young for a period of 5 weeks and is unassisted by the male in rearing them." YOUNG

George Linklater, of Des Barats, once saw a female out in May with 4 young ones that she was teaching to hunt. This was in Algoma.

Since the young need the mother's care all summer, we must believe that but one brood is produced each year. Here the positive evidence gives out, but help for future observations may be found in Cocks's notes on the development of an English Marten brood that he reared in captivity. His main facts are as follows:[10]

They were born April 7, 1882. They numbered 3 (2 males, 1 female), and at first all were *quite white*. On the 10th one examined was 6 inches long; of this the tail was 1¾ inches. On the 14th the white fur was grizzled. On May 6th, they were yet blind, but some time prior to the 20th, that is, at a little over four weeks, their eyes opened. On the 29th, the

[9] See Note 7.

[10] See Note 8.

mother first fed them flesh meat. On June 23, at seven weeks of age, they first left the nesting box and came to the ground. In autumn they were full grown.

One pair which bred in captivity lived till their seventeenth year.

HABITS

This is the most arboreal of all our Weasels. It delights in climbing from crotch to crotch, leaping from tree to tree, or scampering up and down the long branches with endless power and vivacity. One cannot long watch a Marten, even in a cage, without getting an impression of absolutely tireless energy. For hours it will race up and down, leaping from perch to wall, to ground, to perch, to wall, to ground, to perch, over and over again, doing endless gymnastic feats, giving countless surprising proofs of strength, with bewildering quickness, all day long, without a sign of weariness, without a quickening of its breath. It must travel many hard miles each day in this way, yet it is complained that in confinement they suffer for lack of exercise.

Active as a Squirrel is an old adage, and yet the Squirrel is commonly the prey of the Marten.

It is remarkable that the Marten should follow the Red-squirrel in all its range, but hardly anywhere encroach on the territory of the large Gray- and Fox-squirrels—species which seem to afford special inducements to the active destroyer, for their numbers are great, it can follow into their holes, and their weight is so nearly that of its own that it would have no handicap in the leaps from tree to tree.

Daniel Hayward, of Oxford County, Maine, says:[11] "It is impossible for Gray-squirrels to exist in the same locality with the Sable, as their sizes are about the same, and the Squirrels easily become their prey. They will also outstrip the Red-squirrel, and capture him if he confines his retreat to the trees instead of entering holes too small for the Sable to follow. He will pounce upon and overpower a partridge or Rabbit, but usually takes the latter by the long chase, which seemingly

[11] Shooting and Fishing, Vol. 19, Nov. 26, April 16, 1896, p. 537.

FIG. 218—Attitudes of Martens.
Drawn from life. Chiefly in Philadelphia Zoo, 1899.

lasts often for many miles, which speaks well for his endurance and persistent running abilities."

Dr. Merriam, while admitting the beauty and grace of the Marten, says, however:[12] "Its disposition is sadly out of harmony with its attractive exterior. Mr. John Constable has related to me a most interesting and vivid account of an affray that he once witnessed, in company with his brother, Mr. Stephen Constable, between a Marten and a great Northern Hare. The Marten, generally so meek and docile in appearance, assumed the savage mien and demeanour of a fierce tiger, as it attacked and slew the luckless Hare—an animal several times its own size and weight—and even after the poor Hare was dead the Marten's fury did not abate, and he angrily jerked and twisted the lifeless body from side to side, as if to wreak vengeance, for sins never committed, upon the defenceless body of his victim. So intent was he upon this deed of carnage that he was utterly oblivious of the human spectators, who put an end to the scene by driving a bullet through his obdurate pate."

There is much evidence on record to show that Martens taken young are easily tamed and soon acquire an attachment for the one who feeds them, but never become quite docile. Captain Craine's Marten was fond of being petted by her owner, but continued savage to all strangers. On the other hand, many observers testify that in the north-west one often sees tame Martens following the Indian children, who catch them young and bring them up as pets; these are as tame and playful as kittens.

BOTH DIURNAL AND NOCTURNAL

Several authorities maintain that the species is nocturnal and many that it is diurnal. My own experiences incline me to the latter view or both. Obviously, it could not hunt Squirrels by night. But it does hunt by night to some extent. No doubt at this time it preys on Mice, but it is so wonderfully clever at hiding that it is rarely seen.

D. Hayward, of Maine, says:[13] "The Marten is an animal

[12] Mam. Adir., 1884, pp. 52–3.

[13] See Note 11.

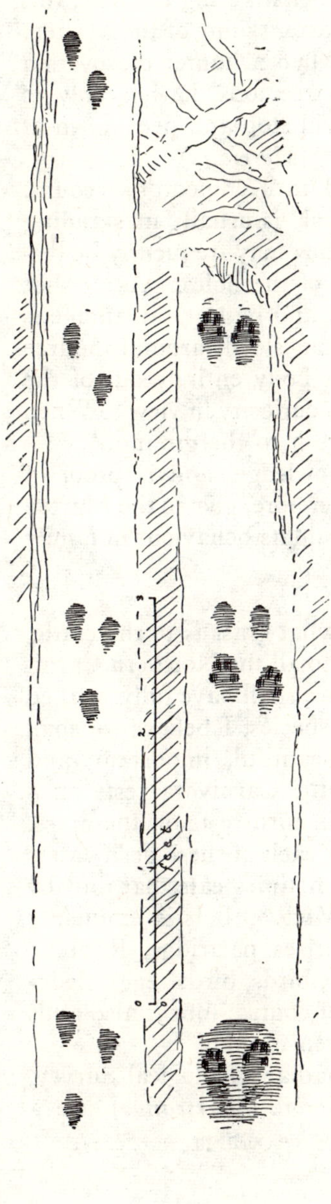

FIG. 219—Marten tracks.
Drawn in Bitterroot Mountains, Idaho, Sept. 26, 1902.

not very often seen at large. I have caught as many as 11 in one day, but in all my experience in the woods have only seen 3 alive and at liberty. This has been thought to be accounted for by their being nocturnal in their habits, but such is not strictly the case. I have often seen their tracks made in the daytime."

Roderick MacFarlane tells me that he has several times found Martens taken in his traps on the same day as they were set, in fact but an hour or two afterwards, although the animal was rarely seen. But he also adds:

"I have known Martens come inside our fort at night to get at white-fish hung on stages, and some have actually been trapped in such places."

I conclude that the creature hunts chiefly by day when the sun is low, but occasionally by night.

CURIOSITY

A Marten in a cage is a picture of life and energy alert to everything. Its keen eyes are quick to take in each new sight and its ears acock for any new sound. It twists its head on one side and peers and sniffs with evident curiosity at every strange object. All the hunters agree that this hankering to know, this itch to see, smell, and find out, is a master failing of the Marten.

"Mr. Constable tells me that when a hunter discovers a Marten climbing about among the tree tops, he has only to whistle and the inquisitive animal will stop and peer down at him, affording an excellent shot."[14] (*Merriam.*)

Similar observations are found in Kennicott's account. "The retreat of the Martens [he says][15] is usually in standing hollow trees, and * * * in winter they may frequently be discovered sitting with their heads out of the holes. As, if shot in this position, they would fall back and be lost, advantage is taken of their inquisitiveness by walking slowly around the tree and inducing the animal to draw its body entirely out of the hole, in order to keep the object of its curiosity in view. When quite out, a well-directed shot brings it to the ground."

Curiosity is a beginning of knowledge, and is proof of intelligence. These observations, therefore, give to the Marten a higher place among its kin than would its behaviour in family matters.

FOOD

What moss is to the Reindeer, what grass is to the cattle, the Mouse millions of the North are to all the Northern Carnivores from Bear to Blarina. When we shall have fully worked out the life-history of each of these species, I believe we shall learn that the whole of that vast beautiful, important, and specialized production that we call the Carnivora rests on a broad simple basis or Muridæ that in turn rests on the grass, that rests on the earth. We shall for each of these flesh-eaters write, 'it sometimes eats this and sometimes eats that, but by far the greatest bulk of its food is Mice.' This is eminently true of the Marten. Its diet comprises partridge, Rabbits, Squirrels, Chipmunks, Mice, Shrews, birds, birds' eggs, fledglings, frogs, toads, fish, and insects of course; but it also adds reptiles, nuts, berries, honey, and even carrion.

A. P. Low, Director of the Canadian Geological Survey, tells me that the rowan berries (*Sorbus americanus*) are a

[14] Mam. Adir., 1884, p. 53.

[15] Quad. Ill., 1859, p. 243.

PLATE LXXIX.—THE EXPLOIT OF THE PET MARTENS.

These were owned by G. A. Paddock. They were extremely playful but mischievous when allowed their liberty.

winter staple of the species, and many of these animals die of starvation in the 'off' years of the rowan.

The Marten, then, like everything else, is omnivorous, but I am satisfied that Mice constitute a larger proportion of its food than any other on its bill of flesh fare.

It is interesting to note that the species never ventures near the habitation of man in the settlements, so that it is unknown as a depredator of the hen-yard. In the North, also, it usually shuns the vicinity of the trading posts. In this it differs from all of its near kin in America as well as from its cousin, the Marten of Europe.

Each fresh investigation adds more creatures to the list of those that lay up for a rainy day. It is surprising to find that most of the Weasels store food, when they have a surplus. Linklater says that the Marten habitually does so. It eats as much as it can of the new catch, then buries the rest. This is true storage because, by setting a trap at the buried piece, you are sure to get your Marten next day. STORAGE

The Fisher and Lynx are supposed to be the chief enemies of the Marten. They are able to take it by open onslaught, therefore, doubtless, in the great struggle they count for little as destroyers. The really dangerous foes are likely to be much less dramatic and obvious. We shall probably find that the Marten's numbers are kept down by something as intangible as it is unexpected and irresistible, possibly a parasite, a disease, starvation, some evil habit of the race, or, as MacFarlane suggests, an inexplicable periodic non-fertility of the females, a non-fertility that I should not be surprised to find a consequence of over-abundant food, which is often as bad as starvation in its effect on the reproductive organs. ENEMIES

An Adirondack acquaintance, Bert A. Dobson, was favoured with an interesting peep into Marten life, and told me of it as follows: November 14, 1900, he had posted himself in a ravine on a Deer path near Moosehead Pond, to wait for a chance as the Deer were running on rut. In the tree below he saw several bluejays shrieking and mobbing something which turned out to be a Marten that had gone up to sun itself in an old crow's nest. Every time the jays swooped the Marten would rise up and hiss like a cat. Dobson shot it through the head from his stand.

SANITATION

Captain Craine's Marten and all those observed by me in captivity, have habitually kept one corner of their cage for the dung-pile. I shall welcome the discovery that this rudiment of sanitation is found with this species when wild, for hitherto the investigations have revealed little of redeeming virtue, but rather tended to show that this externally lovely creature is in disposition absolutely the most unlovely reprobate of all the sanguinary group to which it belongs.

DISEASE

According to Hutchins,[16] "this animal is sometimes troubled with epilepsy." A. H. Cocks also informed me that the British Marten, Stoats, and Weasels in his collection are subject to fits.

TRAPPING

This fur-bearer is unsuspicious and greedy, and therefore easily trapped. Three different ways are practised. First, the old-fashioned deadfall, made on the ground of logs and sticks, so slightly and so simply that a trapper can make and set half a dozen of them in an hour. The principle of this is shown in Figs. 220–221. This is the quickest and most humane way to kill the animal. The objection to it is that the Wolverine, Fisher, and Wolf will follow a trapper for weeks, destroying his traps for the bait or destroying the catch itself. Half a dozen to a mile on a 20-mile round was a usual line of traps.

The second method is the steel trap set in a little cupboard on the side of a tree. This cruelly starves the animal to death, unless the weather is cold enough to freeze it or the trapper comes often on the rounds. When using this kind the

[16] F. B. A., Vol. I, p. 51.

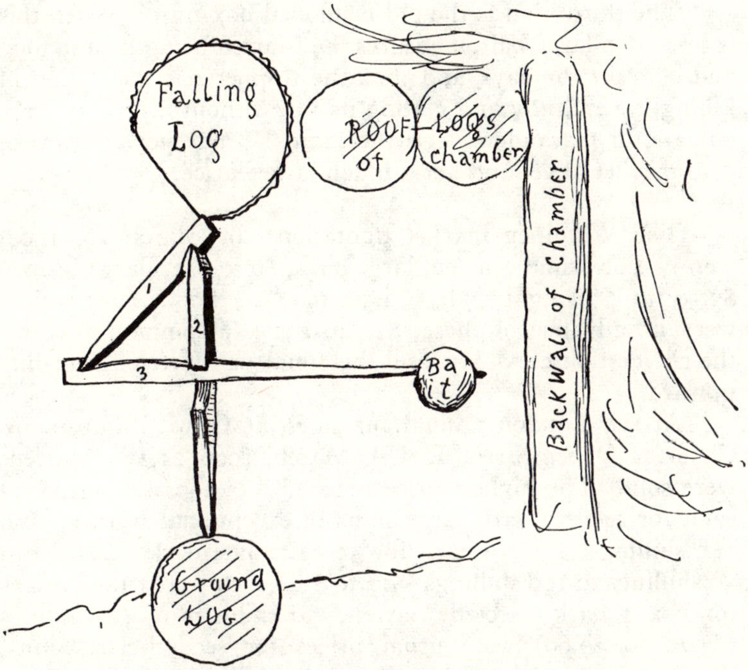

FIG. 220—Section of deadfall, showing trigger set.

FIG. 221—Front view of deadfall set for Marten.

trapper is much pestered by Red-squirrels and Canada-jays, which tug at the bait and get caught.

The third kind is the old-fashioned box-trap. When this is set with a large bait, it ensures the Marten being kept in food and comfort for days, and thus the trapper has the option of killing it mercifully and getting its fur, without injury by trap, climate, or marauders. Furthermore, he may nowadays take it to market alive, and get a much larger price.

FUR

The Winnipeg market quotations for March 26, 1904, were: For prime Marten, large dark, $6 to $12; large brown, $3.50 to $7; light pale, $2.50 to $5. New York prices were in advance of these; $20 to $30 is a common price for the choicest skins to-day, and the trend of Marten is steadily upward.

At the London annual fur sales, at C. M. Lampson & Co.'s (64 Queen Street, E. C.), March, 1906, 21,136 Martens were sold. The highest price reached was 290 shillings ($70) each for 24 very dark large skins of exceptional beauty; and 140 shillings ($33.60) each for 58 extra fine dark skins. But 50 shillings to 100 shillings ($12 to $24) each were ruling prices for No. 1 dark selected Marten; 30 shillings to 40 shillings ($7.20 to $9.60) were usual prices for second-class skins; third-class skins brought about 20 shillings ($4.80), from which they graded down to fourth-class, of which 92 were sold at 5 shillings ($1.20) each.

Russian Sable, however, brought about three times as much, and one lot of 8 extra large No. 1-A, colour black with silver, brought the astonishing price of 980 shillings ($235) each skin. This is probably record price, but these 8 were the pick of 12,007 skins.

During the eighty-five years, 1821 to 1905 inclusive, the Hudson's Bay Company collected 7,006,554 skins of this species, an average of 82,418 for each year. The lowest was 25,524 in 1829; the highest, 177,052, in 1854. The average for the ten years, 1895 to 1905, was 63,926.

Poland's lists show that during the seventy-one years, 1821 to 1891 inclusive, 2,611,500 skins were taken by the other American companies, an average of 36,781 each year. So that the average annual catch of Marten for fur is about 119,000.

Marten-farming.

The supply of Marten has not very seriously diminished, but it has fallen far behind the demand, and the price of the pelt is steadily going up. Realizing that so fine a fur will always find a market, several persons have made attempts at Marten-farming, but so far without success. The animals are hard to get alive, have not hitherto bred in captivity, and are so murderously quarrelsome among themselves that if half a dozen Marten be put in a large cage, only one, the strongest, will be left alive in a very few weeks. Two of my trapper friends, Staley and Leeds, of Idaho, found this out to their cost when they turned several choice specimens loose in a large barn; although they had food, room, and nesting places in abundance, one only survived the first month. These trappers also caught an old male and a young female and put them together in an 8-by-10-foot cage during my sojourn, September, 1902. They quarrelled day and night, and a week later the female was killed by her companion.

The first lesson to be drawn is clear. This wholesale big-cage-plan has not succeeded with any species, and is least of all likely to do so with Martens. Separate cages, judicious management with selection of the gentler individuals, would doubtless solve the breeding problem and open the way to successfully raising them for their fur, just as similar methods have succeeded with the Fox.

While no one so far as I can learn has ever bred the American Marten in captivity, its near kinsman, the British Marten, has been successfully managed by A. H. Cocks, of Henley-on-Thames, England, and a careful study of his results published in the "Zoologist" (1881, p. 333; 1883, p. 203;

1897, p. 270, & Proc. Zoöl. Soc., December 4, 1900), and given to me personally when I visited his collection, should be of the greatest interest to the fur-farmer.

CAGES

In the light of these experiments, together with the results obtained in various Zoological Gardens, I should advise separate cages of ¾-inch mesh galvanized wire, each cage about 6 feet wide, 6 deep and 5 high, with a floor, preferably, of cement. The north side, with a part of the east and west sides and the portions of roof belonging, should be of boards to shut off the wind; the rest open to the sun. Each cage should have a good-sized doorway (2 feet by 2) leading to the cage on each side; all should have a large door opening on a wired alley, itself a complete cage, to prevent escapes. Each should contain a few large branches or trunks and in a quiet sheltered corner a well-roofed nesting box (12 by 12 by 24 inches long). The hole into this should be about 3 inches across, turned away from the front, and should have a sliding door that may be operated quickly and quietly from the outside when it is decided to shut the animal in. The sleeping box should be movable and easily drawn out from the outside, as this facilitates the handling and transfer of the animals. The floor, if of boards, should be treated with boiled linseed oil and afterwards covered with sawdust. Clean food and water pans should be arranged, and every provision made for continuous and perfect cleanliness. The plan shown in the Skunk-farming article will answer for Marten, if we leave out the two large pens and roof each cage with fine meshwire.

FOOD, ETC.

They should have one meal a day. This may be partly raw meat, fowl, or fish, but variety is healthful, and they will be found to eat table scraps and even fruit at times. Overfeeding is as dangerous as underfeeding, and much more likely to prevent breeding. One animal, one cage, should be the rule. Everything should be done to 'gentle' them, and they should early be accustomed to handling.

About the first of January it is well to look for signs of the females coming in heat. These, according to A. H. Cocks, in the articles cited, are usually two or three short straws laid across each other in various parts of the cage. They are mouthed and slobbered over by the female, but apparently are not musked. These, he believes, are left about to attract the attention of a possible mate. BREEDING

The male should now be introduced and left for a week or ten days. This is the critical time, as for slight cause he may kill his partner. Such reprobates should be blacklisted, if possible, and never again used. Selections of the gentler males would in time produce a less murderous race. It is a good plan to make the mating cage of extra size, with many dens and corners, so that the female may have every chance to escape if the male should have one of his murderous impulses. She should be left alone again and always as quiet as possible.

In about three months the young arrive and should not be disturbed. When nine or ten weeks old they begin to appear outside the nest; at four months they may be weaned; at six months they are fully grown and should be put into separate cages. They will be ready to breed at nine months. Castration of the surplus males would doubtless improve their size and coat; probably the second or third year will prove best for marketing the fur, and it is at its best between November and Christmas. It is barely possible that skilful breeding may produce a race that would have progeny twice a year, and thus the returns would be doubled. YOUNG

Although a Marten's pelt does not fetch more than a tenth of what a Silver-fox's does, there are several offsets. The Marten can do with a tenth of the space, a quarter of the food, and is much easier to keep clean, so that the labour of running a Marten ranch is comparatively slight, and many more individuals can be handled by one person. If Russian Sables (which are Siberian Martens) can be got for stock, the returns for the same labour may at once be multiplied by three. FUR

XLI.

Pekan, Fisher, Blackcat or Pennant Marten.

Mustela pennanti Erxleben.

(L. *Mustela*, a weasel; *pennanti*, in honor of Pennant, the English naturalist.)

Mustela pennanti ERXLEBEN, 1777, Syst. Regn. Anim., I, p. 470.
TYPE LOCALITY.—Eastern Canada.

FRENCH CANADIAN, *le Pekan*.
CREE, OJIB., & SAUT., *Oo-djeeg'*.
CHIPEWYAN, *Tha-cho* (Big Marten).

The name 'Pekan,' first recorded by Charlevoix (1744) and popularized by Buffon, 1765, is the Abenaki name, adopted without change (*Rasles*, *Abenaki Dict.*). It is used chiefly in books, but has some currency among the French Canadians.

In Trumbull's Natick dictionary the name given is '*Pekanê*.' C. G. D. Roberts tells me that in Micmac it is called '*Pekwahm*,' and Tappan Adney that the Melecite is '*P'gumpk*,' or sometimes '*Pekonk*.'

'Fisher,' the usual name, is a sad misnomer, as the animal does not fish. DeKay considers[1] that, probably, it was so styled on account of its singular fondness for the fish baits used in trapping. 'Blackcat' is a name often applied in Ontario.

"Wejack, the appellation under which Hearne mentions it, is a corruption of its Cree or Knisteneaux name, *otchœck*, and the word 'Woodshock' has a similar origin."[2]

[1] Zoöl. N. Y., 1842, I, p. 32.

[2] Richardson, F. B. A., 1829, I, p. 53.

In addition to the generic characters which it shares with the Marten, the Fisher has the following: SIZE

Length, about 36 inches (915 mm.); tail, 14 inches (356 mm.); hind-foot, 4 inches (102 mm.). The female is smaller.

Bachman gives[3] 8½ pounds as the weight of a young male. WEIGHT M. Hardy, after weighing many, found them 8 pounds to 12½ pounds.[4] B. R. Ross says the largest he ever caught was 18 pounds.[5]

In general its colour is grayish-brown or brownish-black, COLOUR lighter on the sides, browner below; darker, sometimes quite black, on snout, ears, feet, and tail; and on the head, neck, and shoulders so much tipped with whitish that it has a grizzly gray appearance; the ears have pale linings; like all of the group, it varies greatly in intensity of colour; the claws are whitish horn-colour.

When in its natural surroundings, the Fisher suggests a big black cat with bushy tail, or else a black Fox, according as it is seen in the trees or on the ground.

Two races are recognized:

pennanti Erxleben, the typical form.
pacifica Rhoads, with larger skull and upper molars, also some colour differences.

LIFE-HISTORY.

The Fisher is found in the great pine and spruce forests RANGE from Maine to latitude 62° on the Mackenzie River and west to the Pacific Ocean.

In Manitoba it is rare. W. R. Hine saw one killed on the Assiniboine, near Headingly, some years ago, and had another from the Seine beyond Point du Chêne. I found it in the pine forest about Rat Portage, and have seen one or two brought from the region between Lakes Manitoba and Winnipeg.

[3] Q. N. A., 1849, I, p. 309. [4] Shooting and Fishing, April 13, 1899, p. 526.
[5] Can. Nat., 1861, VI, p. 24.

The Indians tell me that it is found on the eastern part of the north coast of Lake Winnipegosis, and on the south side of Dawson's Bay, but not elsewhere in that country. A. Henry reports:[6] "November 21, 1800, few Fishers along upper Red River."

ENVIRONMENT

It is essentially a forest animal, living on the ground or in the trees, but is not known to burrow, or habitually take refuge in holes underground. This, no doubt, is a weak place in its endowment; it must disappear with the forests. Only those forest animals that are also undergound species are able to hold their own against the axe of the pioneer. Good examples of this are seen in the Woodchuck, the Red-squirrel, and the Chipmunk.

Although not aquatic, it seems to prefer the neighbourhood of swamps, especially if these be among large timber. In the Bitter-roots of Idaho I found it following the bed of a stream that was nearly dry. The trappers on the Ottawa tell me that it lives much more on the ground there than in the trees, and that it follows regular runways.

INDIVIDUAL RANGE

George Linklater, the Des Barats guide, believes that the home-range of the individual Fisher is smaller than that of the Marten. This hardly comports with the family habit, as their range usually corresponds closely with their size.

Madison Grant tells me that one of his British Columbia guides had a number of experiences with a large Fisher that appeared in his locality about every ten or fifteen days, apparently absent during the interval. No doubt, it had a very small home-range while about the camp, but during the two weeks elsewhere it probably changed its home locality half a dozen times. These observations tend to show that it has a certain beat, as have the Otter and Mink, a circle of perhaps a dozen miles around, along which it has dens, that it frequents in turn.

[6] Journal, 1897, p. 155.

MAP 50—RANGE OF THE FISHER AND ITS TWO RACES.

Mustela pennanti Erxleben

Founded chiefly on records by J. Richardson, B. R. Ross, Audubon & Bachman, E. W. Nelson, C. H. Townsend, R. MacFarlane, J. A. Allen, S. N. Rhoads, L. M. Turner, A. P. Low, O. Bangs, C. Hart Merriam, J. Fannin, and E. T. Seton.

ABUNDANCE Nowhere in all its extensive range can the Fisher be styled abundant, yet Hardy says:[7] "It is by no means a rare animal in Maine, being about as plentiful back from the settlements as Otter. When in the fur-trade I used to buy 175 to 200 skins annually. While these were not all taken in Maine, I think, from the best data I could get, that the annual catch of Maine was 150 to 300. The catch varies greatly in different years, just as that of Sable does, as some years both take bait better than others."

MATING The mating is believed to take place about the first of March, and I find among trappers a prevailing opinion that the species pairs.

NEST The animal is so scarce, and its nest so well concealed, that the latter has but rarely been found. It is usually made in a hollow tree at considerable height from the ground, 30 or 40 feet, but has been found in logs and rocky crevices. Although a Fisher is larger bodied than a Fox, it can readily enter a hole but 4 or 5 inches in diameter.

YOUNG The young are born about the first of May. They number from 1 to 5, 2 to 3 being usual. The only litter I ever saw consisted of 3.

In May, 1899, a nest was discovered by some Indians in the woods north of Lake Winnipeg. It was, as usual, in the hollow of a standing tree and about 40 feet up. The young, 3 in number, were bought by N. E. Skinner, in whose possession I saw them at Winnipeg on August 7 of that year. They were then about half grown and exceedingly playful. They showed plainly their Marten affinities, and yet had a close resemblance to a family of Silver-foxes, consequently I was not surprised to read in Kennicott's account that the species has in some parts been called the 'Black-fox.'

While quite small the young appear to quit the nest and follow the mother, like young Martens. Professor Spencer F. Baird tells[8] of an old female and one well-grown young one

[7] See Note 4.

[8] Aud. & Bach., Quad. N. A., 1849, Vol. I, pp. 312–13.

that were treed in the Peter Mountains, 6 miles above Harrisburg, Pa., about the first of February. There was a most desperate fight before "the old one was killed, after having beaten off the dogs, to whose assistance the hunters were obliged to come."

The young one was taken alive; the old one proved a female. She was "no doubt the mother of the one that was captured, and probably died in hope of saving her young."

So far as is known, the young do not usually continue so long with the mother; and the father takes no part in rearing them.

But one brood is produced each year.

SOCIABILITY

Like most Weasels, this is neither sociable nor gregarious. I never heard of anything like a social gathering of Fishers. They have, therefore, but few methods of intercommunication.

VOICE

Its vocal powers are fairly developed. Those from which I made the sketches, uttered an occasional snarl at each other. "A Maine trapper writes: 'Their noise is like a child when it cries in a mournful tone, and again it makes a short, sharp whistle.'"[9]

Bachman describes[10] a male that was treed, as "showing his teeth and growling at the same time * * * elevating his back in the manner of an angry cat," and another as "very spiteful, growling, snarling, spitting when approached," and emitting "a rather strong musky odour."

HABITS

The Fisher is a true Marten, endowed with all the tricks, activity, and the peculiarities of the race. It is probably our most active arboreal animal. The Squirrel is considered a marvel of agility, but the Marten can catch the Squirrel and the Fisher can catch the Marten, so that we have here a scale of high-class agility, with the Fisher as superlative. L. War-

[9] J. G. R., of Bethel, Me., Forest and Stream, June 24, 1886.
[10] Quad. N. A., 1849, Vol. I, pp. 310 and 312.

field, after much experience, says[11] this animal is capable of "jumping from tree to tree like a Squirrel, clearing a distance of 40 feet on a descending leap and never failing a secure grip." And there are several records of Fishers leaping to the ground from a height of 40 feet.

In descending a tree it often comes down head first. But for the Monkeys and some others, we might believe it a rule that no creature is truly at home in the upper world till it *can come down head first* when it likes.

Though so active in the tree tops, it is equally at home on the ground, and is so indefatigable and long-winded that it is known to run down Rabbits and Hares in open chase. If only it could swim and dive well, it would be the most wonderfully equipped animal in the world.

It has much of the blind pertinacity of the smaller Weasels. When I was at Rat Portage, in October, 1886, an Indian brought in a superb Fisher, fresh killed. He saw the animal chasing a Hare. The Hare, with the pursuer close behind, circled about him. He saw the Fisher several times, but could get no shot until the very moment when it sprang on the Hare; then he fired and killed both animals with the same charge.

COURAGE

Its courage, too, is of a high order. In my early days I more than once was told of Fishers—or Blackcats, as they were called in Ontario—which attacked boys and dogs that had disturbed them. I never saw one of these attacks, but they were generally believed in, for all the hunters and trappers entertain great respect for the prowess of this remarkable animal.

WOLVERINE-LIKE HABIT

Though a Marten in most things, the Fisher has many tricks in common with the Wolverine. According to Hardy,[12] one of our best authorities on the species, Fishers often follow up a line of Sable traps to destroy them, stealing the bait, or eating any Sable caught in them. "Usually," he says, "they tear down the boxing or take off the covering from the log

[11] Quoted in Merriam's Mam. Adir., 1884, p. 49.

[12] See Note 4.

FIG. 222—Life studies of Fisher.
The old specimens from Maine; the young from near Winnipeg.

traps, and reach in above the fall and spring the trap. I have had a Fisher remove a piece of water-soaked wood from the top of a trap, set for him, which would weigh a great deal more than he would, then spring the trap. If log traps are set in hollow trees or stubs, they cannot fool with them, and most hunters consider them surer than steel traps."

These details are corroborated by Dr. Merriam, but he adds[13] that the Fisher is "said to be less objectionable than the Wolverine in one particular, *i. e.*, it leaves the traps where it finds them, while the other blackleg lugs them off and hides them."

B. R. Ross's account[14] of the immense Fisher that he caught at Rivière d'Argent (Slave River Delta, 15 miles from Fort Resolution) might easily refer to the Wolverine. He says: "For about two weeks it had been infesting my Marten road, tearing down the traps and devouring the baits. So, resolved to destroy it, I made a strong wooden trap. It climbed up this, entered from above, and ate the meat. A gun was next set, but with no better success. It cut the line and ran off with the bone that was tied to the end of it. As a *dernier ressort* I put a steel trap in the middle of the road, covered it carefully, and set a bait at some distance on each side. Into this it stumbled. From the size of its footprints my impression all along was that it was a small Wolverine that was annoying me, and I was surprised to find it to be a Fisher. It showed good fight, hissed at me much like an enraged cat, biting at the iron trap, and snapping at my legs. A blow on the nose turned it over, when I completed its death by compressing the heart with my foot until it ceased to beat. The skin, when stretched for drying, was fully as large as a middle-sized Otter and very strong, in this respect resembling that of the Wolverine."

COON-LIKE HABIT

In the Northern States and in southern Canada the ranges of the Coon and Fisher overlap. In this common strip the Coon hunters quite often tree the Fisher, for the haunts and

[13] Mam. Adir., 1884, p. 48.

[14] Can. Nat., VI, 1861, p. 24.

behaviour of the two animals are so much alike that the oldest hunters cannot tell from the race, or from the dogs, which animal has been started.

Some interesting observations on this head are contributed by Bachman:[15] "Whilst residing [he says] in the northern part of our native State (New York), thirty-five years ago [about 1814], the hunters were in the habit of bringing us two or three specimens of this Marten in the course of a winter. They obtained them by following their tracks in the snow, when the animals had been out in quest of food on the previous night, thus tracing them to the hollow trees in which they were concealed, which they chopped down. They informed us that as a tree was falling the Fisher would dart from the hollow, which was often 50 feet from the ground, and leap into the snow, when the dogs usually seized and killed them, although not without a hard struggle, as the Fisher was infinitely more dangerous to their hounds than either the Gray- or Red-fox."

When caught in a steel trap, the Pekan has frequently been known to foot itself—that is, gnaw off the imprisoned limb—realizing that it is better to go through a long life halt and maimed than to depart at once on four feet for the Happy Hunting Grounds. Our Maine trapper says:[16] "I once caught one with only two legs; he had been trapped twice before and lost a leg each time."

Although it ranks among our most nocturnal animals, it is known to hunt occasionally by day. Bachman tells of a Fisher that he saw hunting a Gray-squirrel in broad daylight. Many other observers, including myself, have observed it travelling or hunting when the sun was up.

SWIMMING

Knowing its unaquatic reputation, I was surprised to learn from George Linklater that he has often seen this animal swimming rivers and lakes; in fact, it does not hesitate to do so when across seems the easiest way round.

[15] Q. N. A., 1849, Vol. I, p. 310.
[16] J. G. R., Bethel, Me., F. & S., June 24, 1886.

The following, related to me by George Martin (of Paul Smith's, Adirondacks, N. Y.), illustrates at once the savageness of this animal and its power as a swimmer:

In June, about 1870, Paul Smith was guiding a New York sportsman named S———. One night they went out to jack Deer on Spitfire Lake, near Paul Smith's hotel. They noticed a wake in the water and, thinking it was made by a Deer, gave chase. It proved to be a large Blackcat or Fisher, swimming the lake, here a mile wide. S——— insisted that he could catch the animal alive. When they came near, he seized it and threw it into the boat. It attacked him savagely. He drew his hunting knife, but in his excitement he did not notice that the sheath was still on it. With this he fought the Blackcat, stabbing furiously, hitting the boat chiefly, and wondering at the toughness of the Cat. He was getting badly mauled, when Smith threw the beast out of the boat and killed it with his paddle. It was an old female. These animals are often known to swim rivers and lakes in the Adirondacks.

SPEED, ETC. The track of the Fisher is much like that of the Marten, but proportionately larger. On the ground it travels, as do most of the Weasels, by the succession of square-tracked jumps. Bachman describes one running after a Gray-squirrel,[17] with every prospect of overtaking it, but also states that on the ground it appeared to have far less speed than the Fox.

FOOD As already noted, the name Fisher is not happily bestowed, as the animal is probably less of a fisher than any other of our large Weasels. Its food list is most comprehensive and includes every kind of bird, beast, fish, frog, or reptile that it can secure as well as all manner of eggs and fruit—with meats and seeds for variety, its choice staples being Hares, Squirrels, Mice, frogs, and fish when it can find them. Of the last, it is extremely fond, and yet it is not known to go a-fishing. It is said to catch and feed on its little cousin, the Marten, which is evidence of most amazing agility. Audubon and Bachman

[17] Q. N. A., 1849, Vol. I, p. 311.

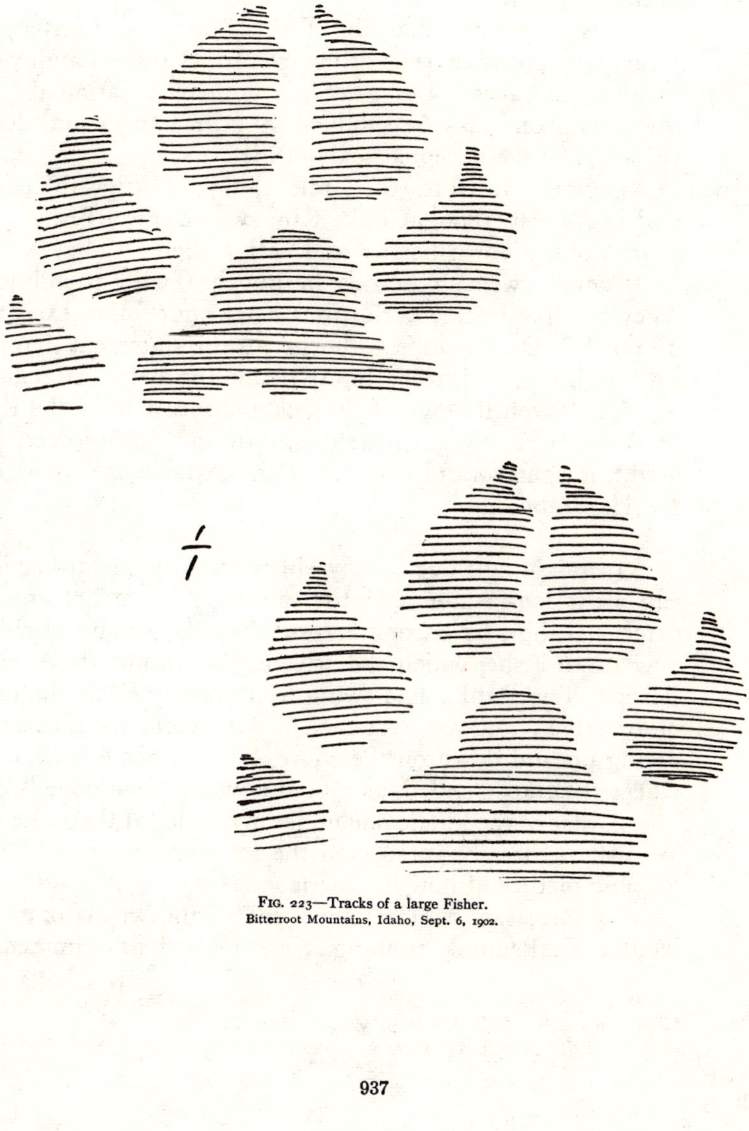

FIG. 223—Tracks of a large Fisher.
Bitterroot Mountains, Idaho, Sept. 6, 1902.

record[18] the killing of an Ontario specimen thus: "A Fisher was shot by a hunter named Marsh, near Port Hope, who said it was up a tree in close pursuit of a Marten, which he also brought with it."

Ross testifies[19] that the Fisher, like the Marten, lives principally on Mice, thus bringing it back to the standby of all carnivorous races. But Mice are not always attainable, and the valiant one may be subdued by grim hunger and descend yet lower in the accepted scale of diet.

OMNIVOROUS

George Crawford (the Indian guide, Mittigwab) tells me that in August, 1896, at Lake Kippewa he saw a Fisher on the shore pulling down berries and eating them. They were, he said, small sweetish berries in bunches with round leaves. They are like huckleberries, but black, not found except well up north. Dr. Coues is authority for the statement that it will stay its hunger with beechnuts if nothing better is at hand.[20]

The favourite food of the Pekan appears to be the Rabbit or White-hare. What little migrating the species does, is, no doubt, irregular wandering in search of woods or regions where the Hare abounds.

RABBIT-RUNNER

In pursuing these it may either stalk them cat-fashion or run them dog-fashion. Hardy says:[21] "I have known one to catch a Rabbit by cutting across when the Rabbit circled. I once saw a Fisher which had driven a Rabbit into the Alleguash River. The Rabbit had swum to a gravel bed in the middle of the river, and sat crouched down, while the Fisher kept racing up and down on the shore, but did not take the water where the tracks ended, as a hound would have done."

A case of the kind came under my notice at the Lake of the Woods, as already noted, and the following incident[22] gives a graphic picture of how it is done.

"I once saw a Hare come out of the woods onto Lake Mollychunckemunk, running at great speed, and, immediately

[18] *Ibid.*, p. 313.
[19] Can. Nat., 1861, VI, p. 24.
[20] Fur-bearing Anim., 1877, p. 70.
[21] See Note 4.
[22] J. G. R., Bethel, Me., F. & S., January 14, 1886, p. 484.

after, a Fisher on his track. They followed down the lake about a mile, when the Hare commenced to circle, quite large at first, and continually making the circle smaller, the Fisher always keeping inside the circle of the Hare, and so gaining quite a distance at every round; or rather, not having to run so fast to keep the Hare on his speed, the Fisher seemed to take it very leisurely, until the circle became so small as to end at a point, and the Fisher was there as soon as the Hare, and made short work of him. I had followed down the lake as fast as I could, in hopes of getting a shot, and so had an excellent chance to see the whole manœuvre, but the Fisher saw me, dragged his prey ashore, and fled into the woods."

The dietary of the Fisher expands on investigation. Linklater and many others of my trapper friends say they have often heard of Foxes killed by this animal. It runs the Fox like a hound, following all day, till the latter gets tired and takes refuge in a hole, where it is easily dispatched. FOX-KILLER

Half as heavy again as the Fox, and twice as much of a fighter, is the common Raccoon, yet it stands in awe of the Fisher, and Dr. Coues says:[23] COON-KILLER

"It may not be generally known that the Pekan successfully assaults an animal as large as the Raccoon; indeed, that the abundance of the latter in some districts depends in a measure upon the rarity of the former. The following letter, addressed to Professor Baird, in 1857, by Peter Reid, of Washington County, N. Y., sufficiently attests these facts:

"'Raccoons are more numerous here now than they were at the first settlement of the country, or for some time subsequent. Thirty years ago they were so seldom found that many boys fifteen or eighteen years old had scarcely seen one. Before the increase of their numbers I once witnessed a circumstance that satisfied my mind on this score. Whilst hunting, early one winter, I found the carcass of a freshly killed sheep, and by the tracks around it in the light snow perceived

[23] Fur-bearing Anim., 1877, pp. 73–4.

that a Fisher had surprised a Raccoon at a feast. A hard chase had ensued, the Raccoon tacking at full speed to avoid his pursuer, the Fisher outrunning and continually confronting his intended victim. I saw where at length the Fisher had made an assault, and where a bloody contest had evidently ensued. The Raccoon, worsted in the encounter, had again broken away, and the chase was resumed, but with diminished energy on the part of the Raccoon; the animal had been soon overtaken again, and a still more desperate encounter had taken place. The Coon had failed fast, and it had at length become merely a running fight, when both animals had entered a swamp where it was impossible for me to trace them further, but I have no doubt the Coon was killed. I have witnessed similar engagements between the Mink and Muskrat, the Weasel and the House-rat, always ending in the death of the assaulted. The Fisher has been nearly extinct in these parts for about twenty-five years, and this, to my mind, accounts for the great increase in numbers of the Raccoon.'"

LYNX-KILLER

Lewis and Clark[24] ascribe a similar habit to the Fisher of Oregon; and we have already seen that this doughty desperado can sometimes fight off a number of dogs. Yet one step farther: The Indians say that a Fisher will kill a Lynx. Linklater, my principal informant on the subject, never saw a case, but was inclined to believe it, as he had had much experience with both animals and knew their relative merits.

DEER-KILLER

Now since the Pekan's prowess is demonstrated and its fighting fame proclaimed, we are prepared for the following from the pen of Manly Hardy:[25]

"In spite of their small size and light weight, Fishers not only kill Deer, but can and do kill those of the largest size. When I first heard of this I doubted it, but know now that they often do it. A year ago last fall, my old friend, Louis Ketcham, was following the track of a large buck near the head of

[24] Aud. & Bach., Quad. N. A., 1849, I, p. 313.

[25] See Note 4.

Nahmakanta Lake. In going along the side of a high granite ledge he saw where the buck had fallen, and there was blood on the snow. After stumbling along a few rods, it had fallen again, and there was more blood. This was repeated several times, and then he saw where the buck had struck a Fisher which had been clinging to its neck and biting it, and had knocked it several feet to one side. The Fisher was evidently badly hurt, as Louis said it dragged its hind-legs, making a track in the snow like an Otter, and had crawled into a crack in the ledge. On going back he found that the Fisher had been on top of the ledge, where the Deer path led along close to it, and had sprung down upon the Deer and was trying to bite the jugular vein. I have known of instances where they have been successful in doing this."

The list of immunes keeps shrinking. There are now but few of the wild things left that, by reason of their size and strength or other gifts, can afford to regard with indifference the Blackcat crawling near. There are indeed two creatures that through ages of security have been led to think themselves exceptions to the rules—the Skunk and the Porcupine; but the Fisher itself is an exception to all rules.

PORCUPINE-KILLER

As far back as 1829, Richardson wrote[26] of the Pekan: "Its favourite food is the Canada Porcupine, which it kills by biting on the belly." Every northern naturalist since has borne similar testimony. Hardy says:[27]

"Their food consists of Porcupines largely. * * * It seems to swallow the quills of Porcupines without any injury. I have examined many hundreds of Fisher skins where there were quills lying flat against the skin, usually either on the back of neck or lower part of the back, but I never saw any signs of their causing any sores or suppuration, as they do in a dog. While I had skins of Fox, Raccoon, and Wildcats, which have been picked up dead, with their necks just filled with Porcupine quills, which evidently had caused their death, I have never seen a quill sticking in a Fisher; and the same is

[26] F. B. A., 1829, I, p. 53.

[27] See Note 4.

true of Bears, which also eat a great many Porcupines. In eating Porcupines they do just as Bears do, turn them over on their backs and eat out most of the meat, leaving the skin nearly entire."

Abundant corroboration is found in "Mammals of the Adirondacks."[28] "I was informed both by an agent of the Hudson's Bay Company and by the trappers themselves [says Merriam] that Porcupines constitute a large and important element in the food supply of the Pekan. Nap. A. Comeau, of Godbout, who secured for me a large and handsome male of this species, tells me that its intestine contained hundreds of Porcupine quills, arranged in clusters, like so many packages of needles, throughout its length. In no case had a single quill penetrated the mucous lining of the intestine, but they were apparently passing along its interior as smoothly and surely as if within a tube of glass or metal.

"Mr. Comeau could not discover a quill in any of the abdominal viscera, or anywhere in the abdominal cavity, excepting as above stated. A great many, however, were found imbedded in the muscles of the head, chest, and back and legs, and it was remarked that their presence gave rise to no irritation, no products of inflammation being discovered in their vicinity. In examining the partially cleaned skeleton of this specimen I still find some of the quills in the deep muscles and ligaments about the joints. A knee, in particular, shows several in its immediate neighbourhood. One is deeply imbedded in the dense ligament alongside the patella; three lie parallel to and close against the tibia, and two can be seen between it and the fibula.

"It is probable that all of these quills entered the body of the animal while engaged in killing and devouring the Porcupine, for those swallowed seemed to have caused no trouble after having fairly entered the alimentary canal. Therefore, there remains no question whatever that the Fisher feeds upon the Porcupine, but I do not agree with Corporal Warfield in the belief that the quills often prove fatal to it."

[28] P. 49.

George Linklater, for many years a chief trader for the Hudson's Bay Company at many different posts, has handled hundreds of Fisher pelts, but *never saw one without some Porcupine quills in it.* The skin is discounted in proportion to the quill damage it shows. He never saw but one Fisher that was seemingly hurt by a Porcupine. He found this—an old one—eating at a Deer carcass. It could scarcely crawl and was full of quills. He thinks it would have died in a few days. This was in December. Another man of the northern woods, Chief Mittigwab, tells me that he never saw a Fisher skin without many Porcupine quills in it, but they do no harm, never fester, and always work out. He has seen them dropping out of Fishers' pelts, but never saw them in their flesh, the back is usually full. Then, adding a final and truly Indian touch, he said: "Fisher's liver given to a dog will force all the quills out of him."

The storage habit is well developed in this species. When it kills an animal, it eats as much as it needs and then buries the rest. If you place a trap at the cache, you are sure to get your Fisher next day. This is well known among the Ottawa trappers and a usual method of catching the Pekan. STORAGE HABIT

It is not generally believed to store food for longer than one or two days, but MacFarlane quotes[29] Colin Thompson as authority to show that for winter consumption the Fishers provide quantities of "hips" in advance.

The pelt of this fur-bearer is cased. The market quotations at Winnipeg for March, 1904, were: Prime, $4 to $9. FUR

During the eighty-five years, 1821 to 1905 inclusive, the Hudson's Bay Company collected 377,338 skins of this species, an average of 4,439 for each year. The lowest was 974 in 1829; the highest, 8,917, in 1868. The average for the ten years, 1895 to 1905, was 3,816.

Poland's lists show that during the seventy-one years, 1821 to 1891 inclusive, 305,570 skins were taken by the other Ameri-

[29] Mam. N. W. Ter., Proc. U. S. N. M., 1905, p. 709.

can companies, an average of 4,224 each year. So that the average annual catch of Fishers for fur is about 8,600.

At Lampson's annual fur sales, London, March, 1906, 2,211 Fisher skins were sold. (The Hudson's Bay Company sold 3,010 the same month.) The highest price realized at Lampson's was 52 shillings ($12.48) each for 24 first-class black skins. But 30 to 40 shillings ($7.20 to $9.60) were ruling prices for first-class paler skins; second-class brought about 20 shillings ($4.80), and third-class 15 shillings ($3.60).

Hardy says[30] he once sold a few extra fine dark skins in London for 100 shillings ($24), while some of the coarsest and palest prime skins in the same lot did not bring over 16 to 20 shillings ($3.84 to $4.80).

[30] See Note 4.

XLII.

Wolverine, Glutton, Carcajou, Skunkbear or Quickhatch.

Gulo luscus (Linnæus).

(L. *Gulo*, from *gula*, the throat, given by Storr on account of its supposed gluttony; L. *luscus*, half-blind.)

Ursus luscus LINN., 1766, Syst. Nat., XII ed., I, p. 71.
Gulo luscus SABINE, 1823, Franklin Nar. Journ. Polar Sea, p. 650.

TYPE LOCALITY.—Hudson Bay.

FRENCH CANADIAN, *le Carcajou.*
CREE, *Kin-kwa-har-gay'-o*, or, according to Richardson,[1] *Okee-coo-haw-gew.*
OJIB. & SAUT., *Kween-go-ar'-gay.*
CHIPEWYAN, *Nog-gy'-ay.*
YANKTON SIOUX, *Skay-cha Tung-ka.*

The original individual of the Hudson Bay Wolverine, to which Linnæus gave the name *luscus*, was said to have had but one eye; possibly, however, it was given on account of its reputation for bad eyesight.

The name 'Carcajou' is probably a French corruption of the Canadian Indian name. Richardson believed that both 'Carcajou' and 'Quickehatch' were derived from the Algonquin or Cree name, 'Okee-coo-haw-gew' or 'Okee-coo-haw-gees.' The name 'Skunkbear' is commonly used in the Rocky Mountains because in size, colour, and shape the Wolverine suggests a cross between a Skunk and a Black-bear.

The genus *Gulo* (Storr, 1780) comprises the largest of the Weasel Family and belongs to the Mustelinæ or true Weasel

[1] F. B. A., 1829, I, pp. 42–3.

subfamily, characterized by having partly retractile claws suited, not for digging, but for climbing. They are stout, bear-like animals with bushy tails, hairy soles, short ears, and 4 extra premolars. The teeth are:

$$\text{Inc. } \frac{3-3}{3-3}; \text{ can. } \frac{1-1}{1-1}; \text{ prem. } \frac{4-4}{4-4}; \text{ mol. } \frac{1-1}{2-2} = 38$$

In addition to these characters the Wolverine has:

SIZE Length, about 36 inches (915 mm.); tail, 6 inches (152 mm.); hind-foot, 7 inches (178 mm.); height at shoulder, 12 inches (305 mm.).

WEIGHT George Cartwright gave[2] the weight of a specimen as 26 pounds.

COLOUR General colour a deep blackish-brown, paler and grayer on crown and cheeks; a band of pale chestnut begins on each shoulder and passes backwards along the sides to meet its fellow on the tail; these become nearly white on the rump in some specimens; the throat and chest are more or less spotted with yellowish-white, which sometimes forms a large irregular patch; claws, whitish horn-colour.

Sexes alike.

Quite recently (1903 and 1905) D. G. Elliot has described two new forms of Wolverine, as follows:

> *luteus* Elliot, is distinguished by the buff colour of its upper parts. California.
>
> *hylæus* Elliot, is very dark in colour, without buff or gray; has auditory bullæ very large. Alaska.

LIFE-HISTORY.

RANGE Wolverines are found in boreal Asia, Europe, and America. In the last the southern limits are shown on the map.

[2] Sixteen Years in Labrador, 1792, Vol. II, p. 407.

MAP 51—RANGE OF THE WOLVERINE AND ITS THREE RACES.

Gulo luscus (Linn)

Founded on records by J. Richardson, Geo. Cartwright, S. F. Baird, E. Coues, C. Hart Merriam, R. MacFarlane, E. W. Nelson, H. C. Yarrow, E. R. Warren, D. G. Elliot, S. N. Rhoads, *et al.*

The main outlines are fairly correct for the primitive range, but further investigations will include most of the Arctic Islands.

It is unknown in south-western Manitoba and very scarce in the north-eastern half of the Province. I have heard of it about Lake Winnipeg, am told that a few are found on Duck Mountain, and in 1885 William R. Hine had a specimen sent him from Brokenhead River. He considers the Wolverine exceedingly rare in Manitoba. The only one he ever saw alive in the Province was kept at the Albion Hotel by Dan MacDonald. It came from Lake Winnipeg.

At no time was it plentiful here, as is shown by the Red River fur returns of Alexander Henry.[3] In 1800–1 and the years following he got 5, 4, 10, 8, 17, 45, 8, 3.

The species reaches its chief abundance in the Barren Grounds, just north of the limit of trees, but it is plentiful also in the Peace River Valley. From this region William Clark, of Winnipeg, informs me that the Hudson's Bay Company received 1,200 Wolverine skins in 1872 and 1,300 in 1882. This, no doubt, includes all skins sent through the Peace River Department from regions more remote. It is evidently becoming scarce in the southern parts of its range.

INDIVIDUAL RANGE

Audubon and Bachman tracked a Wolverine for about 5 miles over the snow-clad hills of northern New York.[4] In Labrador, Cartwright saw one which carried a heavy trap for 6 miles.[5] MacFarlane writes me of another which followed his trail for 12 or 15 miles. The trappers generally say it will follow them along a line of 40 or 50 miles to steal their trap baits. Hutchins says[6] even 60 miles.

The most remarkable case of all, perhaps, is that recorded by Low in his Labrador experiences.[7]

"In the fall of 1893, a Wolverine carried away a trap from the North-west River, and [still bearing the trap] was taken a few days later in another trap on the Hamilton River, some 30 miles away from the place where it had picked up the first

[3] Journal, 1897, pp. 184, 198, 221, 245, 259, 281, 422, 440.
[4] Quad. N. A., 1849, Vol. I, p. 207.
[5] *Op. cit.*, see Note 2.
[6] F. B. A., 1829, I, p. 43.
[7] Labrador Penin., Geol. Surv. Can., 1896, App. I, p. 315 L.

trap." All of this evidence, while not conclusive, has weight, and hunters agree that the Wolverine is a wide-ranger, covering a region of at least 50 miles across in the winter. In the summer it need not go so far for food and must stay in the neighbourhood of its family.

Mating takes place about the middle or end of March. MATING Miles Spencer is of the opinion that the Wolverine pairs, and that the male assists in rearing the young.[8] Abe Leeds, my Idaho guide, tells me that he has seen a pair of Wolverines roaming together in autumn among the mountains of Jackson's Hole, so it is possible that the species pairs for life, though the analogy of other Weasels is against this conclusion.

The den of one which they secured in Rensselaer County, N. Y., is described by Audubon and Bachman, and from their account we may form an idea of the nursery.[9]

"There was a large nest of dried leaves in the cavern, NESTING which had evidently been a place of resort for the Wolverine * * * during the whole winter, as its tracks from every direction led to the spot. It had laid up no winter store, and evidently depended on its nightly excursion for a supply of food. It had, however, fared well, for it was very fat."

The site chosen for the young ones' home is almost any sheltered hollow in the ground or under rocks. Sometimes the old one digs it out, but oftentimes uses any ready-made convenient hole it can find.

Gestation is supposed to last about 60 days; analogy GESTATION would make it about 100; but there is no conclusive evidence. Indians and trappers report the mating season and the bearing season from two to three months apart. Doubtless seasons vary with latitude. Most of the Hudson's Bay Company traders agree that the young are born in June, but in the Barren Grounds, that is, the Arctic region, they may not come till July.

[8] Low Expl. James Bay, 1888, Geol. Surv. Can., Part J, App. III, p. 77 J.

[9] Quad. N. A., 1849, I, p. 209.

As will be seen later, I have evidence of the Siberian Wolverine, supposed to be identical with ours, producing young in April.

YOUNG

The little ones number 2 or 3, rarely 4, but MacFarlane says sometimes as high as 5. The very low numbers are usually offspring of a very young or a very old mother.

In early life they have been little observed, because the mosquitoes are such an insufferable torment in their haunts during summer that no man goes there except under dire compulsion. The half-breeds and Indians of the far Northwest have often assured me that the young are white at birth, but they could not produce the skin of one. I was glad, therefore, to find in the American Museum of Natural History 2 young ones taken by N. C. Buxton in north-eastern Siberia, on May 7, 1901, which, judging from their size, must have been three or four weeks old. They are male and female.

The male, as measured in the dry skin, is: Length, 16½ inches (419 mm.); tail, about 3¼ inches (82 mm.); hind-foot, 2⅛ inches (53 mm.).

It is clad in a dense wool which is yellowish-white tinged with brownish-gray, on crown, legs, back, under parts, and tail; the face bears a mask of brown. Thus it has all the adult marking expressed in very faint colours.

The female is precisely similar.

On one point all my Indian friends and all the books are agreed—that it is as safe to enter the den of a mother Bear as to face a Wolverine when she is with her young. She is a tigress of ferocity, absolutely fearless, and so strong and quick that a man, even armed with a gun, is taking risks if he come near. The young are suckled for eight or nine weeks and fed at home by their mother till quite late in summer. D. T. Hanbury says[10] that August 13, on the Dease River, near Great Bear Lake, he "shot a female Wolverine as she was swimming across the river. She carried a Ground-squirrel in her mouth, which she evidently had intended for her family."

[10] Northland of Canada, 1904, p. 232.

"In October, when the rivers set fast, the Wolverines reappear in families, the young still following their dam, though now not much her inferior in size. They are full grown when about a year old."[11] (*Lockhart.*) But one brood is produced in a year.

The Wolverine has been made the subject of many marvellous stories. We are told, for example, that it habitually lies in wait up some tree for Deer to pass; it drops on them and rides them to death, then devours the carcass at one enormous meal. As a matter of fact, a Wolverine rarely climbs, it seldom attacks a full-grown Deer, and its appetite is no more than that of any other flesh-eater of its size. Its usual prey is small mammals and carrion. It is notorious for its pertinacious persecutions of the trapper. Every trapper in the fur countries can relate personal experiences of the Wolverine, and Richardson also bears testimony to its cunning and its propensity for mischief: HABITS

"The Wolverine[12] [he says] is extremely wary and shows extraordinary sagacity and perseverance in accomplishing its ends. The Indians believe that it is inspired with the spirit of mischief, and endowed with preternatural powers. Though more destructive to their hoards of provisions than the Wolf or even the Bear, and able to penetrate fences that resist their powerful efforts, it is only about thirty inches long, a foot high at the shoulder, and one foot six inches at the rump, but it is very compactly made. With teeth that do not seem to be peculiarly fitted for cutting wood, it will sever a log equal to a man's thigh in thickness, by constant gnawing. In selecting the spot it intends to breach, it shows as much skill as the Beaver, generally contriving to cut a log near one end, so that it may fall down into some void space, and thus open an entrance into the hoard. The animal works so hard in carrying on this operation that it causes its mouth to bleed, as the ends of the logs and the snow often testify. Once admitted into the

[11] Coues, Fur-bearing Anim., 1877, p. 52.
[12] Arc. Search Exp., 1851, Vol. II, pp. 84–6.

hoard, it has to gnaw the pieces of meat asunder, as they are generally frozen together, and then it proceeds to drag them out one by one, and to bury them in the snow, each in a separate place. As it travels backwards and forwards over the meat, it smears it with a peculiarly fetid glandular secretion, after which no other animal will touch it. In this way one of these beasts will spoil a large cache in an hour or two and wholly empty it in a few nights. The pieces which are carried off are so carefully concealed in the snow, and the Wolverine makes so many tracks in the neighbourhood, that it is difficult to trace out the deposits, and they are seldom found."

"In which case," as Thomas Hutchins says,[13] "they furnish a regale to the hungry Fox, whose sagacious nostrils guide him unerringly to the spot. Two or three Foxes are often seen following the Wolverine for this purpose." All Northern traders grow eloquent on the subject of this animal's diabolic pertinacity and destructiveness.

"The winter I passed at Fort Simpson [writes Lockhart][14] I had a line of Marten and Fox traps, and Lynx snares, extending as far as Lac de Brochet. Visiting them on one occasion, I found a Lynx alive in one of my snares, and being indisposed to carry it so far home, determined to kill and skin it before it should freeze. But how to cache the skin till my return? This was a serious question, for Carcajou tracks were numerous. Placing the carcass as a decoy in a clump of willows at one side of the path, I went some distance on the opposite side, dug a hole with my snowshoe about three feet deep in the snow, packed the skin in the smallest possible compass, and put it in the bottom of the hole, which I filled up again very carefully, packing the snow down hard, and then strewing loose snow over the surface till the spot looked as though it had never been disturbed. I also strewed blood and entrails in the path and around the willows. Returning next morning, I found that the carcass was gone, as I had expected it would be, but that the place where the skin was cached was apparently undisturbed. 'Ah! you rascal,' said I, addressing aloud the

[13] F. B. A., 1829, Vol. I, p. 43. [14] Coues, Fur-bearing Anim., 1877, pp. 52–5.

absent Carcajou, 'I have outwitted you for once.' I lighted my pipe, and proceeded leisurely to dig up the skin to place in my muskimoot. I went clear down to the ground on this side and on that, but no Lynx skin was there. The Carcajou had been before me and had carried it off along with the carcase; but he had taken the pains to fill up the hole again and make every thing as smooth as before.

"At Peel's River, on one occasion, a very old Carcajou discovered my Marten road, on which I had nearly a hundred and fifty traps. I was in the habit of visiting the line about once a fortnight; but the beast fell into the way of coming oftener than I did, to my great annoyance and vexation. I determined to put a stop to this thieving and his life together, cost what it might. So I made six strong traps at as many different points, and also set three steel traps. For three weeks I tried my best to catch the beast without success; and my worst enemy would allow that I am no green hand in these matters. The animal carefully avoided the traps set for his own benefit, and seemed to be taking more delight than ever in demolishing my Marten traps and eating the Martens, scattering the poles in every direction, and caching what baits or Martens he did not devour on the spot. As we had no poison in those days, I next set a gun on the bank of a little lake. The gun was concealed in some low bushes, but the bait was so placed that the Carcajou must see it on his way up the bank. I blockaded my path to the gun with a small pine tree which completely hid it. On my first visit afterwards I found the beast had gone up to the bait and smelled it, but had left it untouched. He had next pulled up the pine tree that blocked the path, and gone around the gun and cut the line which connected the bait with the trigger, just behind the muzzle. Then he had gone back and pulled the bait away, and carried it out on the lake, where he lay down and devoured it at his leisure. There I found my string. I could scarcely believe that all this had been done designedly, for it seemed that faculties fully on a par with human reason would be required for such an exploit, if done intentionally. I therefore rearranged things, tying the string where it had been bitten. But the result was exactly the same for three successive occasions, as I could plainly see by the footprints; and, what is most singular of all, each time the brute was careful to cut the line a little back of where it had been tied before, as if actually reasoning with himself that even the knots might be some new device of mine, and therefore a source of hidden danger he would prudently avoid. I came to the conclusion that *that* Carcajou ought to live, as he must be something at least human, if not worse. I gave it up and abandoned the road for a period.

"On another occasion a Carcajou amused himself, much as usual, by taking my line from one end to the other and demolishing my traps as fast as I could set them. I put a large steel trap in the middle of a path that branched off among some willows, spreading no bait, but risking the chance that the animal would 'put his foot in it' on his way to break a trap at the end of the path. On my next visit I found that the trap was gone, but I noticed the blood and entrails of a Hare that had evidently been caught in the trap and devoured by the Carcajou on the spot. Examining his footprints, I was satisfied that he had not been caught, and I took up his trail. Proceeding about a mile through the woods, I came to a small lake, on the banks of which I recognized traces of the trap, which the beast had laid down in order to go a few steps to one side to make water on a stump. He had then returned and picked up the trap, which he had carried across the lake, with many a twist and turn on the hard crust of snow to mislead his expected pursuer, and then again entered the woods. I followed for about half a mile farther and then came to a large hole dug in the snow. This place, however, seemed not to have suited him, for there was nothing there. A few yards farther on, however, I found a neatly built mound of snow on which the animal had made water and left his dirt; this I knew was his cache. Using one of my snowshoes for a spade, I dug into the hillock and down to the ground, the snow being about four feet deep; and there I found my trap, with the toes of a Rabbit still in the jaws. Could it have been the animal's instinctive impulse to hide prey that made him carry my trap so far merely for the morsel of meat still held in it? Or did his cunning nature prompt him to hide the trap for fear that on some future unlucky occasion he might put his own toes in it and share the Rabbit's fate?"

To this bewildering evidence of sagacity Dr. Coues adds:[15] "This propensity of the Wolverine to carry off traps receives confirmation from other sources. In Captain Cartwright's Journal (II, 407)', a similar instance is recorded in the following terms: 'In coming to the foot of Table Hill I crossed the track of a Wolvering with one of Mr. Callingham's traps on his foot; the Foxes had followed his bleeding track. As this beast went through the thick of the woods, under the north side of the hill, where the snow was so deep and light that it was with the greatest difficulty I could follow him even on Indian rackets, I was quite puzzled to know how he had contrived to prevent the trap from catching hold of the branches of trees or sinking in the snow. But on coming up with him I discovered how he had managed. For, after making an attempt to fly at me, he took the trap in his mouth and ran upon three legs. These creatures are surprisingly strong in proportion to their size; this one weighed only 26 pounds and the trap 8; yet, including all the turns he had taken, he carried it 6 miles.'"

"The hunter [says Lockhart][16] may safely leave an animal he has killed, for one night, but never for a second time, without placing it in a strong cache of logs. The first night the Wolverine is pretty sure to visit the place, but will touch nothing. The next night he is certain to return, and, if he can possibly get at the meat, he will gorge himself, and then make away with the rest, which he cunningly hides, piece by piece, under the snow, in different directions. At every cache he makes he voids his urine or drops his dirt, probably to prevent Foxes, Martens, or other animals from smelling the hidden meat and digging it up. Caches must be made of green wood, and be exceedingly strong, or the animal will certainly break

[15] *Op. cit.*, p. 55.

[16] *Op. cit.*, pp. 50–1.

into them. He has been known to gnaw through a log nearly a foot in diameter and also to dig a hole several feet deep in frozen ground, to gain access to the coveted supply. Should he succeed in gaining entrance for himself and yet be unable to displace the logs sufficiently to permit of removal of the meat, the brute will make water and dirt all over it, rendering it wholly unfit to be used; even a dog will then scarcely touch it.

"To the trapper the Wolverines are equally annoying. When they have discovered a line of Marten traps, they will never abandon the road, and must be killed before the trapping can be successfully carried on. Beginning at one end, they proceed from trap to trap along the whole line, pulling them successively to pieces, and taking out the baits from behind. When they can eat no more, they continue to steal the baits and cache them. If hungry, they may devour two or three of the Martens they find captured, the remainder being carried off and hidden in the snow at a considerable distance. The work of demolition goes on as fast as the traps can be renewed."

"The propensity to steal and hide things [says Coues][17] is one of the strongest traits of the Wolverine. To such an extent is it developed that the animal will often secrete articles of no possible use to itself. Besides the wanton destruction of Marten traps, it will carry off the sticks and hide them at a distance, apparently in sheer malice. Mr. Ross, in the article above quoted, has given an amusing instance of the extreme of this propensity: 'The desire for accumulating property seems so deeply implanted in this animal that, like tame ravens, it does not appear to care much what it steals, so that it can exercise its favourite propensity to commit mischief. An instance occurred within my own knowledge in which a hunter and his family, having left their lodge unguarded during their absence, on their return found it completely gutted—the walls were there but nothing else. Blankets, guns, kettles, axes, cans, knives, and all the other paraphernalia of a trapper's tent

[17] *Op. cit.*, p. 51.

had vanished, and the tracks left by the beast showed who had been the thief. The family set to work and, by carefully following up all his paths, recovered, with some trifling exceptions, the whole of the property.'"

How are we to explain this conduct? In ancient days or barbarous countries it would be said that the creature was possessed of a devil and no further explanation considered necessary. This is not quite satisfactory to-day. The Wolverine undoubtedly follows the trapper because it is hungry and sees a chance of securing a bellyful. Having found food, it takes possession of it in a manner of wide usage. As already noted, small boys and Eskimaux take possession by spitting on the object, Squirrels by licking it, Foxes by urinating on it, and Badgers and several Weasels, including the Wolverine, by anointing it with the oil of their anal glands. This is a potent method that carries strong conviction among most creatures that have retained unimpaired the sense of smell. If the Wolverine be not hungry, its provident instinct prompts it to put the possible food away for some day of worse luck, and, acting on the principle 'better safe than sorry,' it brands again in detail with its execrable odour the treasure trove; in so doing, other things, sticks, pots, etc., with an interesting odour of human grease, are accidentally touched with the oil, the convincing holy oil of the anal glands, and so, by a process not without parallel in other worlds, they are converted to its use and receive the honour of a cache into themselves. It is not to be supposed that any part of the procedure is due to malice.

The inordinate sagacity of the species is, as with Wolves, largely fear born of sad experience, stimulated by any suggestion of human touch and assisted by nostrils of marvellous acuteness

TO CIRCUMVENT

Aside from various tricks to decoy it into a trap, there are at least four ways of solving the Wolverine problem. The first is given by Richardson, in his 1851 Journey,[18] thus: "Rae, however, made a safe cellar by cutting a hole in the ice, cover-

[18] Vol. II, p. 86.

ing it thickly with snow, and then pouring water over all until the frost had rendered the whole a solid mass."

The second method is given by Professor H. Y. Hind in his "Exploration of Labrador":[19]

"As an illustration of the ingenuity of Indians in preserving their packs of fur or provisions during the winter months where the Wolverine abounds, Mr. Anderson, chief factor of Mingan, told me that when he was in charge of the Post of Neepigon, north of Lake Superior, an Indian came to him to get some provision, but did not bring his furs.

"'Where did you leave your furs?' he enquired.

"'Made a cache of them,' said the Indian.

"'But, man, the Carcajou will get them; there are plenty in your hunting grounds,' replied Mr. Anderson.

"'No, no; no fear; I'll frighten the Carcajou, I think, if he tries to get my pack.'

"'How did you make the cache?'

"'I wrapped the furs in birch-bark, and tied the bundle at the end of a large branch twice as high as myself from the ground.'

"'Well, that will not keep the Carcajou away. He will climb the tree and jump at the pack and bring it down with him.'

"'No; I think not,' said the Indian, with a smile. 'I fastened two of my little dog-sleigh bells to the pack with a bit of sinew. When the Carcajou comes crawling down the branch to get at the pack, he will ring the little bells, and then you know how quick he'll jump back again and run off. I have tried this trick before, and it never failed me. No fear; the Carcajou will not get my furs."

Somewhat on similar lines is an effectual method that I learnt from J. W. Tyrrell and put in practice during my journey in the far north. The cache is made weather-proof and rain-proof, then left high in a tree, to whose trunk a final finish is given in the form of a complete necklace of cod

[19] 1863, Vol. I, p. 50.

hooks, points downward. Hitherto this has proved very satisfactory.

The fourth and only infallible method is by the use of strychnine. There is, however, a wide-spread feeling against this. The Indians believe it to be an unholy practice that will surely draw down the wrath of the Great Spirit. The trappers say that it ruins the fur of the animal poisoned and tends to ruin all the trapping, as one strychnine bait may claim many victims. Nevertheless, the natives use poison for Wolverines, secretly, but whenever they can get it—just as they also lose no chance of getting certain other contraband poisons for their personal use.

When Linnæus called this animal '*luscus*,' or 'half-blind,' he either knew the creature very well or stumbled on a truth, for the Wolverine has notoriously bad eyesight.

Coues thus comments:[20] "It is said that if one only stands still, even in full view of an approaching Carcajou, he will come within 50 or 60 yards, provided he be to windward, before he takes the alarm. Even then, if he be not warned by sense of smell, he seems in doubt and will gaze earnestly several times before he finally concludes to take himself off.

"On these and similar occasions he has a singular habit, one not shared, so far as I am aware, by any other beast whatever. He sits on his haunches and shades his eyes with one of his fore-paws, just as a human being would do in scrutinizing a dim or distant object.

* * * * * * *

"Lockhart writes that he has been twice eye-witness of this curious habit of the Wolverine. Once, as he was drifting down stream in a small canoe, he came within a short distance of one of the animals on the bank; it stopped on perceiving him, squatted on its haunches, and peered earnestly at the advancing boat, holding one fore-paw over its eyes in the manner described. Not seeming to take alarm, it proceeded on a few paces, and then stopped to repeat the performance, when

[20] Fur-bearing Anim., 1877, p. 56.

Lockhart, now sufficiently near, fired and killed the beast. On another occasion, when the same gentleman was crossing the Rocky Mountains, a Wolverine, which had become alarmed and was making off, stopped frequently and put up his paw in the same manner in order to see more clearly the nature of that which had disturbed him."

Bachman remarks[21] of a captive European Wolverine that he observed in Denmark: "He was somewhat averse to the light of the sun, keeping his eyes half closed when exposed to the rays." All evidence shows that the Wolverine of the north is troubled with eyes that suffer in a dazzling light.

Snow-blindness is a wide-spread complaint among the human dwellers of these white wastes in spite of their various contrivances to prevent it. I have often wondered how the animals escape, especially those that travel by day. Proof that they do not always get off easily is found in the following by Mrs. Mary Austin. In describing the terrors of a great snow on the High Sierra, she adds:[22] "Even the Deer make slow going in the deep, fresh snow, and once we found a Wolverine going blind and feebly in the white glare."

In many early accounts and pictures the Wolverine is presented as a plantigrade animal, that is, one that sets the whole foot on the ground, in bear-fashion. Numerous observations on living specimens, as well as a study of its trail, show that it treads on the toes only, is truly digitigrade, as are most of the Weasel Family. Though bear-like in gait and clumsy in build, this animal is neither slow nor sluggish. A captive specimen which I observed galloped nearly all day up and down its cage, its head low, its back high arched, its movements lumbering but vigorous, and seemingly tireless.

STRENGTH Hearne, on his famous journey, had much experience with Wolverines, and writes thus:[23] "As a proof of their amazing

[21] Quad. N. A., 1849, Vol. I, p. 207.

[22] Land of Little Rain, 1904, p. 257.

[23] Journey, 1795, p. 373.

strength, there was one at Churchill some years since, that overset the greatest part of a pile of wood (containing a whole winter's firing, that measured upwards of seventy yards round) to get at some provisions that had been hid there by the Company's servants, when going to the Factory to spend the Christmas holidays. The fact was, this animal had been lurking about in the neighbourhood of their tent (which was about eight miles from the Factory) for some weeks, and had committed many depredations on the game caught in their traps and snares, as well as eaten many Foxes that were killed by guns set for that purpose; but the Wolverine was too cunning to take either trap or gun himself. The people, knowing the mischievous disposition of those animals, took (as they thought) the most effectual method to secure the remains of their provisions, which they did not choose to carry home, and accordingly tied it up in bundles and placed it on the top of the wood-pile (about two miles from their tent), little thinking the Wolverine would find it out; but, to their great surprise, when they returned to their tent after the holidays, they found the pile of wood in the state already mentioned, though some of the trees that composed it were as much as two men could carry. The only reason the people could give for the animal doing so much mischief was that, in his attempting to carry off the booty, some of the small parcels of provisions had fallen down into the heart of the pile, and, sooner than lose half his prize, he pursued the above method till he had accomplished his ends. The bags of flour, oatmeal, and peas, though of no use to him, he tore all to pieces and scattered the contents about on the snow; but every bit of animal food, consisting of beef, pork, bacon, venison, salt geese, partridges, etc., to a considerable amount, he carried away."

When fighting or under intense excitement it emits a strong musky odour. This, as in all Weasels, is produced by the anal glands. It is very strong in the present species and is another justification of the name Skunk-bear. Although it usually avoids man, the hunters generally testify that it can on FIGHTING occasion face him very bravely, and rarely flies from any foe on four legs.

Such is the opinion by Hearne, who, with convincing reserve, says:[24]

"With respect to the fierceness of this animal which some assert, I can say little, but I know them to be beasts of great courage and resolution, for I once saw one of them take possession of a Deer that an Indian had killed, and, though the Indian advanced within twenty yards, he would not relinquish his claim to it, but suffered himself to be shot standing on the Deer. I once saw a similar instance of a Lynx, or Wild-cat, which also suffered itself to be killed before it would relinquish the prize. The Wolverines have also frequently been seen to take a Deer from a Wolf before the latter had time to begin his repast after killing it. Indeed, their amazing strength, and the length and sharpness of their claws, render them capable of making a strong resistance against any other animal in those parts, the Bear not excepted."

One of my mountaineer friends, Abe Leeds, of Idaho, gives me a stirring account of a meeting between two Wolverines and a Cinnamon Bear, over a dead Elk. All three had been feeding there for some days. But the Bear came once when both Wolverines were in possession. Leeds was waiting for the Bear, but the Wolverines attacked the new guest with great fury, and, although the battle was little more than snarling and heavy growling, extraordinarily so for an animal so small as the Wolverine, the Bear went off and left them in possession.

FOOD

In ancient books the Wolverine is credited with being the inveterate enemy of the Beaver and the Reindeer. One might reasonably infer from two favourite and precious pictures of fifty years ago that these two and none other were its habitual and limited diet. All the evidence I can gather, and it is much, goes to show that while it can climb and swim it is not much at home in the trees or in the water. In

[24] *Ibid.*, p. 372.

FIG. 224—Right-side tracks of Wolverine.

Artillery Lake Sept. 4, 1907. (Natural size.) The small pads and occasional thumb-mark distinguish them from Wolf tracks. The series of the lower right corner shows the trail of the individual—a middle-sized female.

other words, both Deer and Beaver are usually safe from its attacks.

Hearne, after living for years in the fur countries, writes:[25] "These animals are great enemies to the Beaver, but the manner of life of the latter prevents them from falling into their clutches so frequently as many other animals; they commit vast depredations on the Foxes during the summer, while the young ones are small; their quick scent directs them to their dens, and if the entrance be too small, their strength enables them to widen it, and go in and kill the mother and all the cubs. In fact, they are the most destructive animals in the country." Richardson's views are in line.[26] It "feeds [he says] chiefly upon the carcases of beasts that have been killed by accident. * * * It feeds also on Meadow-mice, Marmots, and other rodentia, and occasionally on disabled quadrupeds of a larger size. I have seen one chasing an American Hare, which was at the same time harassed by a snowy owl." Coues, condensing many accounts, says[27] they will devour "anything they can catch or steal. Their own flesh is eatable only in the extreme of starvation," but he does not make it clear whether it is the hunter or the Wolverine that must be starving before it will eat Wolverine meat. Hanbury records the species feeding on Ground-squirrels.

The more light we have on the habits of the Wolverine, the more its living prey diminishes in size, and I doubt not that continued investigation will dwindle its main support into Ground-squirrels or Mice, with even these taking second place in its affections to carrion or stolen meat. Nevertheless, a marked and wonderful exception has just come to hand; an evidence of what this creature can do when pushed by the dire extremity of famine. J. Keele, of the Canadian Geological Survey, while travelling on Third Lake, Ross River (an affluent of the Pelly), March 27, 1908, came on a Moose that was floundering in the deep snow. He and his companion shot it before they realized that it was already done nearly to death by a Wolverine that had leaped on its back from a tree.[28]

[25] Journey, 1795, p. 372.

[26] F. B. A., 1829, I, p. 43.

[27] Fur-bearing Anim., 1877, p. 52.

[28] Forest and Stream, December 19, 1908, p. 971.

The name of Skunk-bear is not at all a bad one in describing the fur of the Wolverine. With the size of a small Bear, something of the quality of a Bear robe, and yet with the two paler bands spreading from the nape of the neck along the sides to unite again in the bushy, Skunk-like tail, which further rejoices in a respectable modicum of smell to complete the semi-imitation, it is quite worthy of its trapper's name. **FUR**

During the eighty-five years, 1821 to 1905 inclusive, the Hudson's Bay Company collected 101,426 skins of this species, an average of 1,192 for each year. The lowest was 402 in 1827; the highest, 2,322, in 1879. The average for the ten years, 1895 to 1905, was 736.

Poland's lists show that during the seventy-one years, 1821 to 1891 inclusive, 10,596 skins were taken by the other American companies, an average of 149 for each year. So that the average annual catch of Wolverine for fur is about 1,300.

At the London annual fur sales, held at Lampson's, March, 1906, 757 Wolverine skins were sold. The highest price realized was 34 shillings ($8.16) each, for 64 first-class dark skins, from which they graded down to 7 shillings ($1.68) for third-class skins.

The Winnipeg market quotations on March 26, 1904, were $2 to $6 for prime Wolverine.

XLIII.

The Hudsonian Skunk, Northern Skunk, Black-tailed Skunk, or Prairie Polecat.

Mephitis hudsonica Richardson.

(L. *Mephitis*, a pestilential exhalation; L. *hudsonica*, Hudsonian, *i. e.*, of Hudson Bay Territory.)

Mephitis americana var. *hudsonica* RICHARDSON, 1829, F. B. A., I, p. 55.

Mephitis hudsonica BANGS, 1895, Proc. Bost. Soc. N. H., XXVI, p. 534.

TYPE LOCALITY.—Plains of the Saskatchewan.

FRENCH CANADIAN, *l'Enfant du diable; le Chinche; la Mouffette; la Bête puante.*

CREE, OJIB. & SAUT., *Shee-gawk'*. In this we see the origin of the word 'Chicago,' meaning 'Skunk-land.'

YANKTON SIOUX, *Mah-cah.*

OGALLALA SIOUX, *Mah-kah'.*

CHIPEWYAN, *Nool'-tsee-a.*

HURON, *Scangaresse* (Sagard-Theodat).

ABENAKI, *Segankú* (Rasles). The word 'Skunk' is traced to the last two Indian words.

The true Skunks belong to the Weasel Family (*Mustelidæ*) and to the Digger or Badger sub-division of the group (*Melinæ*). They form the genus *Mephitis* (Cuvier, 1800) and are about the size of a common house cat; have short ears, long fur, very large and bushy tails, are black in colour, with a thin white stripe on the face and a broad one beginning on the nape, forking on the shoulders to reach to the hind-quarters or sometimes nearly to the tip of the tail; they are at least partly plantigrade, and have the fore-claws very large and suited for digging; but,

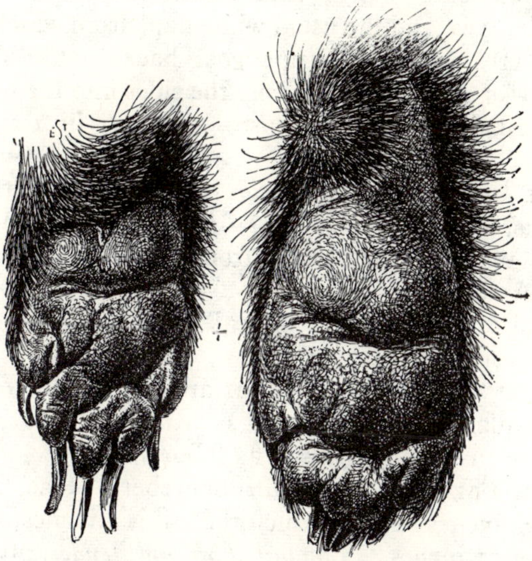

FIG. 225—The right front and right hind-paw of Hudsonian Skunk. Taken at Marshalltown, Ia. (Life size.)

above all, they have greatly developed anal glands which produce the liquid musk that they eject with such notable effect in self-defence.

The teeth are:

$$\text{Inc. } \frac{3-3}{3-3}; \text{ can. } \frac{1-1}{1-1}; \text{ prem. } \frac{3-3}{3-3}; \text{ mol. } \frac{1-1}{2-2} = 34$$

In addition to these generic characters the Hudsonian Skunk has:

Length, about 28 inches (711 mm.); tail, $10\frac{1}{2}$ inches (267 mm.); hind-foot, $3\frac{1}{4}$ inches (82 mm.). SIZE

WEIGHT A large male which I weighed in the Yellowstone Park, July 29, 1897, was 8½ pounds; another, a winter specimen taken in Iowa, weighed 7¾ pounds.

COLOUR General colour black, with a thin stripe down the face between the eyes, and the usual white nape from which a broad white, or creamy white, stripe goes back to the shoulders, where it forks and continues along the sides into the tail, which is of black hairs with white bases, and *ends in a blunt black brush.*

Their *black and white colour*, with their size, their slow movements, and their *immense bushy tail*, usually held aloft, will distinguish the true Skunks from any other animals found in North America.

Brown or cream-coloured freaks of most kinds of Skunks have been found.

At least 8 species of *Mephitis* are recognized; of these, 3 enter Canada. Taking Howell's "Revision"[1] as a starting point, these may be diagnosed as follows:

Canada Skunk (*M. mephitis* Schreber).—Size, large; that is, about 24 inches long, of which the tail is 8 inches; the hind-foot is about 3 inches. The *tail short and slender*, mixed black and white; all its hairs are white at base; *tip, white.* The side stripes from shoulder are narrow, but usually reach the tail. Markings constant; skull large and massive.

Two races[2] of this are recognized, the typical *mephitis*, or Northern form, and the Eastern Skunk, *M. mephitis putida* (Boitard), which differs mainly in being smaller with longer tail; that is, in length about 22 inches, tail about 9 inches, hind-foot about 2⅝ inches.

Northern Plains Skunk (*M. hudsonica* Rich.).—Size, very large; that is, length about 28 inches or more, of which the tail is about 10½ inches; the hind-foot 3¼ inches. The tail is of medium length, *heavy and ending in a blunt black brush;* skull heavy, with a long palate; *zygomata broadly spreading.*

[1] N. A. Fauna, No. 20, August, 1901.

[2] Howell makes them species; I follow Rhoads in making them races.

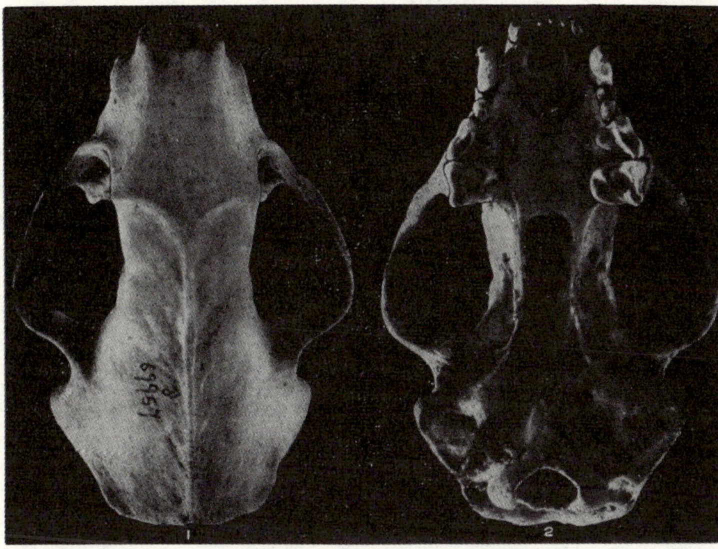

PLATE LXXX.—SKULL OF MEPHITIS HUDSONICA ♂ (NATURAL SIZE).
Cut from A. H. Howell's "Revision of the Skunks," plate VI. N. A. Fauna, No. 20, 1901, Biological Survey, U. S. Dep. Agr.

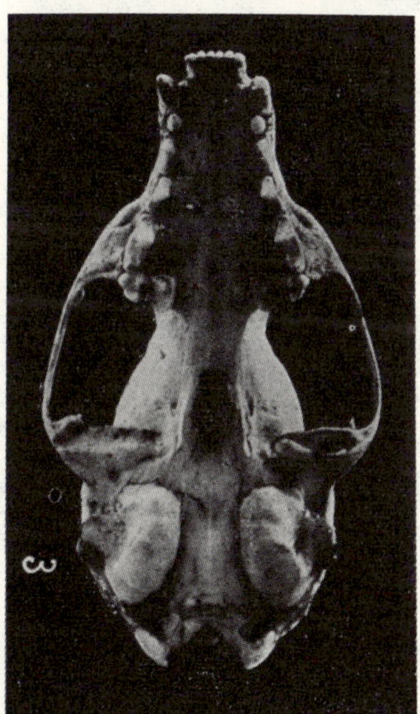

PLATE LXXXI.—SKULL OF MUSTELA AMERICANA (NATURAL SIZE).
Cut from W. H. Osgood's "Yukon Region," p. 44. N. A. Fauna, No. 19, 1900, Biological Survey, U. S. Dep. Agr.

PUGET SOUND SKUNK (*M. occidentalis spissigrada* Bangs).—Much like *hudsonica*, but with longer tail and well-marked cranial characters; skull much narrower, etc.

The Skunk of the Manitoba prairies is the Great Plains Skunk, but it is quite likely that in the wooded north-eastern

FIG. 226—Head of Hudsonian Skunk ♂, from Iowa. (Life size.)

part of the Province we may find the true *Mephitis mephitis* (Schreber) or Canada Skunk.

LIFE-HISTORY.

The present species is found in every part of south-western Manitoba, but is scarce in the pine forest to the north-east. Its greatest numbers are found in the broad pond and poplar belt from Dufferin to Dawson. RANGE IN MANITOBA

Its favourite localities are the edges of the woods and marshes, where sunlight and cover mingling provide it with abundant food, as well as warmth and shelter. It likes the dense forest less than it does the open prairie. The traveller on the Souris Plains is sure to meet with some Skunks or have ENVIRONMENT them visit his camp by night, but the great pine forest is almost skunkless.

HOME-RANGE

The home-range of each individual is doubtless very small. The creature's powers and mode of life preclude the possibility of its roaming far afield. I have often followed its tracks in the early or late snow, from the den through many places and adventures in quest of food, then back to the den, to learn that at no time did it go more than two or three hundred yards from home. In warm weather, because more active, it may go farther, but this is doubtful, because food is more plentiful then and it is still less forced to travel. I believe a half-mile radius would reach the Ultima Thulé of its ordinary wanderings.

ABUNDANCE

In the dry part of the pond and poplar belt of Manitoba, it would be safe to estimate the Skunk at 1 to every square mile. In the prairie region, it is probably a fifth as numerous, and in the pine forest the number may be again divided by 5. This would give us a Skunk population of some 20,000. To approach the problem from another side, the Hudson's Bay Company exports about 10,000 Skunk skins each year. Judged by area, about half must be of this species, and one-tenth of these, again, come from Manitoba, but the free traders get as many as the Company, and half at least of those killed are destroyed by farmers who do not skin them. So that 2,000 each year is not too high an estimate of those killed in Manitoba by man alone, besides which are many enemies that will surely double the casualty list. The average number of young seen with the mother in the fall is 2 or not more than 3; the litter at birth is double as many. This shows that the destruction of the very young by Coyotes, Foxes, Badgers, owls, eagles, etc., is very high, and also that, since winter hardships are still ahead, the Skunks cannot double their number in a year; 50 per cent. drain is all I believe the species can stand in the most favourable food localities. But the Skunks in Manitoba are far from decreasing under the esti-

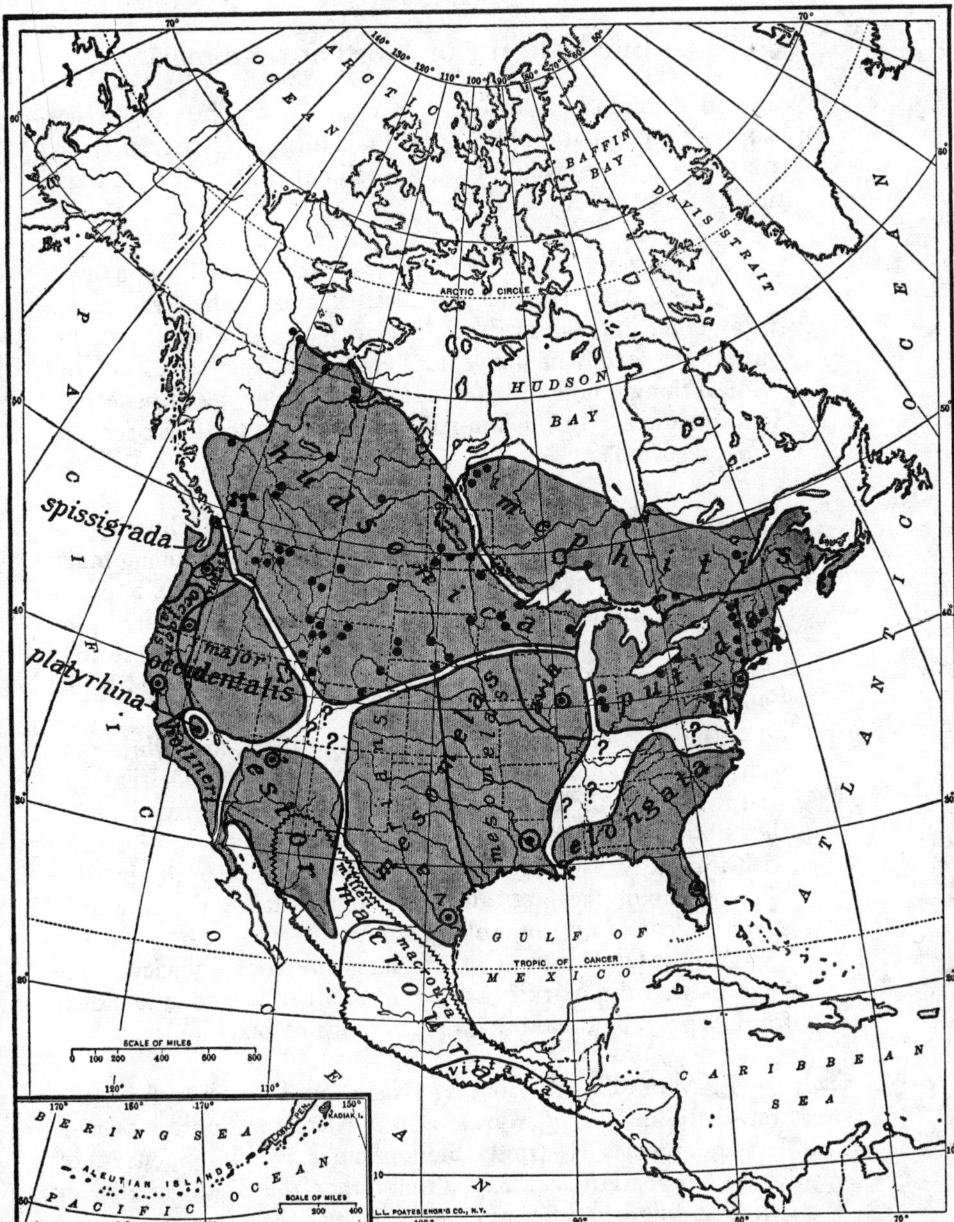

MAP 52—RANGE OF THE LARGE SKUNKS OF THE GENUS MEPHITIS.

This map is founded chiefly on A. H. Howell's Revision, N. A. Fauna No. 20, 1901. Spotted on it are all the records he gives for the species found in Canada, except *occidentalis* which barely enters British Columbia. Additional records by E. A. Preble, J. Alden Loring, and E. T. Seton are marked. The zig-zag line is range of the Hooded Skunk (*M. macroura*); this is left untinted to avoid obscuring the range of *estor*.

Mephitis mephitis (Shaw) with 2 races,
Mephitis hudsonica Rich.,
Mephitis putida Boitard,
Mephitis elongata Bangs,
Mephitis mesomelas Licht., with 3 races,
Mephitis estor Merriam,
Mephitis occidentalis Baird, with 5 races,
Mephitis platyrhina (Howell),
Mephitis macroura Licht., with 3 races.

mated drain of 4,000 adults per annum. This would presuppose a population of not less than 10,000. Between 10,000 and 20,000, then, would be a fair estimate of the Skunks living in Manitoba to-day.

SOCIABILITY The species is said by many trappers to go occasionally in droves of 6, 8, or 10. Kennicott cites[3] a case where there were 15 together in a winter den. This is true, but these droves are simply the family of the year. They stay together all fall and winter, though now full grown, inhabiting one nest and seeking food together. In this limited sense only is the Skunk sociable and gregarious.

VOICE, ETC. It is a very silent animal, but it utters a low 'churring' or scolding sometimes, when it is disturbed, without being much excited, and I have heard one growl as it seized on a grasshopper. It sometimes expresses anger or defiance by stamping with its front feet, and its loud sniffing at curious or strange objects is an expression of interest fully comprehended by its fellows.

The monstrous bushy tail and the black-and-white pattern are no doubt direction or recognition marks that are well known to the live creatures of its region. Day or night, they notify all the world: "I am a Skunk; no one can hurt me with impunity." The wild folk do not fail to profit by this blazonment of the fact, and thus, incidentally, the Skunk is saved a deal of unprofitable exertion.

Abbott H. Thayer, I learn, takes the contrary view. He believes that the markings of the Skunk are intended to hide it from its prey, by breaking up its breadth of black.

PAIRING In Manitoba, the mating season appears to be from the first to the middle of March, and most of the hunters believe that the species is strictly monogamous. Miles Spencer, of Fort George, Hudson Bay, thinks[4] that the Skunk mates in October, but I find no other supporter of this view.

[3] Quad. Ill., 1858, p. 249. [4] Low. Expl. James Bay., 1888, App. III, p. 77 J.

If the Skunk digs its own habitation, it usually selects a dry place on some hillside, but sometimes uses the burrow of a Badger or a Muskrat, or it enlarges the den of a Ground-squirrel to proper dimensions. It is also very ready to use a place under an out-house or farm-building; indeed, any kind of a hole will appeal to the Skunk, provided it be large enough, dry enough, and near enough to food. Kennicott says:[5] "Those who have opened the burrows on the prairie say that it digs a hole 5 to 10 feet in extent and a foot or two below the surface; at the end a large chamber is excavated, and in this a nest of soft grass is placed. The burrows which I have observed were always on high ground, and usually in sandy soil; they were never at the edges of watercourses and ponds, like those of the Mink. In rocky regions its residence will be found in the crevices of the rocks. * * * I have occasionally known it to take refuge in fallen hollow trees." DEN

A. S. Barton writes me that the Skunk abounds in the flat marshy country about Boissevain, Man., and that the unusual conditions there have resulted in a new kind of Skunk architecture. "I found," he says, "a number of Skunk dens on the open meadows one year. They were made like the houses of Muskrats, but much smaller, on dry land, and of fine grass. I thought them the work of some abnormal Musk-rat till I poked a stick in one and provoked the occupant to fire off his unmistakable scent."

I have no evidence on the period of gestation. It is likely to be nearly the same as in the Mink, that is, 42 days. The young are born about the end of April or early in May, and number usually 4 to 6, but have been known as high as 10 in a litter. YOUNG

At birth they are about the size of a Mouse, are naked, and yet show plainly, in two shades, the pattern of the livery they are destined to wear. Indeed it is easier to follow the plan of markings now than at any other time. Eyes and ears alike are closed for some days after they enter this world of sights and sounds.

[5] *Loc. cit.*, p. 248.

M. L. Michael, of the Skunk-farm once existing in Monroe County, Pa., gives the following interesting illustration of the mother Skunk's devotion.

"One night," says he,[6] "I brought a female and her 7 babies, two or three weeks old, and enclosed them in a wooden box. In the morning they were gone. The mother had gnawed through the corner of her prison. Knowing that the young, unable to walk, had been carried by their mother, I called a dog trained to trail them, which at once led off. I

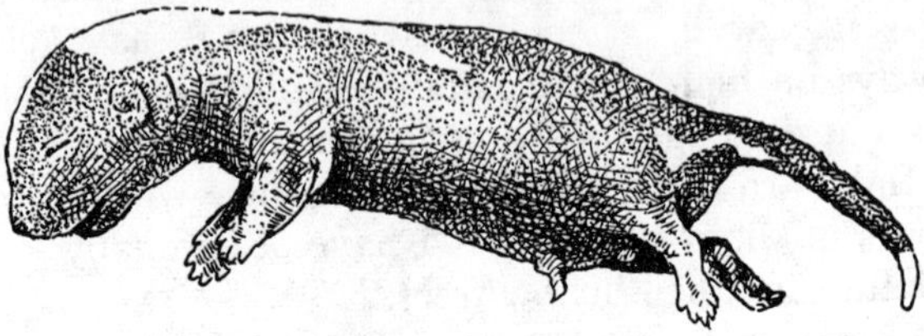

FIG. 227—Young of *M. putida* just before birth (life size). Weight 15 grammes. Cos Cob, Conn., May 12, 1908.

followed closely. A mile away the mother was located in a burrow. My men dug, and there we found her with her 7 children. By noting her tracks across a freshly harrowed field, we discovered that she had made 4 trips, bearing necessarily 2 each trip, except once, when she took but 1." Thus she travelled 7 miles that night.

About the time their eyes opened many young Skunks were brought by the Indians to Hine's taxidermist shop at Winnipeg. Though no larger than half-grown rats, they would at once, when frightened, assume their traditional attitude of defence, and go through all the motions of receiving an enemy and repelling him with the musk. But the musk itself was lacking. When they were about a month old, however, it began to be secreted, and henceforth grew in strength and quantity till at three months the Skunklets were fully equipped and usually had to be destroyed for their over-readiness to prove its power.

The young remain in the den all spring, never going more than a few yards away from home, and live on milk. About

[6] Recreation Magazine, November, 1901, p. 362.

midsummer they begin to follow their mother abroad like a litter of little pigs after the old sow. They are such a pretty playful lot and she such a loving and assiduous guardian that the group realizes the ideal of family life, excepting perhaps in one particular—the father is not present. After many inquiries among hunters and naturalists I am forced to believe

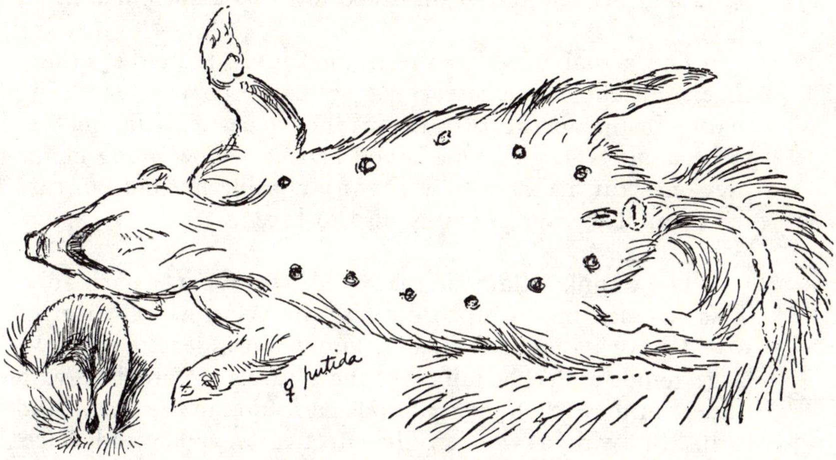

FIG. 228—Mastology of Skunk.
Illustrated by specimen of *M. putida*. Cos Cob, Conn., May 12, 1908.
This one weighed only 3 lbs., yet contained 6 young near birth. The aggregate weight of the young was about 3 oz.

that the mother alone is active in caring for the brood, at least when they are very small.

Their growth is rapid. In Ontario, I once found the young (*putida*) one-third grown and travelling abroad on June 21, and the other two-thirds grown by the first week of July.

Soon after this the group is increased, it would seem, by the return of the father, for I have several times seen a large male Skunk travelling with them, in the fall, and am inclined to think that he returns to his family as soon as the mother gives him permission. Now the reunited party wander about their own range, not caring in the least whether or not they reach home at bedtime, which is dawn.

By October the young might pass for adults, but careful comparison shows them to be a little smaller and much lighter in weight than the old ones.

They continue with their parents all fall and all winter. During the coldest weather they roll up together in their underground home, sometimes the one in which the litter was born, and become torpid till called forth by a spell of warmer days.

The actual breaking up of the family is in the springtime, and the immediate cause seems to be the rearousing mating instinct or, at least, the instinctive desire of the mother to be alone when the next brood arrives. The young of the previous year are now fully grown and able not only to care for themselves, but probably also to breed.

THE WEAPON The Skunk is famous the world over for its 'smell-gun.' This has nothing at all to do with the urine, as vulgar error would have it. The fluid is a liquid musk secreted by two large glands under the tail. All the Weasels are provided with these, but they reach their glorious perfection in the Skunk and furnish it with a wonderfully effective weapon of defence. The glands are situated on each side of the anus; the duct from them is ordinarily hidden away within the rectum, but can be protruded for service.

"The secretion is a clear limpid fluid of amber or golden-yellow colour, has an intensely acid reaction, and in the evening is slightly luminous."[7] (*Merriam.*) It has several other properties of interest. Those who have never smelt it may realize some of its power if they imagine a mixture of perfume musk, essence of garlic, burning sulphur and sewer gas, intensified a thousand times. It is so strong that under certain circumstances it can be smelled for miles down wind. I remember one summer evening at Carberry, Man., being greeted with the powerful odour in great and sudden force; next day I found that at that time a Skunk had been defending himself against a dog on the open prairie, one and a half miles to windward of me.

[7] Mam. Adir., 1884, p. 76.

PLATE LXXXII.—A SKUNK FAMILY.
(*M. putida.*)

And woe to the unhappy creature that is made the target of this battery. If it reaches his eyes it may cause blindness, at least for a time; in his nostrils it acts as a choking irritant. The smell alone is powerful enough to upset most stomachs, and in some cases causes convulsions, fainting, and even death. Certain individuals are much less powerfully affected than others, but, as a rule, men, dogs, and wild creatures with one accord prefer to let the Skunk alone. They will endure a terribly hard pinch of hunger before inviting a volley from the 'Smell-cat's' famous 'breech-loader,' which, by the way, is also a 'repeater,' for it contains not one round, as some have supposed, but enough for nearly a dozen discharges, depending somewhat on the size and age of the Skunk, as well as the time that has elapsed since last it was justified in protecting itself. OFFENSIVENESS

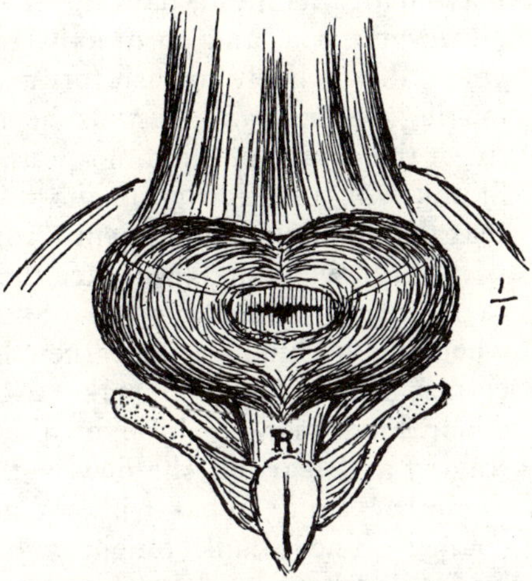

FIG. 229—Anal scent-gland of *M. putida* ♀ dissected and raised to expose the rectum (R). Life size, but a very small example.
Cos Cob, Conn., Oct. 12, 1908.

Like the rattlesnake, it usually gives fair notice, and acts only on the defensive. Let the Skunk alone and it will let you alone. When approached by an enemy, it usually makes off, ambling deliberately, and evidently unwilling to provoke attack. If the enemy follow and overtake it, as a man may easily do, it turns and faces about, and seems to say, "all right if you *will* have it, come on." But it still gives you three fair warnings—which is almost scripturally correct; the first by facing about and stamping, the second by raising and spreading the tail, all but the tip which hangs downward. The third final and dreadful warning is, when the tip rises up and spreads HABITS out. That white flag nailed to the mast does not mean 'surrender,' but clear the deck for action. Then you look out! Stand perfectly still! make no sudden move; it may not yet be too late! The Skunk, especially if an experienced old fellow, may change its mind, haul down the fighting signal, mast and all, forgive you, and go quietly away. A young hunter or a young dog is likely to rush forward at the beautiful, innocent-looking animal, and, just as he is about to seize it, is sure to get the charge of blinding, poisonous spray, after which the Skunk will turn and go on its way rejoicing, quite confident that that enemy is permanently routed.

Very few animals will face the *Mephitis,* it is accustomed to deference. But I knew of a Skunk that made a sad mistake when, on finding a nest of new-born kittens out in a fence corner, it sat down to make a comfortable meal off them. Their loud mewing brought the old cat at racing speed. Raging and fearless, she flew at that Skunk. What can face a mother cat? What will not a mother cat face, for her young? The Skunk fought with all its weapons, muzzle-loader and breech-loader, long arms and short arms, but was badly defeated and escaped, not to be wiser—Skunks do not seem to learn discretion—but probably to die. How well I remember that old cat, smelling to heaven, and blinking her bloodshot eyes so hard as she silently endured the torment and stench. We had a profound reverence for her heroism, but we could not endure her person, and, for many weeks afterwards, she was energetically invited to tarry in the wilderness.

The distance to which the spray can be thrown usually is 4 to 6 or, in exceptional cases, perhaps 10 feet. If the Skunk is approached on the windward side, the distance is greatly reduced. I have more than once seen persons draw near the Skunk with little fear, because the animal was face on, and tradition hath it that the gun points backwards and is only used in Parthian fight. But this turns out at once to be a grave error. The Skunk throws its brush forward and to one side, bends down its back, protrudes the anus, and the pipe of the gland then shoots the dreadful liquid towards the foe, with fearful precision. As a matter of fact, I think the Skunk rarely fires straight backward. Its gun can be trained in any direction or at any elevation, but it prefers to aim at a foe that it can clearly see.

Usually it is careful to keep its coat and tail clear of the fluid; it never deliberately shoots without elevating the tail and clearing the deck. This has given rise to the belief that it *cannot* shoot with the tail down. Some even maintain that a Skunk may be safely lifted by the tail, as such handling puts the gun out of gear. I doubt not Skunks have suffered their captors to lift them by the tail without retaliating, but I am satisfied it was choice, not incompetence, that restrained them.

Although Skunk musk is so potent as to nauseate many animals, to choke some and blind others, it is a curious fact, as every hunter knows, that a hound while running a Rabbit, may run into a Skunk and be so soaked with the musk that he flies in agony to the nearest stream to roll in the mud and wash his burning eyes and nostrils, then, within ten minutes, though still stinking unspeakably, he will take up the faint Rabbit trail again. I do not know how to explain this.

It is often said that clothing once skunked will smell forever. This is a mistake; the odour is strong and durable but can be destroyed. The usual method is to bury the tainted garments. But a better and simpler way is to send them to the cleaner—provided he will accept them—and there the benzine method may be relied on to destroy all traces of the 'child of the devil,' as our French neighbours call the Skunk.

INOFFENSIVENESS

While I was living in a shanty at Yancey's in the Yellowstone Park, in 1897, a family of Skunks of this species made their home under the floor. They came out every evening to pick up scraps about the door or climb into garbage pails in search of eatables. They would even venture into the house when the door was left open. But no one molested them, even the dog refrained, so that the summer passed without offence.

Late one evening, I caught a couple of them in a box-trap in order to keep them till the light was better, that I might take their pictures. The Skunks watched me setting the trap, then waddled into it without loss of time, and were caught. However, among the revolver-carrying mountain men I could not find any one brave enough to help me in carrying the box full of Skunks over to the place where they were to be photographed. I might have had trouble, but that my wife volunteered, and the process of immortalization was duly carried out next day at close range. More than once the Skunks gave warnings No. 1 and No. 2, but I apologized by remaining still, and signal No. 3 was not hoisted, nor had our pleasant acquaintance any rupture.

Near Toronto, July 5, 1888, I had an interesting adventure with the Ontario species (*putida*).

Two men in my employ called my attention to an old Skunk and two well-grown young ones that were walking across a field. I told the men to do as I did, then took a stout board and ran after one of the 'smellers.' He ambled off but, finding me close on him, he faced about and made ready for action. I approached holding the board in front of me. When I was 7 or 8 feet away he fired over his own head. I jumped, and the shower reached the spot whereon I had stood. Before he could get another charge ready I rushed at him, pushed his tail down, with the board laid on his back, then, slipping a hand under each end, I caught him by the neck and the tail, and carried him in triumph to a box. The men tried to do the same, but both got badly 'skunked,' especially the one who attacked the mother; she made good her escape, but the other young one was put in a box with his brother. This adventure cost me two suits of clothes.

AS PETS

These Skunks I gave to Dr. W. Brodie. He kept them in his yard in Toronto, and for some months had no reason to regret it, until one day a neighbour's tom-cat conceived the brilliant but unhappy thought of dining on one of the 'smellers.' The results were many, the cat was temporarily blinded, and the neighbour brought the police, so that ultimately we were compelled to dispense with our Skunks as pets inside the city limits.

I have had several tame Skunks, some of them in full possession of their powers, and rarely have had cause to rue the adoption of such unfragrant pets. Still there was always danger of strange dogs rushing in unwittingly and provoking a round of the irresistible 'scatter-gun.'

One of these Skunks was killed by a meal of very strong cheese; this was always thought to be a rare tribute to the strength of that particular brand.

His final death struggle culminated in a grand discharge of his battery, a parting salute to the earth and his friends, and the cheese.

It is quite possible to disarm a Skunk, not by kindness, but by a surgical operation as "performed by Dr. J. M. Warren, of Boston, in the year 1849. It consists in making an incision through the skin directly in front of the anus and in snipping the ducts of the glands and the basis of the nipple-like papillæ, which projects into the gut just within the sphincter. Adhesive inflammation follows and permanently occludes the ducts at the point of division. Therefore, although the glands themselves are left *in situ*, the animal is forever after incapable of ridding himself of their contents."[8] (*Merriam.*) **DISARMING**

My own experiences with tame Skunks have been slight compared with those of Dr. Merriam, therefore I quote from his account:[9] "Skunks, particularly when young, make very pretty pets, being attractive in appearance, gentle in disposition, interesting in manners, cleanly in habits—rare qualities indeed! They are playful, sometimes mischievous, and manifest considerable affection for those that have the care of them. I have had, at different times, ten live Skunks in confinement.

* * * * * * *

From some of them I removed the scent bags, but the greater number were left in a state of nature. None of them ever emitted any odour, although a couple of them, when half grown, used to assume a painfully suggestive atti-

[8] Mam. Adir., 1884, pp. 78–9.

[9] *Ibid.*, pp. 73–5.

tude on the too near approach of strangers—so suggestive indeed that their visitors commonly beat a hasty retreat.

* * * * * * *

Two summers ago I was the happy master of the cleverest young Skunk that I have thus far chanced to meet. For a name he received the title of his genus and we called him 'Meph' for short. By way of precaution I removed his scent sacs, and he made a rapid and complete recovery after a few days of temporary indisposition. While driving about the country in the performance of professional duties, he usually slept in my pocket. After supper I commonly took a walk, and he always followed close at my heels. If I chanced to walk too fast for him, he would scold and stamp with his fore-feet, and if I persisted in keeping too far ahead, would turn about, disgusted, and make off in an opposite direction; but if I stopped and called him, he would hurry along at a short ambling pace and soon overtake me. He was particularly fond of ladies, and I think it was the dress that attracted him; but, be this as it may, he would invariably leave me to follow any lady that chanced to come near. We used to walk through the woods to a large meadow which abounded in grasshoppers. Here 'Meph' would fairly revel in his favourite food, and it was rich sport to watch his manœuvres. When a grasshopper jumped, he jumped, and I have seen him with as many as three in his mouth and two under his fore-paws at one time. He would eat so many that his over-distended little belly actually dragged upon the ground, and when so full that he could hold no more, would still catch and slay them. When so small that he could scarcely toddle about, he never hesitated to tackle the largest and powerful beetle known as 'horned bug,' and got many smart nips for his audacity. But he was a courageous little fellow, and it was not long before he learned to handle them with impunity, and it was very amusing to see him kill one. Ere many weeks he ventured to attack a Mouse, and the ferocity displayed in its destruction was truly astonishing. He devoured the entire body of his victim, and growled and stamped his feet if any one came near before the repast was over.

"His nest was in a box near the foot of the stairs, and before he grew strong enough to climb out by himself he would, whenever he heard me coming, stand on his hind-legs, with his paws resting on the edge of the box, and beg to be carried up-stairs. If I passed by without appearing to notice him, he invariably became much enraged and chippered and scolded away at a great rate, stamping, meanwhile, most vehemently. He always liked to be carried up to my office, and as soon as strong enough would climb up of his own accord. He was very sprightly and frolicsome, and used to hop about the floor and run from room to room in search of something to play with, and frequently amused himself by attempting to demolish my slippers. I have often given him a bit of old sponge with a string attached, in order to keep him out of mischief. During the evening he occasionally assumed a cunning mood, and would steal softly up to my chair and, standing erect, would claw at my pants once or twice, and then scamper off as fast as his little legs could carry him, evidently anxious to have me give chase. If I refused to follow, he was soon back to try a new scheme to attract my attention."

Skunks can bite when necessary and are credited with having occasionally transmitted hydrophobia, much as any other carnivore might. When they fight among themselves they are said to observe an unwritten law to abstain from using the musk. Evidently it would be wasted in such a combat. It would be like two ducks splashing each other. A Skunk fight is said, then, to be strictly one of tooth and claw. I never witnessed one, but the following detailed case represents several that I have heard of. FIGHTING

About the end of February, 1903, at Welch, Minn., Lee R. Gridley, of Appleton, Wis., was out with another trapper. They were following a Skunk track, and presently came on the animal fighting desperately with another of its kind. The Skunks struggled for a minute or two in silence, and *neither of them used his musk*. The trappers came up and killed one of the combatants; it was a male. The other escaped—doubtless it also was a male—but was so injured that it never came out of the hole into which it had crawled at the approach of the men.

SLIGHTLY AQUATIC On several occasions I have known this animal to show a surprising readiness in taking to water.

On May 23, 1882, while travelling near Turtle Mountain, Man., I surprised a Skunk in the open; he turned to face me, but stones were plentiful and my range was longer than his, so he ran off. I followed and, each time he faced about, I drove him on with handfuls of pebbles till he was routed and fled with unusual haste. Too much so, indeed, for he ran onto a long spit of land that projected into a small lake. At the margin he hesitated, but a new shower of pebbles urged him forward, and he took to the water, swimming for a low island in the middle, fifty yards away. When he got there it turned out to be nothing but floating weeds. This was a sad disappointment; he turned to swim back to the shore, but stones showered in the water stopped him. He was forced to cross the lake, here one hundred yards wide, and very cold. When he reached the bank I was there to meet him. But he was much numbed and lay almost lifeless. Now I took pity on him and pulled him out; he made no attempt to defend himself, but tamely submitted. I carried him to a warm sunny nook, and there left him to recover in peace. This he no doubt did, for I now suspect that he was not so far gone as he seemed, but, finding that his customary defence had failed, was skilfully playing 'possum.

The Skunk is not usually said to be aquatic, but on October 2, 1883, I saw 5 Skunks dabbling in the mud along a pond near Minnedosa, Man., and Miller Christy, in his paper on the "Mammals of Manitoba," says:[10] "One evening last June I assisted in the extermination of a family party—of Skunks—consisting of an old one and six young ones, which were taking a bath at the edge of the lake. The Skunks seem to be fond of the water, as on another occasion I remember shooting one from a boat as he was near by bathing."

[10] Nat. Hist. Journal, May 15, 1885, York, Eng.

PLATE LXXXIII.—SKUNKS FIGHTING FOR A PIECE OF MEAT, WHILE THE FOX JUDICIOUSLY HOLDS ALOOF. THE COMBATANTS DID NOT USE THEIR MUSK.

Scene described by L. W. Walker of Yellowstone Park.

The late W. G. A. Brodie informed me that once, near Toronto, when his dog had discovered a Skunk (*putida*), the latter availed itself of the first opportunity to rush into the Don River, some fifty yards away. The dog followed and, after a prolonged and partly subaqueous struggle, the Skunk floated up dead and the dog returned to the shore perfumed in the usual way.

Similarly Preble relates[11] of the Keewatin Skunk (*mephitica*): "While paddling up the channel between Windy and Pine Lakes, on September 12, we saw a Skunk swimming across the stream, a hundred yards in front of our canoe. On seeing us he redoubled his exertions, but we overtook * * * him just as he reached the shore."

It is quite settled now that by far the largest part of the Skunk's food is grasshoppers, crickets, insects, and Meadow-mice. Ground-squirrels are the next on the bill of fare, with eggs when it can find them. Frogs and crayfish enter largely into the list and snakes provide it an occasional meal. Kennicott records[12] that he knew of a Skunk running Gray-rabbits into their holes and there devouring them. At rare intervals it discovers the hennery and, accustomed to the respect of all the world, enters into possession without a doubt that all this was meant for itself. Eggs and chickens, also hens that happen to be roosting too low, are very much to its taste. Commonly, however, the farmer has the opportunity of executing summary vengeance in the morning, for the Skunk, with its usual effrontery, is frequently found curled up asleep in the nest that it rifled for the midnight feast. FOOD

The Skunk, then, is insectivorous and carnivorous, indeed nothing of animal nature comes amiss, be it flesh or fish, bug or carrion. But its powers are limited; it is as ill-adapted for running down Hares as for catching salmon in a whirlpool, or chasing Squirrels in the tree tops, so that practically it is an insect-eater. And however good (or bad) its intention may be,

[11] N. A. Fauna, No. 22, 1902, p. 65.
[12] Quad. Ill., Pat. Off. Rep., 1859, p. 249.

it is a flesh-eater only at intervals. It is credited with eating fruit. I never saw one do so, but have seen plenty of berry seed in what I took to be Skunk 'sign.'

TRAPPING

This animal has so long enjoyed immunity from attack through the terror of its armament, that it has neglected modes of defence that its ancestors undoubtedly employed. Like the rattlesnake, it has lost its speed, its ability to climb a tree, and its keen wits. In truth, it has become slow and stupid; satisfied with itself and utterly unsuspicious. Foxes and Wolves have a sort of inborn knowledge and distrust of gins and springles, no matter how carefully they are concealed. Nothing seems capable of inspiring the Skunk with such helpful discredence. It will go blundering right into the most obvious of traps, even after seeing a brother taken there the night, yes, an hour, before; yes, even if itself has already been caught therein; pitfall, deadfall, steel, or box-trap, it is all the same to the Skunk, in it goes. When caught in a steel trap it may be easily and safely dispatched by a plan that Dr. Merriam sets forth in his "Mammals of the Adirondacks."[13]

HOW TO KILL A SKUNK

Not by shooting it through the heart or blowing its head off—such a death is usually accompanied by a tremendous discharge—but by one sharp, heavy blow across the back. This paralyzes all the muscles below the point of injury, and without muscular action no musk can be vented. It is quite easy to approach a trapped Skunk if one moves slowly and stands still as soon as it shows alarm by raising its tail or trying to escape.

When caught in a box-trap, box and all may be sunk in water, for a drowned Skunk rarely smells, but the quickest, safest, surest, and most odourless way is that set forth above.

ENEMIES

The Hare is the most harmless of creatures. None fear it; it kills none; therefore all kill it. It has no friends. The fear of the Skunk is on all flesh; therefore none kill it. It has no

[13] Pp. 80–2.

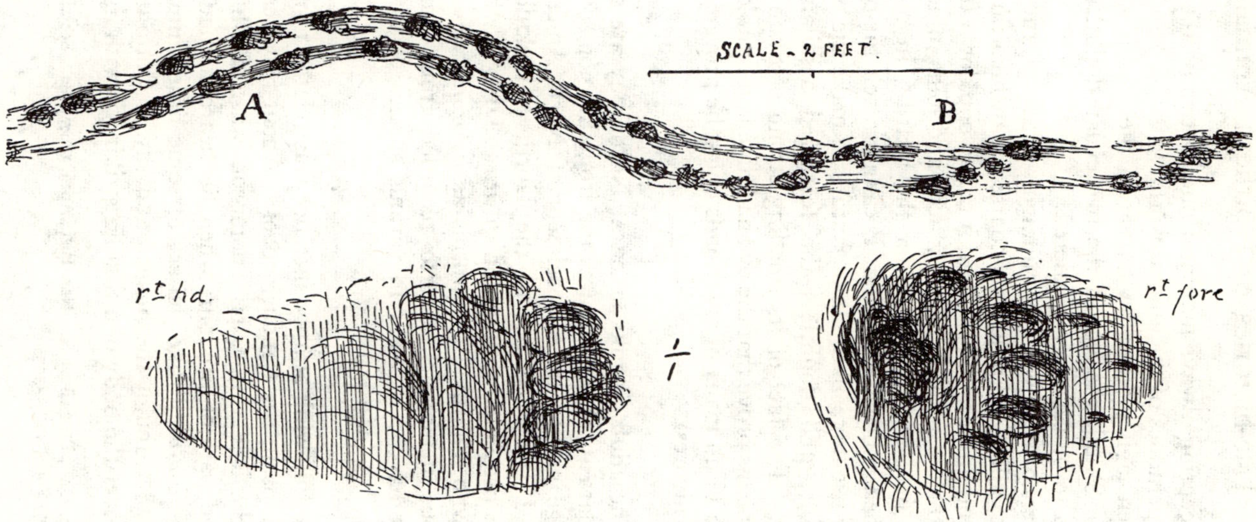

FIG. 230—Tracks of Skunk.

In snow ¾ inch deep. Upper figure the trail; at A, walking; at B, galloping.
Lower figures are life size and show right hind and right fore-foot track. The hind shows no claws and rarely the heel.

foes, excepting always man and one or two of the cutthroats and desperadoes of the animal world.

When facing death by starvation the Fox is said to consider the Skunk the less of the two evils; doubtless this is a question to be carefully pondered.

The horned owl, the midnight pirate of the woods, is known to kill the Skunk. Of course the owl has an advantage over all other foes. It is silent, it can swoop down from above, seizing the Skunk unawares by neck and loins, much as I did when I held my captive under the board. In this way the quadruped is nearly helpless. It cannot reach the owl with its musk or use its teeth or claws, but it can make the whole place intolerable, and doubtless the feathered assailant is often repelled. The fact of its smelling strong of Skunk does not by any means prove that the owl had dined off Skunk. On such evidence I and many of my friends might be proven mephitivorous carnivores.

DISEASE

"The adult Skunks taken at North Bay are all infected by the parasite that disfigures the frontal regions of the skulls of a large proportion of specimens of North American Mustelidæ. I submitted one of the North Bay skulls with the parasites preserved in formalin *in situ* to Dr. W. McM. Woodworth, who identified the worms as *Filaroides mustelarum*, a viviparous nematode hitherto recorded from Europe only, where it has been found in various species of Putorius and Mustela."[14] (*Miller.*)

STRANGE INSTANCES

The following strange instance was related to me by Will H. Thompson, the famous archer: About fifty years ago his father, the Rev. Griggs H. Thompson, was travelling through a wooded part of Missouri when he heard a loud "*qu-a-a-a qu-a-a-a qu-a-a-a*," like the cry of some little animal in pain. He peered through the bushes and saw a Cottontail Rabbit leaping over the body of a Skunk, striking it with its hind-feet,

[14] Mam. Ont., Proc. Bost. Soc. Nat. Hist., April, 1897, p. 42. See also Amer. Nat., March, 1897, Vol. 31, pp. 234-5.

and uttering the squealing he had heard. The Skunk was dead, but evidently killed within a few minutes. Its skull was broken. It seemed impossible that the Rabbit should have done it, but there was nothing to show who did, or why the Rabbit should be fighting the body.

The Skunk is regularly eaten by Indians and trappers. FLESH Provided the animal met sudden death and was not too old, the flesh is said to be white, tender, and well-flavoured.

All the Northern species of the group are standard fur-bearers. FUR Their pelts are cased; they are prime from November 1 to April 1.

During the fifty-eight years, 1848 to 1905 inclusive, the Hudson's Bay Company collected 302,564 skins of this species, an average of 5,216 for each year. The lowest were 0 in 1849 and 1,263 in 1848; the highest, 12,583, in 1889. The average for the ten years, 1895 to 1905, was 9,425.

Poland's lists show that during the thirty-four years, 1858 to 1891 inclusive, 9,765,442 skins were taken by the other American companies, an average of 287,218 each year. So that the average annual catch of Skunk for fur is about 390,000.

The Winnipeg market quotations, March 26, 1904, were 25 cents to $1.

At the London annual fur sales, held at Lampson's in March, 1906, there were sold 445,051 Skunk skins, chiefly from the United States. The highest price realized was 11 shillings ($2.64) each for a superb lot of 233 A1 black skins. Inferior skins went as low as 2 shillings or 3 shillings (48 cents and 72 cents). The ruling price for first-class skins was 8 to 9 shillings ($1.92 to $2.16).

As Skunks are easily managed in captivity, and very prolific, experiments are being made at Skunk-farming for fur. I summarize below our knowledge of this new industry.

Skunk-farming.

There is no doubt that Skunk-farming can be made to pay in spite of the fact that it has failed a great many times. In nearly every case of failure the cause has been the same—the improper bunching of a miscellaneous lot of Skunks in one large enclosure.

Condensing the experience of numerous observers, I should say that the Skunk farmer needs, first, a large loose pen, about an acre in extent, with a 6-foot fence. Second, a number of well-floored pens, each about 10 by 20 feet, for the breeding Skunks; walls 2 feet high would be enough to keep the Skunks in, but it is desirable to keep dogs, cats, and owls out. Chicken-wire over the top of the pens does this very well. It should be high enough to allow head room for the keeper.

Of course, the Skunk is a powerful digger; therefore the walls of the unfloored pen should go 3 feet underground, and at the bottom should have an underhang, either of stone or galvanized mesh wire, extending 2 feet in and 2 feet out, to prevent digging under from either side.

A hollow log or other den should be in each breeding pen and a number of them in the general pen; sometimes the Skunks may be allowed to dig their dens in the loose pen; there is, however, some danger of disaster by a cave-in, if the ground has been disturbed recently. A good plan is to sink a box, or stone-built vault, on some dry knoll.

A bottomless wooden box is easier to make and manage, but it rots in a year or two.

The good plan for a small Skunkery—and no one should begin with a large one—would be on the same lines as that suggested for Mink, p. 898.

The main runs are all the better if much larger, and should have a varied surface; the more plants, grass, etc., the more insects for the Skunks to hunt out and eat.

These animals are omnivorous and should have a greatly varied diet. Table scraps are excellent, but chicken offal, dog-biscuit, milk, oil-cake, mush, johnny-cake, fish, fruit, and insects are acceptable and wholesome food. One meal a day is enough; it should be given in the evening. Two meals are allowable, but should not together exceed the one-meal quantity. FOOD

How much food should each Skunk have? This is a matter to be determined by experiment. If the Skunk commonly leaves good food, you are giving it too much; if it gets thin, you are giving too little. HOW MUCH

A weigh scale is a very good help in determining the latter point.

A Skunk eats about as much as a common cat, and nearly the same diet—with the addition of insects and fruit.

By advertising in any country newspaper of the Northern States, it will be found easy to get as many live Skunks as desired. From 50 cents to $3 each would be fair prices, according to age and blackness—the less white on the Skunk the more it is worth. The black Skunk is one with white on head and tail only. The farther north the better the Skunk. STARTING

On arrival all full-grown individuals may be turned loose in the general run. If any one seems specially quarrelsome, it should be shut up by itself; also any weak, small, or young should be kept apart. MANAGEMENT

In the month of March or April, according to latitude, the pregnant females are separated and each given a breeding pen to herself.

If the Skunks do not admit of handling, you can put them in a small cage with a mesh-wire floor, and then examine them from the under side; the distended belly with the enlarged nipples and milk-glands will show which are destined to become mothers.

These breeders should be extra well-fed and supplied with some fine hay with which to bed their den.

Mating takes place in March (varying with latitude) and gestation is believed to last about 6 weeks. Thus most of the young will be born about the first of May. The mother must not be interfered with at this time, and handling of the young is likely to make her destroy them.

When about two months old they begin to come out of the nest and eat with the mother, but when four months old they are nearly full grown and may now be introduced to the main run.

As winter approaches their food should be increased; the colder the weather the better the fur they produce. The fur is best about or soon after Christmas, and should then be marketed.

THE SMELL-GUN

Most persons ask at once, but what about the Skunk's smeller? All experience goes to show that the animal never uses its musk except in the extremity of self-defence, and may live a long life in captivity without ever becoming offensive.

MARKETING

'Marketing' is the word that covers the unhappy process of killing the beautiful fur-bearer for its pelt. To kill a Skunk the wrong way is to court disaster. The merciful way is by a lethal chamber with illuminating gas or by drowning. The trapper's method, mentioned in the article on the Skunks, is possible, but not one that any tender-hearted person is likely to try on his hand-raised Skunks.

The oil that is rendered out of the fat is said to be of high market value on account of its medicinal properties. It is certainly an excellent lubricant.

The bodies, if used to feed the breeding stock, should be thoroughly boiled with vegetable food or some other meats.

GENERAL HINTS

Never keep more than 50 or 60 Skunks to the acre, otherwise you get crowding, ground-poisoning, and deadly disease.

The soil in the breeding dens should be turned or otherwise refreshed every few weeks.

Cleanliness everywhere all the time is essential.

Skunks, if helped, will keep themselves as clean as cats, and their musk will never be smelt if they are not forced to use it in self-defence.

A diet of all meat, especially raw meat, will kill every Skunk on the farm.

Overfeeding of any kind is as bad as underfeeding.

There is great individuality of temper, as well as of colour—always select the black ones and the gentle ones to breed from. From time to time there will appear fierce, quarrelsome individuals; these should be removed and marketed as soon as possible; never allow them to breed.

Castration of the surplus males will greatly improve the fur.

The wild Skunk pairs, but it is found that one male is enough for a dozen females where they are yarded together.

The chief causes of death to be guarded against are: disease from dirt, overcrowding, wrong food or overfeeding, infanticide by strangers entering the den during the mother's absence, and loss through great horned owls. The last are most dangerous to the young, and these are, of course, safe under the chicken-wire.

The young run from 4 to 9 in a litter. At six months these PROFITS may sell from $1 to $3 per pelt, or, say, the litter bring $10. Forty breeding females is the most one may safely have on an acre, so that under the most favourable circumstances this would bring a gross return of $400, from which we must deduct cost of food, fencing, stock, and care, leaving a very small profit indeed.

Thus it appears that Skunk-farming is not an industry that promises a very large return. It is possibly a paying business if one can handle a stock of 1,000 old ones, but it seems to me that *its chief use is to train fur-breeders for more serious work* with more expensive and immensely more profitable animals, such as Marten, Sable, and Silver-fox, or even Mink.

It is worthy of note that Skunk-farming methods, with cages instead of pens, if applied to black cats of the domestic race, are quite likely to prove profitable; not only are the animals easy to get and handle, but they are much more prolific, and the choice skins, if black and prime, fetch as much as an ordinary Skunk skin.

XLIV.

Common Badger of America.

Taxidea taxus (Schreber).

(L. *Taxidea*, from *taxus*, a Badger, and Gr. *eidos*, like; Latinized into a name applied because of the creature's resemblance to the Old World badger or taxus.)

Ursus taxus SCHREBER, 1778. Saugthiere, III, p. 520.
Taxidea taxus RHOADS, 1894, Am. Nat., XXVIII, June, p. 524.
TYPE LOCALITY.—Usually given as 'Labrador' but almost surely Saskatchewan River.

FRENCH CANADIAN, *le Blaireau d' Amerique; le Brairo.*
CREE, OJIB., & SAUT., *Mit-ten-usk'.*
YANKTON SIOUX, *Ho-cang.*
OGALLALA SIOUX, *Ho-ka'* (=shaggy or bristly).

The genus *Taxidea* (Storr, 1780) comprises large animals of the Weasel Family (*Mustelidae*). They have thick, heavy bodies, very short tails, short legs, front feet immensely powerful, with long claws and developed for digging; ears, very short, and the following teeth:

$$\text{Inc. } \frac{3-3}{3-3}; \text{ can. } \frac{1-1}{1-1}; \text{ prem. } \frac{3-3}{3-3}; \text{ mol. } \frac{1-1}{2-2} = 34$$

To these generic characters the Badger adds:

Length, about 28 inches (711 mm.); tail, 5 inches (127 mm.); hind-foot, $3\frac{7}{8}$ inches (97 mm.). SIZE

WEIGHT The following Badgers I weighed at Clayton, N. M., in 1893:

Female, taken October 26, was 10½ pounds.

Female, taken November 2, was 14 pounds 5 ounces.

Female, taken December 28, was 16¼ pounds (excessively fat).

Male, taken December 29, was 14¼ pounds.

Bachman gives[1] 23 pounds as the weight of one he examined in the Menagerie at Charleston, S. C.

COLOUR General colour above, silvery gray, each hair being yellowish-white at base, then blackish with a white tip; neck, crown, and muzzle above, brown; cheeks, chin, and stripe from nose over head to shoulders, white; under parts generally yellowish-white; bar on each cheek, back part of ear, and the feet, dull black; tail, tinged yellowish-brown.

When seen alive it looks like a small Bear that has been flattened somehow, coloured silvery gray, and adorned with black and white marks on the head.

The following races are recognized:

taxus Schreber, the typical form.

neglecta Mearns, differs in being smaller, with longer tail, and with colours deeper and richer than in either the preceding or following.

berlandieri Baird, is distinguished by having the white line continued along the back in some cases to the tail, also by a general buffiness of colour (as compared with the silvery gray of *taxus*), and heavier markings.

infusca Thomas, similar to *berlandieri* but darker.

Life-history.

RANGE The map (No. 53) shows the range of the Badger in the North-west to coincide with the untimbered regions in which

[1] Aud. & Bach., Q. N. A., 1849, Vol. I, p. 363.

MAP 53—RANGE OF THE AMERICAN BADGER, AND ITS FOUR RACES.

Taxidea taxus (Schreber).

This map is founded on records by J. Richardson, S. F. Baird, R. Kennicott, E. Coues, R. MacFarlane, C. H. Townsend, E. A. Mearns, D. G. Elliot, J. Fannin, V. Bailey, E. A. Warren, O. Thomas, E. T. Seton, and C. C. Adams, in Northern Michigan.

It is fairly correct on the north and east, further investigation will change it somewhat on the west, and greatly in Mexico.

This gives the primitive range, but it is little changed to-day, excepting perhaps that the species is now exterminated in most of the region about Lake Michigan.

the larger kinds of burrowing rodents are found in numbers. In the dry poplar country, that is varied with open glades, it is occasionally found far from the prairies. MacFarlane records a specimen from Isle ā la Crosse, in 1889, and 2 from Green Lake, 1889 and 1890 (*Mam. N. W. T., p.* 715). (Spots 2 and 1 on map.)

While travelling on the Athabaska, in 1907, I was shown by the pilot, John MacDonald, a range of hills where 2 Badgers were killed by François Black, one in 1905, the other in 1906.

The exact neighbourhood was Red Willow Lake, 18 miles south-east of Fort MacMurray; the place is spotted (3) on the map. He said that the animal was previously unknown there, but some Plains Indians, who happened to be at the Post, knew these at once and called them '*Mittenusk.*'

The spot (4) in northern Michigan is on the authority of Charles C. Adams (*Ecological Surv., N. Mich.*, 1906, *p.* 130).

MANITOBA

In Manitoba, the Badger is confined to the dry prairie regions. It is very rare in the half-timbered country, and unknown in the thick woods to the north-east.

It seems indeed to be exact complement of the Woodchuck, which, on account of these facts, is known to the Hudson's Bay Company people as the Thick-wood Badger.

ENVIRONMENT

Dry rolling prairies of light or gravelly soil, with a high populative rate of Ground-squirrels, are the Badger's ideal surroundings.

John Atkinson, the Lake Winnipeg guide, writes me that he found a few of these animals in the drier country between Winnipeg and Whitemouth River, also about the gravelly hills east of Gonor.

On the heavy clay prairies of the Lower Red River, however, it is scarce, and, of course, on swampy lands it is unknown.

HOME-RANGE

There is little direct evidence at hand to show the home-range of an individual Badger, but obviously it must be very small. As a matter of opinion, I should say that a Badger may pass his whole life within a mile or two miles of his original home. In Texas, Vernon Bailey found that one Badger had worked all summer in a 20-acre field, and of the species in

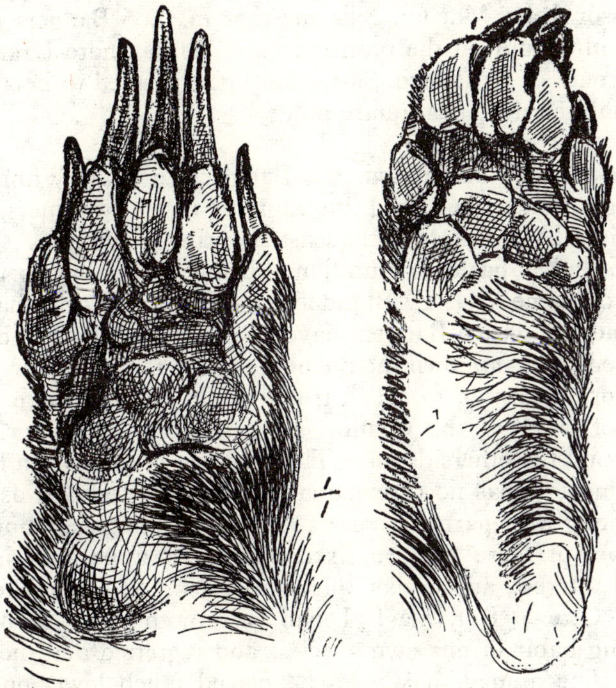

FIG. 231—Right fore and hind-foot of Badger.
Taken in Colorado, Sept. 7, 1901. (Life size.)

general, he says[2] that, when food is scarce, they become great travellers, "sinking a house in the earth wherever sleeping-time overtakes them."

In early days there was at least one Badger for every square mile of high, dry prairie in Manitoba and perhaps one-third as many on the heavy clay prairies of the Red River Valley, which would give a Badger population for the Province ABUNDANCE

[2] N. A. Fauna, No. 25, pp. 184–5.

of some 20,000. To-day their numbers are much reduced by trapping, poison, and the destruction of Ground-squirrels, their food supply, as well as by the disturbance of ploughing the land, for the Badger is a shy prairie animal, and is likely to disappear when all the open country is under cultivation. Professor John Macoun tells me that in 1906 Badgers were yet so abundant on the prairie around White Shore Lake, 40 miles south of Battleford, Sask., that there seemed to be about 10 inhabited dens per square mile.

SOCIAL AMUSEMENT

So far as I have seen, the Badger is a solitary animal, leading a somewhat sordid life, minding its own business, but confining that business to the least elevating of pursuits. The British Badger has the reputation of being remarkably sociable and frolicsome. G. E. Blundell, of Bristol, tells me that in England the native Badgers have a sport which he has often observed. At sundown the members of the family repair by a well-worn pathway, to a low trunk or stump, and there play a sort of 'King of the Castle' game, each one trying to climb up, or pull the others down. They indulge in this for an hour at a time. It has no connection with the sex feelings, as old and young take part as soon as the latter are strong enough. The fact of there being a fixed place and apparatus is of special interest, and ranks this amusement with the sliding of the Otter. But, alas! I had seen nothing of such an engaging habit in our own species, and regretted that though such a fine animal, it was to be placed much lower on the scale of development than its congener, and was glad indeed to find later from Paul Fontaine's account of the Badger[3] that on bright moonlight nights he had often watched them for hours gambolling and playing like dogs. Thus, as in every case, the more we learn of the animal the more claim it has on our sympathy and interest. To the casual glance the wild animal is a fierce, elusive creature, occupied chiefly with eating and running away. It is only on getting gently nearer that we realize the other half of

[3] Great Northwest, 1904, p. 40.

PLATE LXXXIV.—BADGER STUDIES FROM LIFE.

its life, the side which shows love for the mate, its young, and the pleasant society of its own kind.

INTER-COMMUNI-CATION

The Badger has many sounds that it uses in expression. Unfortunately, only those used in battle have been recorded. The hiss, the grunt, the growl, and the low husky snarling, which seems to show that the snarler is a little afraid of the one it is snarling at. As recognition marks, or signal service apparatus, the black and white face-spots are no doubt important, for they announce its species to all the wise world that can see; but there is another contrivance highly developed in the species, that is, the anal group of glands. Just how it is used is not known, because this belongs to the gentle side of the Badger's life, and all our observations so far have been that of the bitterly hostile.

MATING

Little is known of the mating habits in this species. I am inclined to believe that, like all the higher mammals, it is monogamous, and that, as in the highest, the male, sometimes at least, stays with the female all summer and helps to protect and feed the young.

All the evidence I have been able to gather is given here. Professor John Macoun tells me that in Saskatchewan, where Badgers still abound (1906), he commonly saw 2 adults at each den door during the first week of August, but never 3. In each case the bigger one, presumably the male, remained sitting head out of the hole, with its bristles up, and uttering a sound of menace. He saw no young.

The following incident also goes to show that the species pairs and the male continues with the female all summer. It was related to me by Russell Brown, of Sunnyside, Wash. While haying late in June, 1902, his dog was attacked by two full-grown Badgers. He went to the rescue with a fork. On killing the assailants, they were found to be male and female, but he saw nothing of any young.

R. W. Cowan, ranchman, tells me that near Calgary, Alta., during the month of September, he more than once has seen 2 adult Badgers trotting along the trail together. At such times they are more ready to fight than to turn aside.

All of these observations point to an all-season association of the pair.

Collateral support is found in the ways of the British Badger, which is known to be a model husband and father. Although it is dangerous to rely on such oblique light, for the British Badger is not now even in the same genus with this, I cannot refrain from giving Sir Alfred E. Pease's remarks on the nesting of that amiable species. (*Monog. Badger.*)

"He is fond of company; he is monogamous, and clings closely and faithfully to his own wife. With Badgers, as with the human race, the sexes are not precisely equal in numbers, and often, from the force of circumstances, a Badger has to remain a celibate, but he is not a bachelor by choice. He may become a widower, but in either case he will travel far to seek a partner to share his shelter and his lot. It is not altogether rare to find an old solitary dog Badger, who has loved and lost, or taken in late age to a hermit's cell; but he, as often as not, when he failed to secure the companionship of the gentler sex, has found some other male to share his home, when they live comfortably *en garçon.*

"Nor do the married pair shun the society of their kind. I have often seen large Badger 'sets' almost as full of Badgers as a warren is of Rabbits. One evening, near my house, I waited an hour of midge-plagued time to watch the Badgers come out from a small 'set,' and was rewarded by seeing a procession of 7 full-grown Badgers emerge from a single hole, and I had them all in full view for something like twenty minutes. As this was in July, they could hardly be one family."

It is an open question whether the hibernating Badger mates in spring like the hibernating Ground-squirrels, or in the fall like the hibernating Bear. Paul Fontaine states[4] positively that "they pair in autumn, before they hibernate." This we know to be the case with the British Badger, so that the evidence is strong, though not conclusive.

[4] *Ibid.*

GESTATION

The gestation of the species is unknown; cannot be guessed at until we know the exact time of mating.

DENS

The residential burrows of the males and unmated young have not been investigated. It is probable that each Badger makes a burrow every twenty-four hours during the summer while in search of food. As there is no certain way of distinguishing these prospect shafts from actual residences, the labour

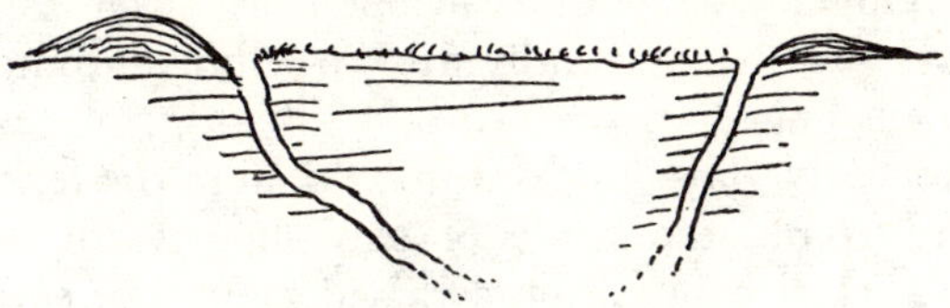

FIG. 232—Badger hole, 6 feet deep.
Carberry, Mass., July 25, 1892.

of digging out all, in order to run the facts to earth, has hitherto proven too serious for the investigators. It is not by any means certain that the 'foot-loose' Badger does stick to any one den in his home-range.

But the female has a different way of life. Early in the spring, accompanied, we believe, by her mate, she prepares a complete nest of grass in a well-drained hole, two or three feet below the surface, and here, in late May or perhaps early June, are born the young. They number 2 to 5; 3 is probably the usual number.

YOUNG

I cannot learn that any one ever saw a Badger travelling and accompanied by its young. Apparently they remain in the home den until big enough to dig for themselves, which may mean till a year old. Senator J. N. Kirchhoffer, of Brandon, tells me that on the Souris he once saw an old one and 2 young ones together at the den about the end of September.

SPEED, ETC.

This creature has bartered its speed for strength to dig. A man can easily overtake it if by rare chance it is surprised from its home, but that does not mean that it is caught, for in loose soil it can dig so fast as to escape into the ground before the foe can come near. It is supposed to be at home nowhere but underground, and I never expected to see one go aloft, so was much surprised one day to see a caged specimen climb readily to the roof of its cage by hooking its claws in the wire netting; and another in the Winnipeg Zoo that easily and often climbed a low branching tree in its enclosure.

HABITS

The Badger is a winter-sleeper. A 'seven-sleeper,' the country folk say. It generally appears above ground as soon as the snow is gone.

In the early days of Manitoba, before the fence and the plough had come, the traveller saw, hourly, on the sunny mornings, a whitish bump on a raised mound of earth not far from the trail. As he approached it, the white bump might develop a sharp and movable point at one end, the point would sway in the wind, then the white thing disappear into the earth, showing that the bump was simply a Badger taking his morning sun-bath. On the Souris Plains Badgers were thus seen a dozen times a day.

They rarely go far from their holes, and when they do, they are much alarmed by discovery, and go shuffling about to each promising place in search of a road to the friendly shelter of mother earth.

I overtook one once on the open plains in Arizona. He skurried about but could find no hole, so faced about, and as he made short leaps towards my companion I caught him by the only safe handle, his rough, strong tail. But possession seemed to satisfy the hunter's instinct, and once we had conquered him we freed him and left him in peace.

On another occasion, in June, 1897, on the Upper Yellowstone, I met a Badger waddling over the prairie. I had a camera with me and, meaning to get a picture, ran after him. To my great surprise, he came rushing towards me uttering a loud snarling. Fully believing in my ability to avoid his attack, if indeed he really meant to make one, I continued to run, when, just as we were within thirty feet of each other, he fell tail-first into a shallow badger-hole that he had not seen, and I fell head-first into another I had not seen. We both were greatly surprised, quite shocked indeed, but he recovered first. He scrambled out of his pitfall, ran ten feet nearer to me, then dived down his home-hole, towards which he had been making from the first.

Those who know the Badger of Europe have little idea of the life of the prairie species. The former seems to live much like a Skunk, trotting about at night, above ground, seeking its food in the woods and thickets, retiring to an underground home to rest during the hours of daylight. But the prairie Badger spends the greater part of its life underground, where it digs, feeds, sleeps, and multiplies much like a Mole. It rarely comes out during the day, except to bask in the sun by its doorway, and then is ready to plunge below at the slightest alarm.

Deep underground, this animal is safe from violence. It is hopeless to dig it out, for it can burrow like a Mole; it succumbs to nothing but a few barrels of water sent suddenly after it. This shuts off our diver's wind, and forces it to the surface to breathe and meet its adversaries. On one occasion I ran after a Badger on the prairie, and just as I neared him he plunged into a hole that was but three feet deep. I seized his tail as he was digging it deeper and tried to haul him out, but he braced himself with both fore-feet and defied my best efforts.

Water was at hand, but a couple of bucketfuls thrown in merely caused him to swell out his body till it plugged the hole, and no water whatever passed him to get near his head. A spade handle, however, pushed between him and the wall let the flood down with a sudden 'gulch,' and the Badger was forced to turn about and wage unequal fight.

AS A FIGHTER

The Old World Badger has long been famous as a fighter, and the prairie species seems no whit behind its cousin. It is so strong that a man cannot pull it out of its hole if once it gets fairly braced. It is so protected by its thick, loose-fitting hide that a bull-dog may be holding it by the scruff of the neck without in the least shutting off the Badger's wind or preventing its operating with teeth and energy on any or all four quarters of its preoccupied assailant. Its jaws are so strong that it usually leaves a gash at each bite, and its courage such that it *never* surrenders, no matter how numerous or strong its assailants; it dies fighting to the last. A fifteen-pounder would be a large one, and any dog twice the weight would surely be worsted by the Badger.

In my journal for 1892 I find this note: July 27. This morning at Carberry a Badger was pitted against four large dogs on the open prairie, and he beat them all off, escaping almost unhurt. The Badger was aggressive as soon as a dog came near him. He continually uttered a sort of hissing, also grunted like a pig. As a matter of fact, it was a bloodless battle consisting of little but noise. The dogs seemed afraid to close in. When taken back to the stable the Badger drank a large quantity of water. Where does he get it when home?

WINTER SLEEP

As late as November 4, 1884, I found a Badger active above ground. The fresh snow was plentifully marked with its tracks, showing where it had gone about sniffing at all manner of Ground-squirrels' holes, seeking those that gave token of inmates sleeping below. In New Mexico, this animal is active all the year round; but in Manitoba, as soon as the ground freezes, it goes below and sleeps through the winter without any store of food other than its fat, until, in April, it is again aroused to life.

FOOD

The species is carnivorous, strictly so, as much so as any animal is ever strictly anything. The bulk of its food is, no doubt, Mice and Ground-squirrels. I have often seen places where a Badger had ripped open the long surface burrow of the Striped Ground-squirrel or had sunk twenty or thirty prospect holes at intervals to strike the deeper burrow, and I have no doubt that its labours were rewarded with a meal.

In the November of 1884, as noted above, I followed a Badger's tracks in the fresh snow to learn that the night before he had sunk a number of burrows to the depth of five or six feet. In each case he had reached the winter den of a Richardson Ground-squirrel and doubtless had devoured the sleeper, for its nest lining and its grain stores were scattered about.

Alexander Henry gives the following curious note in his Journal on Red River, 1799, p. 158: (Coues. ed. 1897).

"This afternoon I saw an extraordinary race—a Badger in pursuit of a Skunk. I wished to see what would be the consequence, but one of my men killed both with a club before I thought of preventing him."

FRIENDSHIPS

Many of our quadrupeds are known to form curious, apparently platonic, friendships with totally different creatures. The British Badger has frequently been found living in good-fellowship with a Fox, and, on several occasions, our own comfortable species has been accredited with a similar partnership, which shows that he is much the same all-round good fellow as his British cousin twice removed. The first cases were friendships with Coyotes and were recorded by A. H. Hawkins, the surveyor.

"During the progress of my survey," he says,[5] "in southern Alberta I noticed on two occasions a Badger and a Coyote travelling in company. The same thing was observed and reported by the men who did my mounding on three different occasions, all of which were in different localities.

"The men reported having seen the animals travelling in company in Township 1, Range 13, West of 4th Principal Meridian. The first time that I saw them together was in Tp. 6, R. 17, and the second time in Tp. 7, R. 17, W. 4th. This last time I had the best view. Seated one day eating our noon lunch, I noticed two animals coming towards us and drew the attention of my men to the fact. We remained perfectly quiet, so that they came within 20 to 30 feet of us before seeing that we were so near. The Coyote travelled ahead, and the

[5] Ottawa Nat., May, 1907, p. 37.

Badger followed along as fast as he could, right at the heels of the Coyote.

"I could see no reason nor could I explain it in any way satisfactory to myself, and, although I asked several people in the West about it, the occurrence is still a mystery to me."

Some similar cases have been reported to me by G. A. Rimington, of Penrith, Eng. Several times, near Calgary, in 1907, he saw a Badger and a Coyote associated and travelling together. In these cases it seemed to be a partnership affair, which was probably involuntary on the part of the Badger. No doubt the Coyote knew very well that the Badger would dig out Ground-squirrels, some of which would bolt and thus give the Coyote a chance to share in the spoils.

In exactly the same way the Badger is followed by hawks, etc., in California, as graphically described by Mary Austin in "The Land of Little Rain."[6]

But the most remarkable case of all is a friendship between a Manitoba Badger and a lost boy. This was related to me by George Fraser, a native of Manitoba, and corroborated by his mother, Mrs. Fraser, of Kildonan, and Archbishop Matheson.

In 1871, a little seven-year-old boy, named Harry Service, wandering from his father's house at Bird's Hill, near Winnipeg, was lost for two weeks. When found, he was living in a den with a Badger. His clothes were torn so that he was nearly naked, and his face was all scratched. He told his parents that he had taken shelter in the hole during a rain-storm, and that the Badger came later and scratched his face. At first they fought, but the child was plucky and would not give up the hole. Later the Badger brought some food and, after another quarrel, allowed the child to eat some of it. In the days that followed the Badger brought him food several times. The beast always entered the den by one of the entrances not used by the child.

When found they were on terms of friendship, and the child cried bitterly when taken from his savage friend. The boy's story, however, was not clear. He said at one time that

[6] 1904, p. 152.

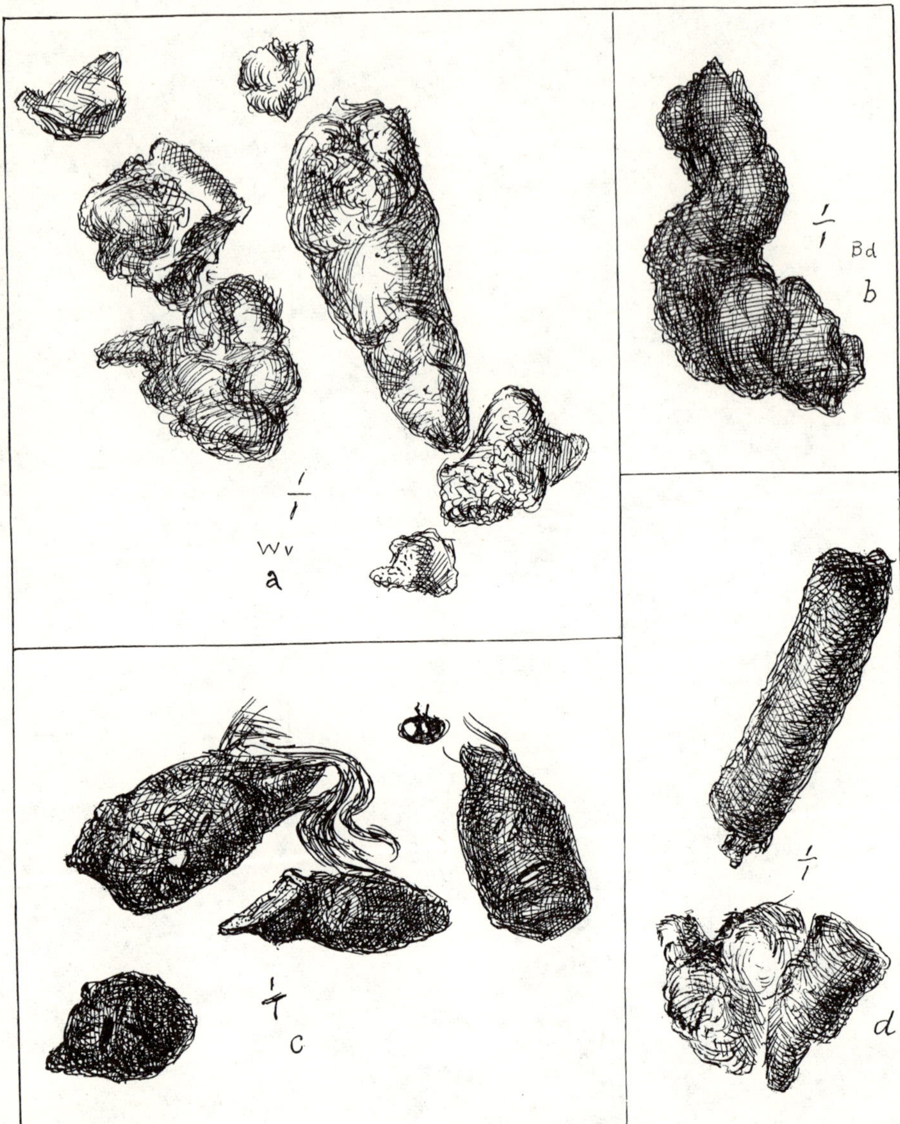

PLATE LXXXV.—SCATOLOGY OF CERTAIN MUSTELIDAE (ALL NATURAL SIZE).

a. Wolverine, October, 1908.
b. Badger, October, 1908.
c. Skunk, October 10, 1908, chiefly remains of grasshoppers and wasps, but in this case also the tail feathers of a small bird, Cos Cob, Conn.
d. Fisher, April 28, 1905.

he lived on mud. His face, mouth, and tongue were black with mud and much swollen. Nevertheless, his description of the Badger was beyond question. He even said it had five toes on one foot and four on the other.

The Badger must be considered a valuable animal, as surely as we reckon the Gophers a nuisance. The only sin I ever heard charged against it is that the holes it makes endanger the limbs of stock and the lives of horsemen on the plains. But this objection scarcely exists in Manitoba to-day, since we no longer have vast open plains. The work of the Badger is now confined largely to the strips of prairie that exist along the road allowances, where it can do but little harm. USE TO MAN

The pelt is prime about October 1. It is of beautiful silvery gray and the hide strong and durable, so it is much in use for chair rugs. In price it ranges from 75 cents to $1.50. FUR

During the sixty-four years, 1842 to 1905 inclusive, the Hudson's Bay Company collected 81,837 skins of this species, an average of 1,278 for each year. The lowest was 289 in 1904 (none at all prior to 1842); the highest, 4,000, in 1891. The average for the ten years, 1895 to 1905, was 2,445.

Poland's work (p. 131) appears to show that double as many are taken by the other American companies, so that the catch of Badgers for fur may be about 7,000.

At the London annual fur sale held at Lampson's, March, 1906, 5,955 Badger skins were sold. The highest price reached was 19 shillings ($4.56) each for 27 unusually fine first-class skins. More usual prices were 5 shillings ($1.20) to 10 shillings ($2.40) for first-class skins. Inferior skins sold for a shilling (24 cents), or even less.

XLV.

Raccoon or Coon.

Procyon lotor (Linn.).

(Gr. *Procyon*, from *pro*, before; *cyon*, a dog—the name of a star group that rises just before the Dog-star. The name was given to this genus by Storr in 1780, probably without special reason. L. *lotor*, a washer, on account of its habit of washing its food.)

Ursus lotor LINNAEUS, 1758, Syst. Nat., X ed., I, p. 48.
Procyon lotor DESMAREST, 1819, Dict. d'Hist. Nat., XXIX, p. 91.

TYPE LOCALITY.—Eastern United States.

FRENCH CANADIAN, *le Raton*.
CREE & SAUT., *Es'-see-ban*.
OJIB., *Es'-see-pan*.
YANKTON SIOUX, *Way-atch-a*.
OGALLALA SIOUX, *Wee'-cha*.

'Coon' is abbreviated from 'Raccoon' or 'Racoon,' which is the Englished form of 'Arocoun,' the Indian name of the creature in Virginia.

The Raccoon Family or *Procyonidæ* comprise middle-sized animals, kin of the Bears, having on each foot 5 well-developed toes with fixed claws, the soles naked, the hind-feet plantigrade; they have pointed nose and ears; tail, rather long and bushy, usually ringed.

The genus *Procyon* (Storr, 1780) has the above characters and has the teeth:

$$\text{Inc. } \frac{3-3}{3-3}; \text{ can. } \frac{1-1}{1-1}; \text{ prem. } \frac{4-4}{4-4}; \text{ mol. } \frac{2-2}{2-2} = 40$$

In addition to the Family and generic characters unmistakable specific peculiarities. SIZE

An adult male killed at Springfield, near Toronto, Ont., June 25, 1888, was $32\frac{1}{2}$ inches (826 mm.) from tip of nose to tip of tail-bone. The head and body from nose to ischium were $23\frac{3}{4}$ inches (604 mm.); the tail-bone, $10\frac{3}{8}$ inches (263 mm.); from tip of front toe to tip of outstretched hind toe, $37\frac{3}{8}$ inches (950 mm.). This was a very fine and fat individual. Hind-foot of another specimen was $4\frac{1}{4}$ inches (108 mm.). A female caught at the same place, fifteen days earlier, was $31\frac{1}{2}$ inches (801 mm.) from tip of nose to tip of tail-bone; the head and body from nose to ischium were $21\ \frac{11}{16}$ inches (552 mm.); the tail-bone, 11 inches (280 mm.). She was still in milk; her 6 teats were as shown in Fig. 233.

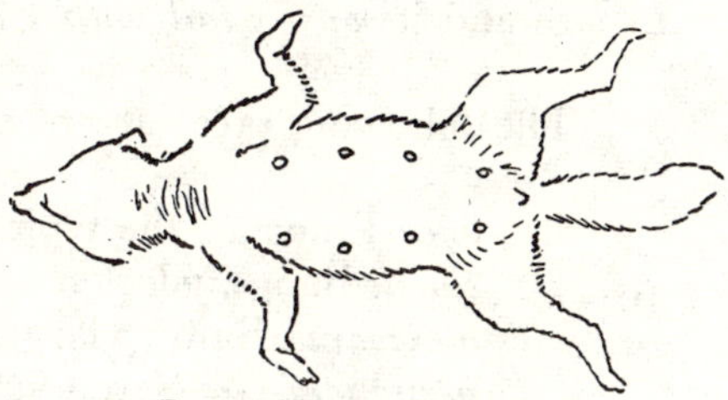

FIG. 233—Mastology of the Coon ♀. Springfield, Ont., June 25, 1888.

The above male weighed 18 pounds; the female barely 10. WEIGHT I consider the male about average size and the female small. Audubon and Bachman give[1] 22 pounds as the ascertained weight of a good size male.

In general the Coon is of a dull brownish-gray, becoming COLOUR yellower on the back, strongly yellow on nape and on tail, and a paler gray on the belly and feet; on all the upper parts, especially along the spine, the long hairs are black tipped, and on the under parts they are white tipped. On the cheek is a black patch that includes the eye, and joins with the narrow blackish stripe that runs from the nose to the dark colour on the forehead. Around this the face is dull white, whitest in a band above each eye; the lower back part of the ear is black, which joins with a black patch on the neck behind the ear—the tip of

[1] Q. N. A., 1849, Vol. II, p. 76.

the ear behind is whitish. The tail, beginning with the tip, has 6 to 7 rings of very dark brown or black on a pale yellowish background. The under-fur is dark brownish-gray and shows much on the under parts; the throat also is dark brownish-gray; the eyes are dark; the whiskers white. The sexes are alike.

Where seen alive, the Raccoon is readily distinguished by its size and its *bushy tail with black rings*.

The following races are recognized:

lotor Linnæus, the typical form.
elucus Bangs, a darker, shorter-haired, long-tailed race.
mexicanus Baird, a large, pale gray race, with long tail.
hernandezi Wagler, a very large dark form, with very narrow rings on tail.
pallidus Merriam, a medium-sized pale gray form, without any yellow suffusion.
insularis Merriam, small and pale.

Life-history.

RANGE

Map 54 sets forth the range of this wide ranger. In the east and north-east it is fairly well ascertained, but must be greatly modified by further investigation in the south-west, west, and south.

IN MANITOBA

In primitive days the Raccoon was quite abundant along the Upper Red River, as is attested by the Journal of A. Henry, the fur-trader, who, about 100 years ago, wrote at Park River House, on Red River, 35 miles south of the Boundary. Here he collected annually 100 to 200 skins. Since then this animal's numbers have increased and decreased several times; they are now at a low ebb. It is still found in south-western Manitoba, but is exceedingly rare, although, according to Herrick,[2] it ranges over the entire wooded parts of Minnesota.

[2] Mam. Minn., 1892, p. 139.

MAP 54—RANGE OF THE RACCOONS FOUND IN NORTH AMERICA.

This map is founded chiefly on papers by Messrs. D. G. Elliot, V. Bailey, R. MacFarlane, W. H. Osgood, C. Hart Merriam, John Richardson, R. Kennicott, L. Adams, J. Rowley, J. A. Allen, G. S. Miller, S. F. Baird, E. A. Mearns, and E. T. Seton.

In its north, south, and east, its lines are tolerably accurate, but in all the Rocky Mountain, Pacific Coast, and Mexican regions, must be modified by future work.

Four species of Raccoon are recognized:

- *Procyon lotor* (Linnaeus), with its 6 races,
- *Procyon psora* Gray, with its 2 races,
- *Procyon maynardi* Bangs,
- *Procyon pygmaeus* Merriam.

D. Nicholson, of Morden, tells me that in twenty-five years' residence he has seen but 2 Coon skins taken in Manitoba; both were from Pembina Valley.

According to the *Boissevain Recorder*, 2 Coon skins were brought into Wooton's store in Manitou from Pembina

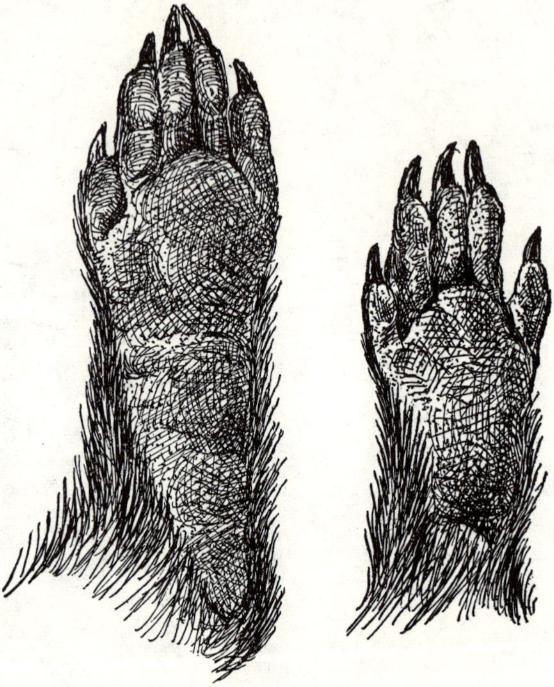

FIG. 234—Paws of Coon, left hind and left fore. (Life size.)

Valley on March 13, 1902. C. C. Helliwell saw one that was killed on the Souris River some years ago and heard of several others.

Apparently they are increasing again in that region, as J. S. Charleson writes me that in the spring of 1907, 3 Coons were found in a hen-house near Blyth, 13 miles south-east of Brandon. One was killed and mounted for Rex McPhee; another, he says, was caught in the previous winter on the Assiniboine,

10 miles above Truesbank, and a number of Coon signs were reported along the river banks near Blyth.

W. R. Hine mounted one taken near Winnipeg—it is now in possession of Sheriff Inkester—and William G. Tweddell tells me that he knew of one being killed in the country north of Shoal Lake.

I saw a very large and dark specimen that was taken on the Upper Assiniboine, near Fort Ellice, about 1884.

In September, 1904, J. J. G. Rosser, of Hudson's Bay Company, at Winnipegosis showed me a coon-skin taken on Waterhen River at the second rapids by an Indian, Francis Katchaway, October, 1903. The trapper did not know what he had caught—said it looked like a cat. None of his people had seen one before. This is the only one ever taken near this post, and is the northernmost record for the Province. Rosser heard of another that was killed at Valley River (Dauphin Lake) quite recently.

Angus Brabant, Inspector for Hudson's Bay Company and former Chief Factor, saw a Raccoon taken at Pine Creek, 60 miles north of Dauphin, Man., Lake Winnipegosis, 1890.

OTHER RECORDS

"William McKirdy, of Nipigon, told me that a few years ago a Raccoon was killed by some Indians near Lake Nipigon and brought to the Hudson's Bay Company's post. Neither Indians nor traders ever had seen the animal in the region before, and to most of the former it was entirely unknown."[3] (*Miller.*)

George F. Guernsey writes me from Fort Qu' Appelle: "December 14, 1906.—Within the last 20 years I have known of 2 Raccoon being taken some 50 miles north of here, in Touchwood Hills, which are heavily timbered with poplar and birch. But they are so rare that the Indian who took one of them did not know what it was; there are none in the Qu' Appelle Valley—not enough timber, for one thing."

A newspaper clipping recently directed my attention to a still farther record. On writing to the person interested, W. H.

[3] Mam. Ont., Proc. Bost. Soc. Nat. Hist., April, 1897, p. 41.

Jaeger, of Edmonton, Alta., I was courteously furnished with particulars: A Raccoon, the first ever heard of in the district, was taken by an Indian about 1903, on Red Deer River, at a point some 75 miles south of Edmonton.

Another, also captured by an Indian, was secured at a place 40 miles south of Edmonton, in February, 1905. The skins were brought in for sale to Thomas Hourston's store at Edmonton.

C. C. Chipman, the Hudson's Bay Company Commissioner at Winnipeg, writes me on December 3, 1906: "There was 1 Raccoon killed in the Peace River Country about fifteen years ago and they did not know what it was. I never heard of any having been killed at Lake Winnipeg or Lake Manitoba."

William McInnes, of the Geological Survey of Canada, examined the skin of one killed at Attawapiskat Lake (Lat. 52° 20′, Long. 87° W.) in the winter of 1893.

These various records are spotted on the map. When extra-limital occurrences multiply, it is usually proof that the species is extending its range.

INDIVIDUAL RANGE

We have little light on the individual range, but it seems much wider than might be expected from such a slow-footed creature. Bailey speaks[4] of Coons in Texas going regularly half-a-mile to a mile from their dens to their hunting grounds, and, of course, back before dawn. Bachman tells[5] of following a pair through the snow, and they made a journey of about a mile, ending where they began. W. S. Williams, of Panther Creek, N. C., informs me that a pet Coon he had, escaped, and within a couple of days was killed while raiding a hen-roost 5 miles from home. Obviously this one had no home.

ENVIRONMENT

This is a creature of woodland edges, preferably hardwood; dense coniferous forests do not please it, one reason being that hollow trees are essential to its well-being. It does occasionally lodge in rocky crannies, even in bank burrows,

[4] N. A. Fauna, No. 25, 1905, p. 193.
[5] Quad. N. A., 1849, Vol. II, p. 81.

but this is exceptional and imposed by the absence of more congenial quarters. No matter what its daytime residence is, its nightly prowling is always close to the water.

Its numbers are much greater than is commonly supposed. In 100 acres of hardwood bush, near Toronto, I got 3 Coons, and had evidence of several remaining. Yet they were considered scarce. This woods was chiefly second growth; heavy timber has proportionately more Coons. Poland's Fur Trade Reports[6] show that for 40 years prior to 1891, 500,000 to 700,000 Coon skins have been marketed each year from North America. As Mexican Coon fur is worthless, we can see by the map that the region paying this tribute without apparently suffering is about two-thirds of the United States, or 3,000,000 square miles. I reckon that an annual drain of 25 per cent. is all that such a species could stand without diminishing, and there is evidence that the Raccoon is rather increasing. NUMBERS

Furthermore, it is probable that not more than half the Coons killed are marketed in London as fur. Therefore, the low annual return of 500,000 would represent an annual kill of 1,000,000, and a total population of 4,000,000 in North America; that is, a pair of Coons to every 1⅓ square miles of their range. A safe estimate, indeed, even though we have included vast farming regions in the Middle States, where the species is now exterminated.

It is a common thing to find half-a-dozen Coons in one hollow tree. It is a rare thing to find a solitary Coon. Therefore, I consider the Coon a sociable animal. But they do not run in bands, except as families, nor are several nests placed together; therefore, they are but slightly gregarious. SOCIABILITY

During approach, its singular black-masked face; during retreat, its yet more singular ringed tail, are label marks that proclaim to friend and foe with equal emphasis that *this gray beast is a Coon.* INTERCOMMUNICATION

[6] Fur-bearing Animals, 1892, pp. xxii–xxxiii.

But its voice is even more serviceable to it. The querulous '*churr*' of a captive Coon squabbling over provender is familiar to all. The growl and snarl of Coons in fight are well known, and the soft '*err-err-err*' of a young Coon, begging for food, has been heard by all who know the Coon as a pet. But it has yet another note, one that has been the cause of much dispute. In the black woods, on still nights, I have often heard it, a long-drawn, tremulous '*Whoo-oo-oo-oo*.' This is the 'whicker.' It is often passed for or confounded with the call of the screech owl. But I think I can tell them apart by the stronger and more squally quality that characterizes the sound of the quadruped; the bird's note is much softer and sweeter, as well as more often uttered.

DEN The ideal den of this creature is some hollow branch high up in a large tree that is fully exposed to the sun. But Coon ideals are as scarce as those of man, and next choice is any available hole in a standing tree or tall stub; failing this, it will use any hollow trunk it can find, preferably standing, but not to be despised when down, and it will even rest content with a cranny in a cliff. So far as I can learn, it draws the line only at a hole in the ground.

The den is not only the nursery, it is the year-round home of the family. There seems little doubt that, like some other species, the Coon maintains a central home-den and several hunting lodges scattered in convenient proximity to favourite and remote feeding grounds, each to be used as occasion seems to warrant.

MATING All the evidence there is goes to show that the Raccoon is a monogamous animal and that the male stays with the female, helping to some extent in the rearing of the brood. When the mating takes place is not known. The analogy of their near relations, the Bears, would fix on autumn as the nuptial time. Coons are undoubtedly noisier then than at any other time, which is a mite of proof for autumnal mating. The fact that the species hibernates is another indirect evidence, as

PLATE LXXXVI.—RACCOON STUDIES FROM LIFE.
By E. T. Seton.

winter torpor is usually associated with long gestation, since the days passed in torpor are scarcely counted in those operations of nature where high functional activity is essential; furthermore, we may argue from this that the gestation will vary greatly in proportion to the length of the individual's winter sleep.

The young are born in April or May, varying somewhat with the latitude, those in the north being later; they number from 3 to 6, 4 being usual. YOUNG

The home-life of the Coon family is nearly ideal. I think, but am not sure, that the father continues to form one of the circle. During May the little ones stay home and are nourished only with milk. In late June they are one-third grown and begin to sit outside the den on bright days, enjoying their sun bath, but ready to seek the home-nest on the slightest hint of danger. HOME-LIFE

A charming picture of young Coon life in Texas is supplied by Vernon Bailey:[7] "While watching for Fox-squirrels [says he] one morning [June 6] in the heavily timbered bottoms, I heard a scratching sound from an old cypress in the edge of the swamp near by, followed by a loud splash. A young Coon, less than half grown, had fallen into the water. At the sound, the old Coon and 2 more young ones came out of a hollow some 30 feet up in the trunk and climbed down to near the bottom of the tree. They came down the tree slowly but steadily, head-first, as a squirrel would have done, with the hind-feet reversed and slightly divergent.

"When the old Coon saw the young one climb out of the water upon the tree trunk, she turned about and ascended the trunk, followed by the 3 young. The one that had fallen, besides being very wet, was slightly hurt and climbed with difficulty. When half-way up, he stopped on a limb to rest and began whimpering and crying.

"The mother had already reached the hole, but, on hearing his cries, turned about and climbed down to him. Taking a

[7] N. A. Fauna, No. 25, 1905, p. 194.

good hold of the back of his neck and placing him between her fore-legs, so that he, too, could climb, she marched him up the tree and into the hollow."

Whether the parents ever bring home food to the young is not known.

In mid-June, about Toronto, I found the young still suckling.

In July the young begin to accompany their mother, or possibly both parents, in nightly excursions to the edges of swamps and streams where they learn the rudiments of frog-hunting, crayfish catching, and many other pursuits that make their life.

In August—the Green-corn Moon—the mother Coon will lead them to the fields where grow the milk ear-rows, and they revel in a feast that is to them what honey is to Bears.

A lively scene took place one mid-August night outside my shanty in the Adirondacks. Two Coon families had met at a certain delectable fishing pool, which was responsible indeed for my presence. Their meeting was accidental and unfortunate, if one might judge by the row that followed, for they squabbled, scolded, and fought for half the night. As nearly as could be ascertained in the gloom, there were 2 old ones in charge of one family and but 1 caring for the other.

All through autumn and winter the family life continues; not even the mating season seems to mar their good-fellowship. Merriam says:[8] "It is unusual to find a Raccoon alone, for they commonly live and travel in small companies, consisting of the several members of a single family. They do not return to the same nest every morning, but often make little excursions in various directions, being gone several days at a time, and taking refuge, about daylight, in any convenient arboreal shelter.

* * * * * * *

"In tracking Raccoons upon the crust I have sometimes observed a family to separate and go in different directions, spending the day in different trees, to come together again on the night following."

[8] Mam. Adir., 1884, p. 94.

So far as I have been able to determine, the young Coons stay with the old folks as long as the latter allow it; and these make their full-grown offspring welcome until their quarters are needed by the new family, which arrives with commendable regularity as soon as the late April showers and the greening hills proclaim that now and truly is the world astir with spring.

The hibernation of the Coon is strictly dependent upon temperature. In the Red River Valley it lasts from mid-November to early March. In the latitude of New York it is shortened at both ends by several weeks. In the Southern States the species dispenses with hibernation altogether. HABITS

The remarks of Alexander Henry on the Raccoon of the Upper Red River give a clear idea of their times and seasons in northern Minnesota. The records in his Journal are thus:[9]

"Park River, September 8, 1800. On the beach, Raccoon tracks are plentiful. (P. 90.)

"October 4. Caught * * * 2 Raccoons, in * * * trap. Caught 5 Raccoons. (P. 112.)

"October 5. My men caught 5 Raccoons in their traps along the beach. (P. 112.)

"October 6. My men caught 3 Raccoons, in traps. (P. 112.)

"October 18. My men have caught 20 Raccoons. (P. 122.)

"October 19. Bring in daily some Raccoons. * * * very fat. (P. 122.)

"November 7. My men took great numbers of fat Raccoons in their traps. (P. 136.)

"November 21. They take no more Raccoons with traps. These animals are lodged in hollow trees where they will remain like Bears until spring without any subsistence. (P. 155.)

"November 30. * * * Some went Raccoon hunting, the weather being warm. They returned in the evening with 7, which they found in one hollow tree. The size of this tree was enormous, having a hollow 6 feet in diameter, the rim or

[9] Journal, pub. 1897.

shell being 2 feet thick, including the bark. Raccoon-hunting is common here in the winter season. The hunter examines every hollow tree met with, and when he sees the fresh marks of the claws, he makes a hole with an axe, and thus opens the hollow space in which he lights a fire to find out if there be any Raccoons within, as they often climb trees in the autumn, and, not finding them proper for the purpose, leave them and seek others. But if they be within, the smoke obliges them to ascend and put their heads out of the hole they entered. On observing this, the axe is applied to the tree; with the assistance of the fire it is soon down, and the hunter stands ready to dispatch the animals whilst they are stunned by the fall. But sometimes they are so obstinate as to remain at the bottom of the hole until they are suffocated or roasted to death. (P 157.)

"March 5, 1801. My men have raised and put their traps in order for the spring hunt, as the Raccoons begin to come out of their winter quarters in the daytime, though they retire to the hollow trees at night. * * * My men begin to take Raccoons which are very lean." (Pp. 171–2.)

The product of all this was 197 Coon skins. (P. 184.)

Now it begins again its season of active life, although it is ever ready to resume its cold-sleep if the return of cold weather should render it desirable. As the ground is still covered with snow, and the Coon does not store up food, it is hard put for a time, and draws freely on the reserves that are afforded by its fat. These are usually exhausted before Dame Nature again provides its daily bread, so that, as we have seen, the spring Coon is a very lean beast.

This is strictly nocturnal if any animal ever is; the darkest hours of night are its favourite time for prowling, which, nevertheless does not prevent enterprising reformers of the race occasionally setting forth on a diurnal excursion, for which they not uncommonly share the fate of unnumbered reformers, and win, without wearing, a martyr's crown.

Although nesting and resting in trees, where it moves about with the slow caution of 'Possum and Bear, rather than the reckless agility of Marten and Squirrel, the Coon travels, hunts, and feeds almost exclusively on the ground.

It may occasionally rob the nest of woodpecker, Squirrel, or other tree-dweller, but such must not be considered its normal habit of life—by far the greatest bulk of its food is taken on or near the ground. FOOD

It is quite omnivorous. Frogs, fish, flesh, fowl, eggs, reptiles, insects, shell-fish, fruit, nuts, grain, vegetables, and sweets are acceptable fare with the Coon; not equally so, but all welcome at all times.

In a wild state, the summer-long main support of the Coon is frogs. In catching them by night it is singularly expert, and when the frog takes refuge in the muddy bottom, the Coon, with wonderfully dextrous, tactile fingers, gropes after it. Leaving the enterprise entirely to its paws, its eyes may scan the woods and shores in a vacant way, but its mind is in touch with the finger-tips, and the frog that escapes them must indeed be worthy to live and father a superior race.

As Merriam says:[10] "They overturn stones and catch the crayfish that lurk beneath, and also gather the fresh-water mussels (*Unio* and *Anodon*) that live on sandy and muddy bottoms. They also catch and devour the hapless fish that chance to get detained in any of the little pools along-shore, but are unable to dive and pursue their prey under water, like the Otter and Mink."

Pennant describes[11] this animal as particularly fond of oysters, and says it "will watch the opening of the shell, dextrously put in its paw, and tear out the contents; sometimes the oyster suddenly closes, catches the thief, and detains it till drowned by the return of the tide."

In the Southern States its coat may change to a less substantial style, but its appetite for all nutritious dainties is the same. Audubon and Bachman detail[12] its watching of "the soft-shelled turtle, when she is about to deposit her eggs, for which purpose she leaves the water and, crawling on to the

[10] Mam. Adir., 1884, pp. 91–2.
[11] Arctic Zoöl., 1784, Vol. I, p. 70.
[12] Quad. N. A., 1849, Vol. II, pp. 76–8.

white sand-bar, digs a hole and places them underneath the heated surface. Quickly does the rogue dig up the elastic ova, although ever so carefully covered, and appropriate them to his own use, notwithstanding the efforts of the luckless turtle to conceal them.

"Sometimes by the margin of a pond, shrouded, or crouched among tall reeds and grasses, Grimalkin-like, the Raccoon lies still as death, waiting with patience for some ill-fated duck that may come within his reach. No negro on a plantation knows with more accuracy when the corn (maize) is juicy and ready for the connoisseur in roasting ears, and he does not require the aid of fire to improve its flavour, but attacks it more voraciously than the Squirrel or the blackbird, and is the last to quit the corn-field. * * * and although it generally visits the corn-fields at night, sometimes feeds on the green corn during the day; we have seen it thus employed during the heat of summer."

Although the frog-pond and the corn-patch supply its choicest foods, the Coon is not averse to a fat fowl. Some individuals, indeed, seem to give way to the chicken habit and riot in the hen-house night after night, killing first a fowl and then a dozen at a time, until they fall into the power of the barn-fowl's proper guardian.

These, however, are abnormal individuals and are not to be considered representative of the race's food-habits. It is possible that, like most Lords of the Forest, its principal revenue is derived from Mice, which are available when frogs and fruit are not.

Summing up its dietary—there is nothing in it, except occasional thefts of corn and fowl, to blacklist this creature on the farm-book; and these are so offset by its usefulness as a fur-bearer and beast of the chase, that most persons are glad to hear that the Coon is rather increasing in America.

WASHING HABIT

The Latin name (*lotor* or *washer*) and German name (*Wash-bear*) record a common habit of this animal. If near the water, it rarely eats a morsel of food without washing it.

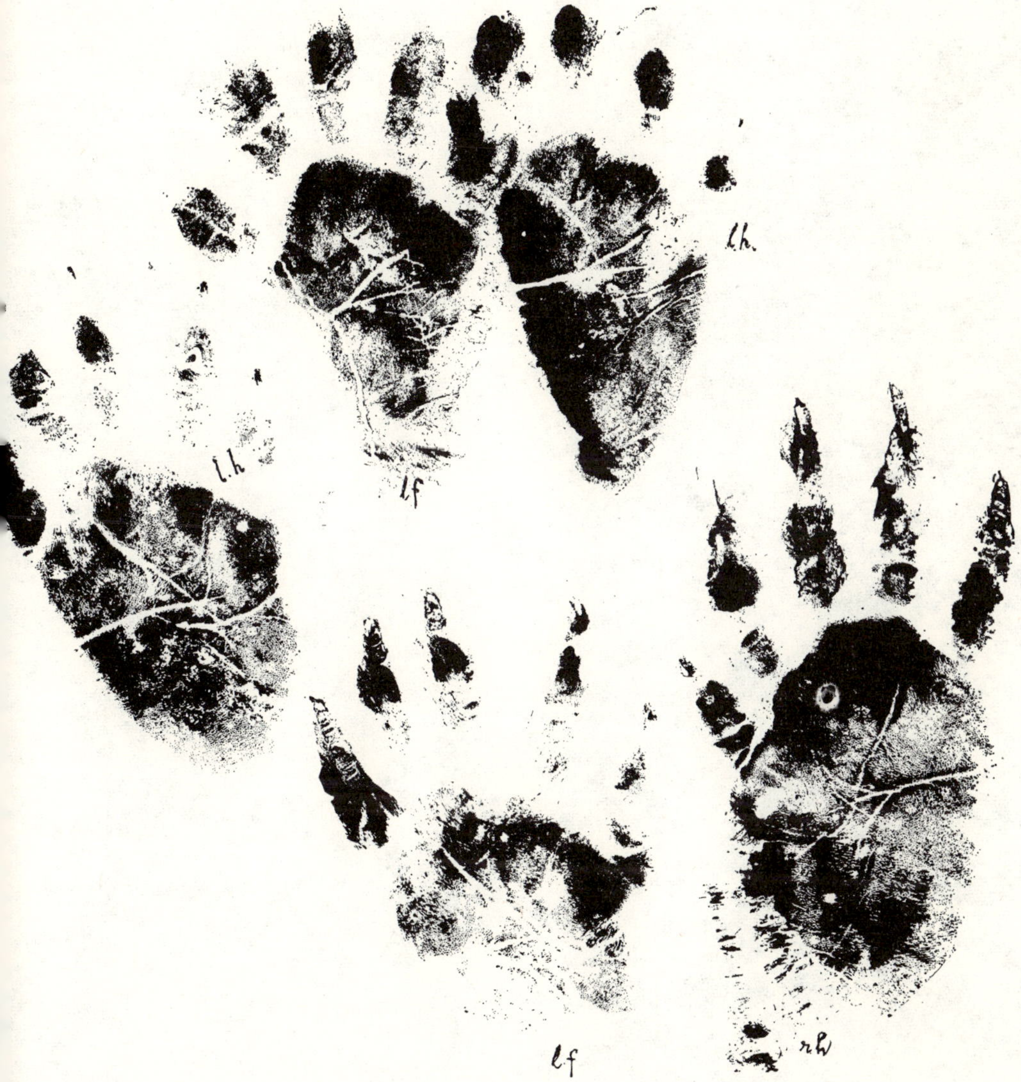

PLATE LXXXVII.—TRACKS OF RACCOON (LIFE SIZE).
Made by driving the animal over fresh black paint then across a clean sheet of soft paper. Secured by Mrs. Grace G. Seton.

This Mosaic habit seems to have arisen from its fashion of groping with busy nervous fingers in the mud for frogs, fish, or insects. Then, having secured some wriggling prey, its first care, before eating, is to clear it of sand and clay, by dabbling it in the open water. Taking advantage of this, many trappers catch Coons by setting in the mud on some favourite frogging point. I shall never forget the sensation I had in my early days when, one morning, on going to a trap set for Muskrat I found, firmly held in it by one paw, a huge and savage-looking Raccoon.

If necessary to reach some desirable food or to escape from an undesirable caller, the Coon will swim fearlessly and well, but ordinarily is not fond of water in which it cannot comfortably wade. SWIMMING

As a runner it takes low rank. I never saw but one running before the hounds in daylight, and its speed seemed barely half of theirs. Moreover, in many nights cooning, I never knew one of these animals to run more than a quarter of a mile before treeing. RUNNING

It is a desperate fighter. I have seen one beat off two large hounds, each of which was over double his weight, and saw another defeat three dogs—a terrier and two hounds. A Bedlington terrier, a famous fighting dog in Toronto, was said to have reached the final pitch of war-glory when, single-handed, he killed a full-grown Coon whose weight was about the same as his own. FIGHTER

The old Raccoon is sullen, dangerous, and untameable if kept captive, but the young, if taken at an early age—that is, before they have begun to hunt for themselves—make, as Merriam says,[13] "intelligent and interesting pets; being easily tamed and evincing considerable affection for their master. AS PETS

"But they cannot be allowed their liberty like tame Skunks, because of their innate propensity for mischief. If

[13] Mam. Adir., 1884, p. 93.

not closely watched, they will slyly enter the house through some open door or window, and are liable to do considerable damage, for their natural curiosity prompts them to examine everything within reach, and anything out of reach of a 'Coon' must be inaccessible indeed. They invariably manifest an insatiate desire to investigate the pantry shelves, and rarely neglect to taste every edible thing that happens to be there. They have a special penchant for sweetmeats, and greedily devour preserves, honey, molasses, sugar, pies, and cakes; and even bread, butter, lard, milk, etc., are by no means disregarded. They remove the covers from jars and pails, and uncork bottles with as much ease and facility, apparently as if they had been instructed in this art from earliest infancy. Doors that latch, as they do in most old country houses, are soon opened, even by unsophisticated Coons, and it takes them but a short time to acquire the method of opening knob doors. Their fore-paws are employed as hands, and can be put to almost as great a variety of uses as those of the monkey—which animal they further resemble in the propensity for mischief-making."

SANITATION The species has progressed but little along the paths of sanitation. Its dung is dropped anywhere, at any time, excepting while in the nest. Like cow, horse, and Bear, the Coon can void as it walks—is, in fact, a peripatetic defecator—but, owing to conditions, this bovine habit is not accompanied by bovine, much less equine, success.

Audubon and Bachman tell of a tame Coon that enjoyed a bucket of water thus:[14]

"After playing for a short time in the water it would commonly urinate in it and then upset the pail."

TRACKS The tracks are shown in Fig. 235; the pairing of the front feet when at full speed agrees with the tree-climbing habit of the species.

[14] Quad. N. A., 1849, Vol. II, p. 79.

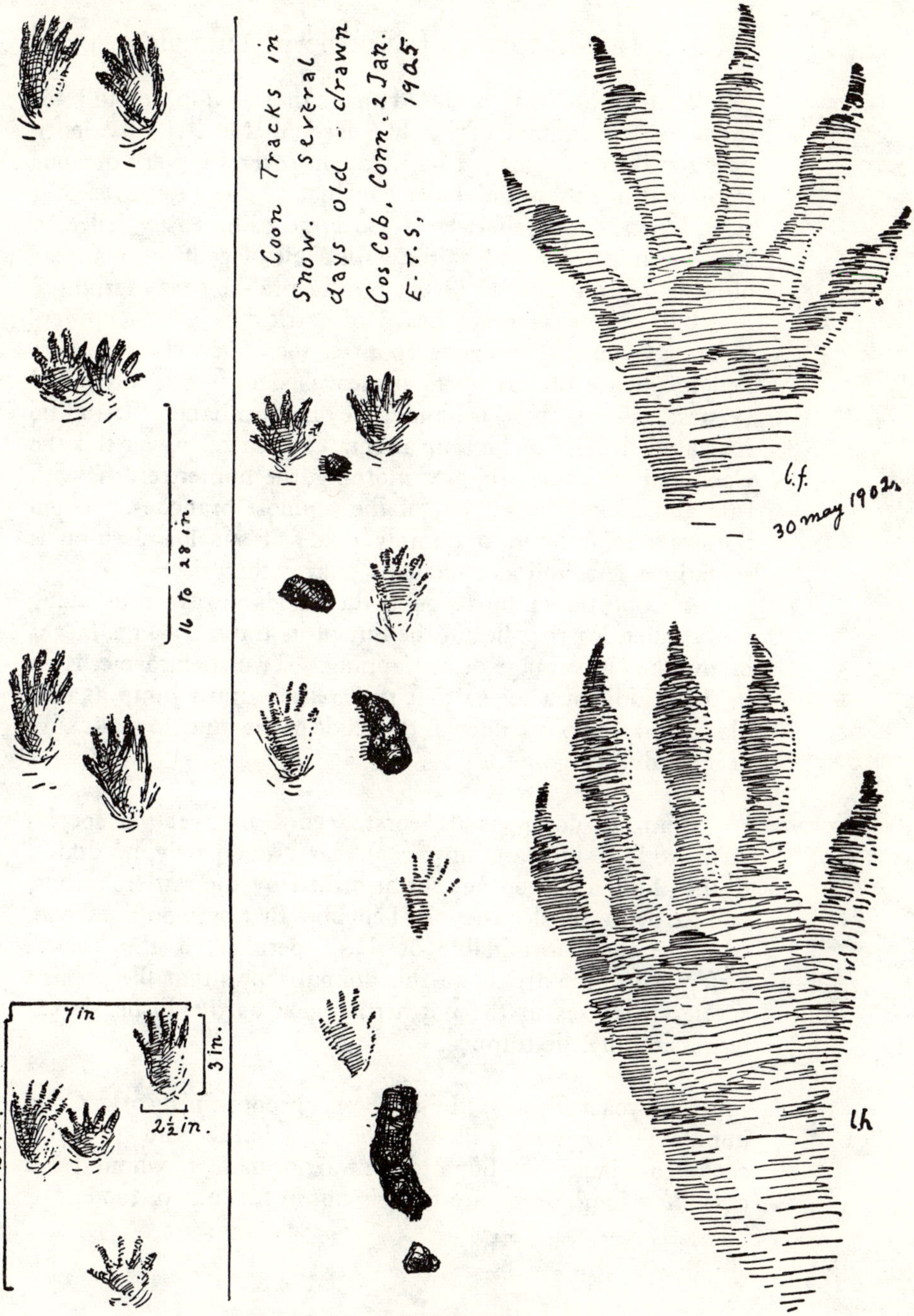

FIG. 235—Tracks of Raccoon.

The two large tracks on the right are left front and left hind of a large individual; they are natural size. The series on the left was made by a large Raccoon running through the snow. The middle series is part of this and shows where the creature walked when he should have stood still.

TRAPPING 'Cute as a Coon,' is an old adage that is supposed to refer to the fox-like cunning of the old Raccoon. My own experience does not bear this out. I have caught several in very obvious traps that a Fox would have scoffed at. Merriam says of a treed Coon:[15] "If the tree is too large to be easily felled, a trap set at its foot and baited with a bit of codfish or an ear of corn, is pretty sure to secure him before the next morning."

I have several times heard of a trick by which hunters secure a Coon that has gone up a tree too large to be cut down and too dense of leafage to be searched out with torch and keenest eyes. If straw is handy, they make a band of it and tie it around the tree as high up as they can reach. Over this the Coon will not come; so, next morning, the hunter returns with rifle and picks him off out of the topmost branches. When no straw is at hand, a dummy made of surplus clothing is sometimes left, and as effectually guards the prisoner.

A Coon travelling through the woods always runs along every fallen log that lies in the line of its travels. This fact is often taken advantage of in trapping. Two sticks are crossed in the middle of a log so that the creature must jump at that place, and a trap is hidden at each side of the jump to make sure of it coming or going.

ENEMIES Man, no doubt, is its worst enemy, and yet the species seems to have increased in recent years; due partly, no doubt, to the abundant food supply furnished by the settlers' crops, and partly to the destruction of enemies that prey on the Coon.

The most formidable of these, perhaps, is the Fisher. Reference to the article on this animal shows that the present species increases in the northern region as the Fisher is exterminated by the trapper.

USES The roast Coon is the supposed proper finish of a Coon hunt. Young ones killed without a battle and properly cooked are palatable, but a tough warrior male, to whom death came in a long, desperate fight is not fit for human food.

[15] Mam. Adir., 1884, p. 94.

The fur is the creature's chief contribution to human comfort. Half a million skins are shipped to London in good years to be manufactured into overcoats, sleigh-robes, and motor-car wraps. For this it is particularly well adapted, being a deep warm fur of the coarser kind on a strong and durable leather. The abundance of the animal keeps the price low, so it has an established popularity. FUR

During the eighty-five years, 1821 to 1905 inclusive, the Hudson's Bay Company collected 167,027 skins of this species, an average of 1,847 for each year. The lowest was 0 in 1829; the highest, 24,783, in 1866. The average for the ten years, 1895 to 1905, was 3,814.

Poland's lists show that during the seventy-one years, 1821 to 1891 inclusive, 27,138,479 skins were taken by the other American companies, an average of 382,091 each year. So that the average annual catch of Raccoon for fur is about 384,000. The record high year was 1883, when 711,071 were received by the traders.

At the London annual fur sales, held at Lampson's, March, 1906, the highest price paid was 26 shillings ($6.24) each for 34 unusually large, dark, fine skins; but this was exceptional. The nearest to this was 12 shillings ($2.88) each for 236 No. 1 large skins. More usual prices were 6 shillings ($1.44) and 7 shillings ($1.68), and many lots of inferior quality were sold at eight pence (16 cents) and 1 shilling (24 cents) a skin.

XLVI.

Grizzly-bear, Silver-tip, Roachback, or Montana Grizzly.

Ursus horribilis Ord.

(L. *Ursus*, a bear; L. *horribilis*, horrible.)

Ursus horribilis ORD., 1815, Guthr. Geog., 2nd Am. ed., Vol. II, pp. 291 and 300.

TYPE LOCALITY.—Montana.

FRENCH CANADIAN, *l'Ours gris ou féroce.*
CREE & OJIB., *Mish-e muk-wa'.*
CHIPEWYAN, *Klay'-zy.*
OGALLALA SIOUX, *Mah-to' shah-kay' han-ska'.*

The name 'Grizzly' or 'gray' was, according to the Century Dictionary, given to this Bear on account of "its usual colouration." The word 'grisly' (grewsome, grim, or terrible) was applied later.

The Bear Family *Ursidæ* is composed of very large, heavy, thickset carnivorous animals of dull colours, not spotted or striped. They have 5 well-developed toes on each foot; their hind-feet are plantigrade; they have fixed claws; very short tails; short, rounded ears, and the true molars with broad, flat tuberculated crown.

The genus *Ursus* (Linn., 1758) has the above characters, with colors, black, dull brown, yellowish-gray, or white, and the following dentition (often wanting in premolars):

$$\text{Inc. } \frac{3-3}{3-3}; \text{ can. } \frac{1-1}{1-1}; \text{ prem. } \frac{4-4}{4-4}; \text{ mol. } \frac{2-2}{3-3} = 42$$

PLATE LXXXVIII.—GRIZZLY BEAR.

Life study by E. T. Seton in 1886. From specimen in Philadelphia Zoo.

In combination with the above Family and generic characters, the Grizzly has very distinctive specific features of size and colour. SIZE

It is greatly to be regretted that full authentic measurements of typical Grizzlies have not been put on record. At best

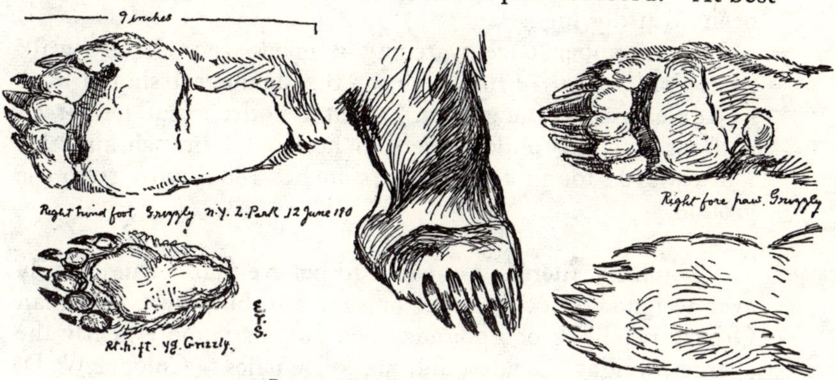

FIG. 236—Life studies of Grizzly paws.

we can go to Lewis and Clark.[1] They killed a large specimen near Porcupine River in Montana. It measured as follows:

Length from the nose to the extremity of the hind-foot	8 feet	7½	inches	(2,630 mm.)
Circumference near the fore-leg	5 "	10½	"	(1,790 mm.)
" of the neck	3 "	11	"	(1,194 mm.)
" of the middle of the fore-leg .	1 foot	11	"	(584 mm.)
Length of the talons		4⅜	"	(111 mm.)

"But this was not the largest Bear killed by the party. They give an account of one which measured 9 feet [2,743 mm.] from the nose to the extremity of the tail, and the talons of another were 6¼ inches [159 mm.] in length."

From various data at hand, I should say that an ordinary male Grizzly stood about 3½ to 4 feet (1,067 to 1,220 mm.) at the shoulder, and a female, about 3 feet (914 mm.).

The hind-foot of an ordinary adult is about 10 inches (254 mm.) long from heel to tip of longest claw; 9 inches

[1] Guthr. Geog., Rhoads reprint, 1894, p. 300.

(229 mm.) would be small, and 12 inches (305 mm.) very large. Of course, the track is larger than the foot.

The creature looks so big when charging over the mountainside or lying stretched at the feet of the victorious and excited sportsman, that all guesses at its weight, etc., have been absurdly high.

I remember once watching a good-sized Grizzly walk past; had I *guessed* his height at the shoulder I should have said at least 4 feet(1,220 mm.). But I noticed that he passed without stooping under a certain horizontal branch and this I afterwards found to be but 35 inches (889 mm.) from the ground.

WEIGHT

Similarly, there is no reason to believe that a true Grizzly ever weighed 1,500 pounds, or that any but the Californian Grizzly reaches 1,000 pounds; 600 pounds is more nearly the average weight of males, and 500 of females. Colonel W. D. Pickett, of Meeteetsee, Wyo., for thirty-five years one of the most successful Grizzly hunters in the West, says that of 40 wild Grizzlies that he actually weighed, the heaviest went less than 800 pounds.[2]

In the Washington Zoo is a large Grizzly from the Yellowstone Park. In September, 1894, he weighed 730 pounds, and has since added considerably to his bulk.

The heaviest weight on authentic record is 1,153 pounds. This is given by G. O. Shields as the weight of an enormous Grizzly that lived eighteen years in Union Park, Chicago. "He was fed to suffocation by the thousands of visitors and in his later years grew so fat he could not walk, could only crawl around."[3] The *estimates* set his weight at 2,000 pounds.

COLOUR

In general, the Montana Grizzly is of a deep brown colour, darkening to brownish-black along the spine, on the limbs, and on the ears; and grizzled or frosted over with a white tipping of the hair on the upper parts of the body. In some individ-

[2] Personal letter to G. O. Shields.

[3] Recreation Magazine, August, 1899, p. 135.

FIG. 237—Life studies of various Grizzlies.

uals the tipping is so reduced that the Bear looks deep brown or cinnamon; in others it is so increased that the Bear at a distance looks dirty white, but, in all, the limbs are dark. Another general feature of colour is an upright bar or patch of dull white behind the shoulder.

FORM The distinctive external characters of the Grizzly's form are its concave facial outline (the Blackbear's is about straight), the maned hump on its back, over the shoulders, and the great size of its front claws, which are twice the length of the hind ones and but little curved.

From the Blackbear group it is easily distinguished by these peculiarities; from the Polar-bear by the latter's white colour, convex profile, hairy soles, and short claws; but to distinguish it from the many Alaskan species of Fish-bears and Brown-bears, by external characters, is not at present possible.

The following races are recognized:

horribilis Ord., the typical form.
horriæus Baird.
californicus Merriam, a very large race with little silver tipping.
alascensis Merriam, a large northern race.
phæonyx Merriam, a large race with dark-coloured claws.

Life-history.

RANGE The original range of the Grizzly is shown on Map No. 55. It is now, of course, greatly restricted, especially in the east. The Grizzly is no longer found on the open plains, and in California it is nearly if not quite extinct.

IN MANITOBA The claim of the Grizzly-bear to be entered on the Manitoba list rests chiefly on the narrative of Alexander Henry, the nephew. He was an expert hunter and trapper, and close observer. He was, moreover, well acquainted with the Grizzly-

MAP 55—PRIMITIVE RANGE OF THE NORTH AMERICAN BEARS.

(Exclusive of the Blackbear group.)

This map is founded on records by J. Richardson, S. F. Baird, R. Bell, A. P. Low, E. W. Nelson, J. A. Allen, C. H. Merriam, Jas. H. Kidder J. W. Tyrell, and many arctic voyagers.

The following are recognized:

Ursus maritimus Phipps. The Polar or White Bear,
Ursus middendorffi Merriam, Kadiak Brown-bear,
Ursus kidderi Merriam, Kidder's " "
Ursus dalli Merriam, Dall's " "
Ursus dalli gyas Merriam, Peninsula " "
Ursus eulophus Merriam, Admiralty Ids. Bear,
Ursus kenaiensis Merriam, Kenai Ids. Bear,
Ursus sitkensis Merriam, Sitka " "
Ursus horribilis Ord., Montana Grizzly with 5 races,
Ursus richardsoni Swainson, Barren-ground Bear,
Ursus phæonyx Merriam, Dark-clawed Grizzly.

bear of the Mountains, and his account, especially in view of the supporting testimony, must be allowed full weight. In the year 1800, he built a trading post on the Red River at the mouth of Park River in North Dakota, about thirty miles south of the International Boundary. The journal he kept of those times

FIG. 238—Life studies of Grizzly.

shows that Blackbear were extraordinarily numerous, and that Grizzlies were occasionally seen on Red River. In his journal for 1800 are these entries:[4]

"October 17. * * * During my absence the hunter killed a large Grizzly-bear[5] about a mile from the fort. He had seen two males and a female, but the latter escaped. My people, having cooked and eaten some of the flesh, were taken very ill, and most of them threw it up. This Bear had been wounded in the fore-leg some time before by an arrow, the iron head of which stuck fast in the bone, and was beginning to rust. Grizzly-bears are not numerous along Red River, but more abundant in the Hair Hills [Pembina Mountains]. At Lac du Diable [Devil's Lake], which is about 30 leagues west, they are very common—I am told as common as the Blackbear is here, and very malicious. Near that lake runs a principal branch of Schian [Cheyenne] River, which is partially wooded. On the banks of this river I am informed they are also very numerous

[4] Alexander Henry Journal, 1897, p. 121.

[5] Dr. Elliott Coues, the accomplished naturalist who edited Henry's Journal, identifies this as the *Ursus horribilis*.

and seldom molested by the hunters, it being the frontier of the Sioux, where none can hunt in safety; so there they breed and multiply in security.

"October 18. * * * My hunter plagued me for a small keg of liquor, having vowed, on killing the Grizzly-bear, that he would make a feast of rum. This is a common custom among the Saulteurs, when they kill any uncommon animal."

Again, on page 145, he writes: On Cheyenne River "Grizzly-bears are to be seen in droves; and it may be called the nursery of Buffalo and Deer. It is a delightful country, but seldom can our Saulteurs kill a Beaver there without falling in with their enemies."

Later, in his returns of fur from Lower Red River Department, 1800–1, Henry enters:[6]

	BLACK-BEAR.	BROWN-BEAR.	GRIZZLY-BEAR.
From Reed River (Roseau River), Man.	52	20	4
From Park River, Dak.	125	23	2

In the following year, though 152 Blackbear and 42 Brown-bear appear, there was but 1 Grizzly taken, and that was brought into Portage la Prairie, Man. (P. 221.)

In 1804–5, he records 10 Grizzlies; 4 from Hair Hills (Pembina Mountains), 4 from Salt River, 2 from Pembina River. (P. 259.)

In the season of 1805–6, the returns show 125 Blackbear, 49 Brown-bear, and 4 Grizzly-bear. Of these Grizzlies, 1 was taken at Portage la Prairie and 3 at Pembina River. (P. 281.) As the furs collected at each post were, for business reasons, classed separately with great care, there was little chance of importations to impair the scientific value of the list.

On August 29, 1808, Alexander Henry, writing at a point on the Saskatchewan some 60 miles below the Forks, that is, near W. Long. 104°, says:[7]

"Tracks of animals are very numerous along the beach, including those of Bears both of the common and Grizzly species."

[6] *Ibid.*, p. 184.

[7] *Ibid.*, p. 480.

This note is of special interest as marking the probable north-easternmost extension of the Grizzly-bear. But its range began to shrink with that of the Buffalo that it preyed on, or both retreated before the mounted riflemen, who now began to appear.

In Brackenridge's time (1814):[8] This Bear was "not usually seen lower than the Mandan Village [near Bismarck, N. Dak.]. In the vicinity of the Roché Jaune [Yellowstone] and of the Little Missouri,[9] they are said to be most numerous."

In 1820, Richardson saw a Grizzly killed at Carlton House,[10] on the Saskatchewan, and intimates that the species was well known to the Indians there, though not apparently farther east. He gives its range as "the Rocky Mountains and the plains lying to the eastward of them as far as Latitude 61°, and perhaps still farther north,"[11] and as late as 1875 Colonel Dodge reports the species numerous in the Black Hills.[12]

HOME-RANGE

The range of the individual Grizzly varies greatly with local conditions. In a rugged mountainous region where food abounds it will not go half a dozen miles from a central point. In the days when it followed the Buffalo herds it probably went ten times as far, for, unlike the Blackbear, it is at home on the plains. But a typical Grizzly in ordinary mountain country to-day will ramble over a home-region at least 25 miles across.

W. H. Wright, after exceptional experiences, says:[13] "The Grizzly will live his life in a restricted area. He will go but a few miles in any direction if there is food at hand, but he will seek the food he wants if it is 20 miles away. A Grizzly, however, nearly always selects a range where he will not have to travel very far to feed."

ABUNDANCE

In ancient days it was common to see a dozen of these monsters in a day's march. Old hunters say that they would

[8] H. M. Brackenridge, Views of Louisiana, 1814, p. 55.
[9] Probably not the present Little Missouri.
[10] F. B. A., 1829, I, p. 25.
[11] *Ibid.*, p. 28.
[12] The Black Hills, 1876, p. 122.
[13] World's Work, August, 1905, p. 6540.

FIG. 239—Grizzly poses. (From life.)

average that number in the spring of the year. From various reports it seems likely that in choicest ranges a Grizzly to a square mile was quite within the bounds of probability.

But its numbers have dwindled with its range. It is gone from the open country. It is found now only in the most inaccessible mountains.

In California, the greatest of these great Bears is near extinction. I spent a month in the High Sierra in 1897 looking for Grizzlies and saw only two fresh tracks in all that time—probably made by the same animal. In 1902, I was told by Dr. James R. Walker, of Pine Ridge, S. Dak., that a few Grizzlies still lingered about the Black Hills. But I doubt their existence to-day (1908).

The range of the species is reduced to one-quarter; its numbers are divided by 20. The day is even now in sight when the Grizzly-bear, as a wild inhabitant of the United States, will cease to exist.

In British Columbia it holds out fairly well as yet. Not many years ago W. H. Wright saw 21 there in a single spring trip.[14]

ENVIRONMENT

The ideal home of this animal is high rolling uplands, where dry, open prairies are interspersed with rocky ridges and densely wooded thickets. Here it finds food in abundance as well as sunning places and shady retreats in which it can wallow in mud and coolness, and defy alike the over-hot sun, the bot, the gnat, and the relentless mosquito.

Even in the days when it roved the wide plains it was usually found in the bottom-lands and places that had a vestige of cover rather than bold and bull-like on the level open.

SOCIABILITY

Notwithstanding many accounts of Bears in droves, etc., the Grizzly is not a sociable or gregarious animal. The groups of 5, 6, or 7 Bears recorded are doubtless accidental and temporary associations of two families.

[14] *Ibid.*

Grizzlies cough, growl, grunt, roar, and sniff, in expression of various feelings. Dr. W. T. Hornaday, writing after years of experience among Bears in zoölogical collections, says:[15] INTER-COMMUNICATION

"I have learned the language of our Bears sufficiently that whenever I hear one of them give tongue I know what he says. For example: In warning or threatening an enemy, the Sloth-bear says, '*Ach! Ach! Ach!*' and the Grizzly says, '*Woof! Woof! Woof!*' A fighting Bear says, '*Aw-aw-aw!*' A baby's call for its mother is '*Row! Row!*' A Bear's distress call is '*Ew-wow-oo-oo-oof!*'"

But the Grizzly, in common with the Blackbear, has another means of sending tidings to others of its race, and that is by the use of bear-trees or sign-posts. The remarks on the sign-posts of the Blackbear apply equally here. So far as I have seen, these Bears register their call in the same way.[16]

Hornaday, however, writing on the subject, says:[17]

"On those trees we saw where several of the rubbing Bears had bitten the trunk high up, tearing the bark open crosswise. We also found, on some, raking claw-marks across the bark. Charlie Smith said that the tooth-marks are always made by the Grizzlies and the claw-marks by Blackbears."

The only difference I have been able to see between the marks of the two species is that while the Grizzly leaves 5 claw marks at each place, the Blackbear commonly leaves but 4, for the reason that its thumb is so short that the claw often misses.

Grizzly-bears mate in midsummer. A pair of Grizzlies in the Central Park Zoo, New York, mated in July. The old Grizzly (Monarch) at Golden Gate Park, San Francisco, and his wife consummated their union on June 19, 1904. MATING

Little is known of their marriage customs in a state of nature, but it is generally conceded that they are not promiscuous, that one male mates with one female, that they continue together for a month or more, then part for good. It is very doubtful whether any of the sex feeling persists after the waning

[15] Language of Animals, N. Y. Sunday Magazine, June 25, 1905, p. 7.

[16] See pages 1060–2.

[17] Camp-fires Can. Rockies, 1906, p. 159.

of the honeymoon. The pair, if they meet again, may accept each other as friends, but not as mates. Whether they reunite on the next breeding season is utterly unknown.

GESTATION In the first of the above-mentioned cases the gestation lasted a little over 6 months, and in the second 6 months and 4 days. This, strange to say, is a month less than in the Blackbear and may be modified by shorter hibernation in the Grizzly.

My notes on the first are as follows: New York, January 17, 1901. The male Grizzly in the Central Park Zoo was brought here in 1891; the female in 1884. Both were full grown when

YOUNG they came. Two years ago they produced a cub. Last year again they mated in July, uniting many times. To-day, about 11.20 A. M., a young one (a female) was born. (Whether there were others is not known.) It weighs 1½ pounds, is 8½ inches long (216 mm.) from tip of tail to end of nose. It is blind and apparently naked, but covered with very fine, short, gray hair, and is of a pale pink or flesh tint. The ears are low and the openings not yet visible. It squeals like a child when it is hungry, and is very restless, nosing about, falling on its back, and screaming in temper. It began to suck its dog foster-mother at 1 P. M., when 1 hour and 40 minutes old. The little one had all the form of a Grizzly—the shape of head, the hump on shoulders, the paws, the lower jaw. The tail, however, was proportionally too long for the adult. The mother had been carrying it about in her mouth. It was somewhat scratched and bruised by her teeth. Of this I made several sketches. It died a few days later.

The second case was in San Francisco. On June 19, 1904, Monarch, the old Californian Grizzly at the Golden Gate Park, united with a female Rocky Mountain Grizzly. On December 23 (6 months and 4 days later), 2 little cubs were born in the cage. They were kept hidden by the mother for several weeks. They are now (March 18) nearly 3 months old and about a foot high at the shoulder. They weigh, I should think, 12 or 15 pounds, are very pretty and

FIG. 240—Montana Grizzly. New-born, life size.
Central Park, New York City, Jan. 17, 1901.

playful. Their ground colour is gray, with dark ears, feet, and patch around eyes. Already they eat meat, fruit, bread, etc. Their mother is very careful of them. She seemed unwilling to let them go out of the den, but they got past her, and she followed. As we fed them it came on to rain very heavily and suddenly; at the same time a loud slam of the bars alarmed her. At once she raked and cuffed the two young ones *under her body*, then, straddling very wide, sheltered them from the rain and guided them back into the den. She often uttered a sort of choppy coughing sound to them.

Sometimes the little ones put their paws through the cage to their father. He would sniff at the paws very loudly and utter a sort of short, quick "*koff, koff, koff, koff.*" It was not a menace, as he offered them no harm, though he had ample opportunity. At any sudden alarm the old mother reared up on her hind-quarters to look around.

The young commonly number 2, rarely 3, but 4 have been noted in one or two extraordinary cases.

Ordinarily, they are born in the mother's winter den, exactly as described in the Blackbear. They are suckled all winter in the den, but begin to eat solid food as soon as they can get it; that is, after they have begun active life in the spring.

Ordinarily, they pass the summer with their mother as sole guardian, but there seem to be some cases in which the father—that is an interested adult male—has joined the party.

Catlin[18] tells of a male and female Grizzly-bear that, accompanied by the cubs, came into his camp on the Missouri near the Yellowstone Fork and, on being molested, attacked him. Evidently it was in the height of summer. Several instances of the kind are on record.

It is commonly agreed that the young Grizzlies stay with the mother till winter, and that all den up during the coldest weather, but it seems uncertain whether on this, their second winter, mother and young lie up in the same den.

It is probable that the Grizzly breeds but once in two years, unless the young are destroyed before midsummer. The

[18] Life Among the Indians, pp. 128–31.

young are able to shift for themselves the second year and are supposed to breed in the third.

At one time it was believed that the species would not breed in captivity, but we are better informed to-day in the handling of cage-animals, and the list of non-breeders grows

FIG. 241—Young Grizzlies, 3 months old; born in Golden Gate Park, San Francisco. The offspring of Monarch.

less. Arthur B. Baker says:[19] "A Grizzly-bear in one zoölogical garden produced, in 12 litters, 22 cubs, but only 1 was reared."

In general, the Grizzly's habits resemble those of the Blackbear. Both are shy, but desperate fighters when cornered. Both are lovers of the twilight, but travel in full day or black night on occasions. They differ in this: the Blackbear rarely quits the woods; the Grizzly often lives permanently in HABITS

[19] Smithson. Miscel. Coll., 1904, No. 1434, p. 178.

the open country. The Blackbear is much at home in the trees, so are the Grizzly cubs, but adult Grizzlies do not climb. It is commonly supposed that they cannot, but Superintendent A. E. Brown, of the Philadelphia Zoo, assures me that he once saw a tame adult Grizzly climb a smooth telegraph-pole to the crosstrees. It had, however, to be greatly urged before it would make the attempt; and it is a fact that the wild adult does not climb. The hunter who succeeds in getting up a tree is as safe from a Grizzly as from a bull.

FOOD Though omnivorous in food habit, the Grizzly is more carnivorous than the Blackbear. Its great strength enables it even to master the range steer as it once did the Buffalo. Some individuals are much more given to meat diet than others, and such have become veritable nuisances on the cattle-range near their head-quarters. On the open prairie the Grizzly is said to feed largely on the prairie-turnip (*Psoralea esculenta*) alternated with Ground-squirrels, Gophers, and other products of the soil. For procuring these, its armed feet are a perfect combination of crane and crow-bar, pick and shovel, rake and forceps.

Like the elephant's trunk, their mighty force in heaving a huge log or boulder is only equalled by their manual dexterity in picking up eatable mites. I have seen a Grizzly use two claws to pick up small objects, exactly as a Chinaman might use his chopsticks. I have noticed further that it never uses two claws when one will better serve.

DENNING Like its kin, the Grizzly does not den up any sooner than it can help; that is, it roams its range as long as it can find food.

The males remain active longer than the females, and in the southern part of the range they do not den up at all. W. H. Wright is of the opinion, however, that in Montana the denned-up Grizzly does not come out until a month or six weeks after the Blackbear.[20]

TRAILS This Bear also has fixed pathways through the woods and over difficult places. These it will use for years, till they are

[20] See Note 13.

PLATE LXXXIX.—"A NARROW ESCAPE."
The pony and the Grizzly.

deeply worn. In going up a bank or over logs, it will put its feet into the same tracks each time till they become a kind of a stairway. The bear-trails in the Bitterroot that are described in the Blackbear chapter were undoubtedly made in part by the numerous Grizzlies that came each year to hook out the running salmon.

It is a remarkable fact that though all animals make and use trails more or less, it is only those of the lower order, such as rodents, that take the trouble to repair or improve their trails. If a tree branch falls across a Rabbit or Beaver trail, it is cut in two; if a twig grows up, it is nipped off. But the Bear, the Wapiti, and the Buffalo take no such trouble. If mere wear will improve their highway, it is improved, but if a tree fall across it or a self-rolled stone should block it, the lazy giant goes round the balk and strikes out a second trail.

Those who form their idea of a Bear's speed from watching a hulking, slouching prisoner, are sure to be amazed at the real thing. For 50 or 100 yards a Grizzly can go as fast as a good horse, and in rough country it can go faster than any horse and keep it up indefinitely. It is well known that in the spring of the year the Indian ponies that have wintered out and are poor, very commonly become the prey of the Grizzly, who can now catch them on the open plain. Townsend[21] tells of a wounded Grizzly that pursued closely a man on horseback, for half a mile, snapping at the horse's heels, and apparently would have captured the object of his wrath but for a timely volley from the man's comrades. SPEED

In view of this, it will be seen how absurd it is for any man to think he may escape from a Grizzly by simply running.

This species is a good swimmer, but seems less ready to take to the water than is the Blackbear. SWIMMING

The strength of a Grizzly-bear is as might be expected from its weight. STRENGTH

[21] Nar. Journ. Rocky Mts., 1839, p. 68.

In the Yellowstone Park, I was witness of a quarrel between a Grizzly and a very large Blackbear (the mother of Johnny). The Grizzly, with a blow of his paw, hurled her some sixteen feet against a pine root and ended all her desire for martial glory. Cattlemen everywhere attest that a Grizzly can *carry off* the carcass of a steer, which means that it can, and does occasionally, drag it half a mile or more. Similar instances are on record where the carcass was that of a full-grown Buffalo. The New Mexican cowboys who described to me the scene which I used in the lassoing of "Monarch," said that on that occasion one blow of the Bear's paw disabled a horse. The various incidents of Grizzly strength that are used in my stories of Wahb and Monarch are gathered from life and authenticated by numberless witnesses.

In addition to its muscular strength, the Grizzly is tenacious of life. It can and will fight with wounds in all non-vital parts, and, for the reason that its vitals are well protected, it is hard to kill. Nevertheless, a shot through the heart will drop it within a few yards, and a shot in the spine or brain will down it finally on the spot. But it will be seen how little chance of success against the Grizzly had the Indian armed with bow and lance, or even with musket and flintlock gun. No wonder, then, that a claw-necklace from a Grizzly of one's own killing was the outward and visible sign of valour extraordinary, and worn with becoming pride. This, it will be remembered, was one of the greatest exploits of the hero Hiawatha. He went to the far Westward, there he clubbed and killed the mighty Mishe-mokwa, and returned with the necklet trophy of claws.

SANITATION

The species has not gone far in the evolution of sanitation. At best it keeps its den clear of excrement. In Colorado, I found that a certain fly propagates in this Bear's dung. The Grizzly is even fonder of a mud-bath than the Blackbear. Its wallows are found wherever the Grizzly roams. They no doubt offer protection from the mosquitoes and also, when hot weather and shedding time arrive together to produce a general sense of prickly heat and discomfort, the cooling, healing mud of the bath furnishes a most delicious relief.

The incidents of the sulphur bath in Wahb and the final scene in Death Gulch are founded on fact. Death Gulch, on the eastern side of Yellowstone Park, was examined by Professor T. A. Jagger, of the United States Geological Survey, in July, 1897. The deadly gas in it turned out, as was expected, to be carbonic dioxide. When Professor Jagger entered the valley, there was the body of an old silver-tipped Grizzly lying in the quietest corner among many bones and bleaching skeletons. It had not been there many hours, and there was every evidence that this was the latest victim of the poisonous gas.

MENTALITY

The mentality of the Grizzly is far inferior to that of dog or Wolf. But its powers of scent and hearing are so exquisite, and its ability to travel fast, far, silently and unseen, so great, that, once it has learned the danger of rashness, it shapes its life behaviour on lines that look like profound sagacity. There are, moreover, great individual variations. The stupid Grizzlies are early weeded out by the hunters, and the ten-times sifted remnant are the wisest of the wise in their kind.

It is common saying in the West that a Grizzly is a most unreliable animal. You never know what it is going to do next, but you may be very sure it is going to be quick about it.

CATTLE-KILLER

Cattle-killing Grizzlies are rare now, but undoubtedly exist. The following instances related to me by Edward C. Russell, attorney, of Helena, Mont., shed interesting light on their habits:

About 1880, Grizzlies were very numerous in the country some 60 miles south of Helena. They used to come in there for the berries, but would occasionally kill cattle. When a beef was thus killed, Russell used to sit up over it in a tree. The Bears used to come sniffing up the wind and when 50 yards down from him could smell him, but when they got nearer the scent seemed to go over their heads. They were puzzled and would go back, get it again, and work up nearer, to lose it once more.

One night a large Silver-tip came to the carcass. The watcher was ten feet up a small cotton-wood. The Bear smelt him and came sniffing up the wind for some time to find him, but each time lost the trace as he got near. Several times this happened. Then, in determination to find his foe, he began to break down the brush around. He would smash down a thicket with great uproar and then remain as still as death to see if the man was flushed; then another, and again a wait. Sometimes he would wait for 3 or 4 minutes without a move. In this way he flattened all the brush for an acre about the hunter and the carcass, but the night was so dark that there was no opportunity for a shot.

Another time a cow was killed by a Grizzly, dragged half-way across the river, and then left lying in the water; 50 yards away was a high cut-bank covered with brush. It was not on the usual trail of the Bears, and as they could not climb up the face, Russell decided to hide there and shoot from it. On getting ready to go, however, he found that he had but one cartridge, and so gave it up. Next morning he learned that the Grizzly had come, but before feasting was careful to break down all the brush on this commanding point, so that had the man been there he would certainly have fallen into the power of the Bear.

PASSING

But the cattle-killer is vanishing just as surely as the buffalo-killing Grizzly is gone. A great shrinkage of the Big Bear's range is seen to-day, and a wonderful change in the Bear himself. All the old travellers from Lewis and Clark to those of forty years ago aver that the Grizzly had little fear of man, and ofttimes claimed and received from him the right of way. But we have lived to see another mind in the Range-king; we see in him now the exemplar of an ancient law—the beasts are shy in proportion to their bigness that is really in measure of man's eagerness to add them to his bag. The Mouse will let you walk up within a few yards, will even run

PLATE XC.—DEATH GULCH.
From photograph by Prof. T. A. Jagger, Jr., July, 1897.

PLATE XCI.—THE OLD GRIZZLY IN DEATH GULCH.
From photograph by Prof. T. A. Jagger, Jr., July, 1897.

over your foot if you are still; the Moose will flee on the slightest intimation that man is within a mile. And the Grizzly, too, accepts the common creed. No longer the arrogant despot of all trails and ranges, he has retreated to secluded fastnesses, to wild and inaccessible regions of thicket and swamp. He is changed in temper as in life, and the faintest whiff of man-scent is now enough to drive him miles away.

And what is it that has made this change?—that has turned the heart of the mountain terror and made him shyer than ever fawn or hare? The educating force was *modern guns*. Repeating rifles have instilled the idea that man is master—omnipotent, merciless—therefore shun the onset they can end in only one way. The fallen monarch is become a fugitive in his own kingdom. In many parts of the country, particularly the south and east, his kind is extinct. In a little while he will have left the United States, or will continue only as a pensioner in the Yellowstone Park. And I, for one, would gladly see the total abolition of all bounty laws on the Grizzly's head. I should welcome a movement to prevent his extinction. His day and sceptre are gone; right well he knows that; he is harmless now, and is, moreover, a magnificent animal, whose extinction would be just such a loss to zoölogy as the destruction of St. Peter's would be to the world of art.

XLVII.

Common Blackbear, American Blackbear, or Cinnamon Bear.

Ursus americanus Pallas.

(L. *Ursus*, a bear; L. *americanus*, American.)

Ursus americanus PALLAS, 1780, Spicil. Zoöl. Fasc. XIV, p. 5.

TYPE LOCALITY.—Eastern North America.

FRENCH CANADIAN, *l'Ours noir, l'Ours d'Amerique.*
CREE, *Kus-kit-tay' Mus-kwa'* (Blackbear); *Sau-wis' Mus-kwa'* (Yellow or Brown-bear).
SAUT., *Mak-a-tay' Muk-wa'.*
OJIB., *Mah-kay-tay' May-kwa'.*
CHIPEWYAN, *Sass.*
YANKTON SIOUX, *Wah-conk-seach Sa'-pa.*
OGALLALA SIOUX, *Mah-to'-wah-hay. See'-cha* (Blackbear), *Mah-to'-ho'-tah* (Brown-bear).

The Blackbear has the characteristics of the genus as set forth already, and is further distinguished by the short-curved claws on its front feet and the straight profile of the skull. Besides these, its peculiarities of dentition, size, and colour are very distinctive.

SIZE A fair-sized and fat she Bear shot in Colorado measured as follows: Snout to tail-tip, 63½ inches (1,613 mm.); tail, 5 inches (127 mm.); hind-foot, 7¼ inches (184 mm.); height at shoulders, 25½ inches (648 mm.). It weighed 227½ pounds.

WEIGHT A large male Cinnamon Bear killed by J. H. Cadham, 25 miles south-east of Winnipeg, weighed 265 pounds; 200 pounds may be considered a medium-sized northern Blackbear; 300 pounds a very large one.

In Florida, however, this animal attains much greater size. Charles B. Cory, Director of Field Museum, tells me

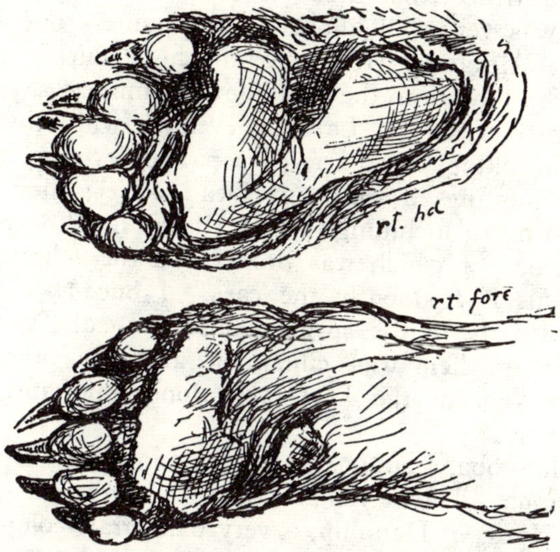

FIG. 242—Paws of a large Blackbear; right hind and right fore. (*Summer.*)

that he killed a comparatively small Floridian female that weighed 350 pounds, a male, 411 pounds, and a large male, 512 pounds.

Audubon and Bachman give[1] the following dimensions of a very large specimen:

From nose to root of tail, 6 feet 5 inches (1,957 mm.); height to top of shoulder, 3 feet 1 inch (940 mm.).

The typical Blackbear of Eastern America is deep, glossy black everywhere, excepting the muzzle, which is more or less brown, and the white spot sometimes seen on the breast. As one nears the Mississippi, various shades of cinnamon brown are found, and in the Rocky Mountains fully a quarter of the Bears are of the Cinnamon variety. This difference of colour, COLOUR

[1] Q. N. A., 1849, Vol. III, p. 188.

however, does not mean a difference of species; they are mere freaks or sports of the black race. A Blackbear may have cinnamon young this year and black the next, or even one of each kind in the same litter. So also a Cinnamon mother may give birth to either Black or Cinnamon young.

This was very well known to the hunters one hundred years ago. Alexander Henry, writing his 'Journal' on Dead River, which flows into the Red above Winnipeg, says:[2]

"August 11, 1808. Late this evening, while the Indians were still drinking, there arrived a party of young men who had been hunting *en canot* up Dead River; they brought some fresh meat, including that of a large Blackbear and her two cubs, one of which was brown and the other perfectly black. This is frequently the case. I once saw a Blackbear killed early in the spring whose two cubs were taken alive; one of them was cinnamon and the other black. Both were kept at the Fort for a long time and became perfectly tame."

In Manitoba, I suppose, about one in twenty Blackbears is a cinnamon. I saw a remarkable specimen in the collection of H. C. Nead, of Dauphin, a very pale straw-coloured Bear with chocolate-coloured face and legs; yet it was clearly of the Blackbear species.

N. E. Skinner tells me that two young Bears were found in a den at Carberry, Man., about 1895. One of them was a cinnamon, the other black with brownish-gray muzzle.

A different colour variety is the albino, or pure white freak. Alexander Henry thus records a case on the Red River:[3]

"October 13, 1800. Two Indians were with him, Nanaundeyea and Grosse Loge; they had made no hunt as yet. One of them a few days ago saw a full-grown Bear as white as snow. His gun missed fire and the Bear escaped. He assured me that it was not the Grizzly, but the common kind."

[2] Journal, A. Henry (1799–1814), pub. 1897, p. 449.

[3] Journal, 1897, p. 118.

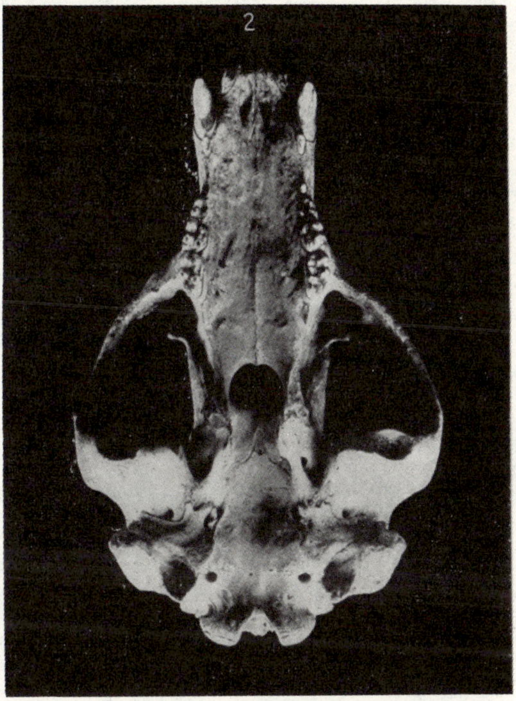

PLATE XCII.—SKULL OF BLACKBEAR.
(*Ursus americanus.*)
Cut supplied by the U. S., Biol. Surv. from N. A. Fauna, No. 21, p. 30.

The following forms are recognized:

americanus Pallas, the typical form.
sornborgeri Bangs, "skull smaller, shorter, broader," etc. (colour unknown).
carlottæ Osgood, a large race with skull longer than in *americanus* and teeth larger and heavier, etc.; said to be glossy black.
eremicus Merriam, a brown-nosed, black-furred race with brown under-fur, slender muzzle, and naked nose-pad very long.
floridanus Merriam, very large and wholly black.
luteolus Griffith, a very large form with large teeth.
altifrontalis Elliot.
hylodromus Elliot.
machetes Elliot.
emmonsi Dall, a brown-nosed black Bear of small size, frosted or silver-tipped, with white on the neck and body.

To complete the list of small Bears found in North America, we need add only *Ursus kermodei* Hornaday, a pure white species recently found in British Columbia. It is related to the Blackbear, but is a smaller form.

LIFE-HISTORY.

The range of the species and its various races is set forth in Map No. 56 with fair accuracy. On the north and west it is limited only by the limit of trees. On the south it will doubtless be found farther than I have yet been able to trace it. RANGE

In Manitoba it is of general distribution, but most plentiful in the poplar belt from Pembina to Pelly.

How are we to form any idea of the primitive and present Blackbear population? The material at hand is far from satisfactory, but such as it is I give it. ABUNDANCE

Hearne says[4] that Blackbears were "so numerous in the country between York Fort and Cumberland House that in [June] 1774 I saw 11 killed in the course of one day's journey."

As many as 30 Bears have been killed in one year in Lewis County, N. Y., about 300 square miles (according to Merriam[5]), and the inference is that this was unusually high—moreover, we know now that it was too much for their numbers to stand.

I reckon that an animal breeding so slowly as the Bear could not stand a greater drain by man than 10 per cent. per annum, therefore Lewis County must have had a Bear population of considerably less than 300. Yet this was considered an abundance.

Throughout the last quarter of the nineteenth century the Hudson's Bay Company exported about 7,000 bear-skins each year, and the other fur companies about the same, an aggregate of 14,000 bear-skins. But we know that thousands are killed when the hides are not worth shipping, and half of those taken are used or misused by the natives, therefore 30,000 will more nearly represent the annual kill on an area of about 5,000,000 square miles. As during the time cited the supply has, apparently, not dwindled, it implies at least 300,000 Bears, one to every 16 square miles.

This I suspect is very near the truth to-day, although there are reasons for believing that in ancient times they were more numerous.

FLUCTUATION

A study of the fur returns shows that the Bear population rises and falls as does that of most species. Roderick MacFarlane calls my attention to this, but cannot satisfy himself of the reason. He mentions, to cast doubt on, epidemics and migration, then adds:[6]

"There are other circumstances also, such as an unfavourable season for breeding, a scarcity of the required food, and the destruction by fire of extensive areas of forest, which would, of course, more or less affect the abundance of these

[4] Journey, 1795, p. 370.
[5] Mam. Adir., 1884, p. 103.
[6] In personal letter.

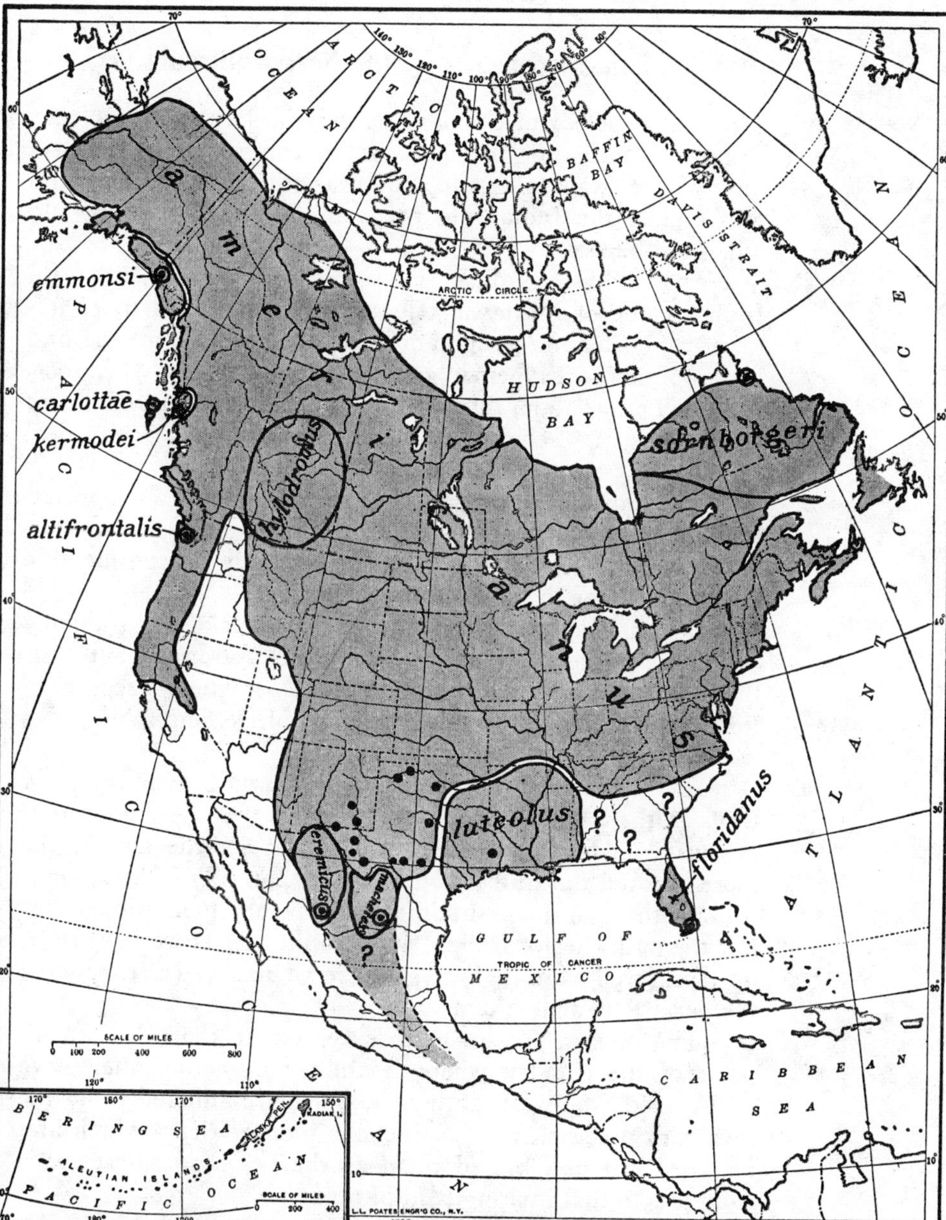

MAP 56—RANGE OF THE AMERICAN BLACKBEARS AND THEIR NEAR KIN.

This map is founded chiefly on papers by Messrs. E. W. Nelson, R. MacFarlane, R. Bell, S. Hearne, J. Fannin, C. H. Townsend, V. Bailey, O. Bangs, W. T. Hornaday, E. A. Preble, W. H. Osgood, C. Hart Merriam, E. T. Seton.

It is fairly correct, except on the south, where further investigations must greatly modify it.

The spots in Texas and New Mexico are V. Bailey's records. The record in Costa Rica is by Geo. K. Cherrie.

The following are recognized:

Ursus americanus Pallas, in 3 forms,
Ursus luteolus Griffith,
Ursus floridanus Merriam,
Ursus emmonsi (Dall.),
Ursus carlottæ Osgood,
Ursus hylodromus Elliot,
Ursus machetes Elliot,
Ursus altifrontalis Elliot.

and other species of animals in certain localities. A very lingering spring, for instance, would compel Bears to leave their winter 'washes'[7] while snow was still on the ground, and thus enable the Indians to track and kill more than would otherwise be possible."

HOME-RANGE

The individual Bear is a wide ranger. Mittigwab, Indian guide of Mattawa, tells me that he several times followed a big Bear on a round of 15 miles from its home. I find it the opinion of trappers in general that adult Bears when footloose will range about this distance. A mother Bear, with very young cubs, is, of course, more of a stay-at-home.

MIGRATION

The hunters generally agree that the Blackbear migrates. Bachman, speaking apparently for the Carolinas, says:[8] "In hard winters [it] is found to move southerly in considerable numbers, although not in company."

It is a common experience to find Blackbears suddenly numerous where a few months before they were rare, but what the nature and extent of this migration is, or whether regular in time or direction, I have not yet been able to determine.

TRAILS

All animals have a tendency to form beaten roads or trails in their home-region—this trail to the water, that to a favourite feeding ground, etc. The heavier the animal the more marked the trail becomes. The pathway of a Ground-squirrel through the prairie grass may be nearly invisible to us, but the pathways of Buffalo and Wapiti are, as we have seen, open and well-worn highways that serve mankind as they serve the beast that made them.

The present species is no exception to the rule. In all parts of the country where Bears are numerous they have well-placed, well-worn pathways, which are adhered to by the race with precision that justifies the pioneer Bears which first searched out and blazed them as the best and safest roads from this to that inviting land of promise.

[7] *Cf.* Beaver, p. 460, foot-note.
[8] Quad. N. A., 1849, Vol. III, p. 196.

There is abundance of testimony for this. L. Allen says:[9] "I have noticed that Bears are methodical in their habits, always following their own trail until their tracks are deep depressions in the ground."

After telling of a long hunt through the snow after three Bears, C. Wasgatt adds:[10] "All this time the Bears had walked in the track made by the leader."

In detailing the methods of a successful Bear trapper, J. B. Burnham says:[11] "Bear-trails in places are well-defined paths. * * * In passing over these trails the Bears step in each other's foot-prints, and if one Bear fifty years ago crossed a log at a certain spot, every Bear that followed is morally certain to have chosen the same place. Moreover, they never deviate from the exact line of their trail if it is in any way possible to avoid leaving it. Knowing these facts, Guy never baits his traps. In setting them he has two considerations to keep in mind—first, placing the trap where a man will not set his foot in it, and, second, where a Bear will.

"On the Twin Pond runway he found a spot where a small spruce tree had grown up directly in the Bear's path. A man would step to one side to pass this if he happened to be following the Bear's route, but the Bears themselves, on account of their conservatism, preferred to go under the low reaching boughs."

To this I can add my own corroborative experience in the Rocky Mountains. Where the Bears abound, the whole country is laid out in trails that are the outcome of necessity in a rough country, and knowledge of the best sources of supplies. They differ from human trails only in that the Bear makes no effort to improve the road; it merely selects the best available. In this particular many rodents are in advance of their betters.

In some regions where I have camped—the Bitterroots, for example—where the forests on the bottom lands are par-

[9] Recreation Magazine, April, 1900, p. 305.
[10] Maine Sportsman, October, 1900, p. 12.
[11] Forest and Stream, January 7, 1899, p. 3.

ticularly dense, and the bear-trails are exceedingly numerous; not only are they the best routes, they are the *only possible routes*. If you are in search of the things those Bears sought—berries, fish, water, or pleasant open hillsides, breeze-swept of flies—your part of wisdom is to follow the bear-trail. It was made during generations of search for these same joys of life, by those who were past masters of the road and had sought out all the most delectable spots in the hills.

The bear-trail differs from the wapiti-trail in several ways: first, it is deficient in head-room, unpleasantly low for a man; second, it always runs *along* a fallen trunk, if such be in the line; third, it crosses a stream by a log in preference to a ford.

It is well to remember further that in a bear-trail there is always danger of a bear-trap.

SOCIABILITY

The Blackbear is essentially a solitary animal. Occasionally a number of them have been seen together, but these gatherings were either for the purpose of mating or were a family of grown-up young ones with their mother. Nine out of ten grown-up Bears will be found leading solitary lives.

INTERCOMMUNICATION

The Blackbear has a long list of vocal sounds to express his feelings to others of his kind. Besides the growl of anger and loud cough of menace, they have whining calls and sniffs of many sorts, also a number of bawls expressing rage or pain.

But a still more curious outburst of intercommunication is recorded in the following extract from Audubon and Bachman:[12]

"At one season the Bear may be seen examining the lower part of the trunk of a tree for several minutes with much attention, at the same time looking around and snuffing the air. It then rises on its hind-legs, approaches the trunk, embraces it with the fore-legs, and scratches the bark with its teeth and claws for several minutes in continuance. Its jaws clash against each other until a mass of foam runs down on both sides of the mouth. After this it continues its rambles."

[12] Aud. & Bach., Quad. N. A., Vol. III, p. 189.

In this connection, Merriam says:[13] "In traversing unfrequented portions of the [Adirondack] wilderness one occasionally meets with a tree whose bark has been scratched and torn at some little height from the ground, in a manner that cannot fail to excite his attention and surprise. This is the work of the Bear, but the object of it is not known. Hunters claim that whenever a Bear passes one of these trees he stops, stands on his hind-legs, and gnaws and scratches it before resuming his journey. The only account of the strange proceeding that I have seen is given by Audubon and Bachman."

But the fact is widely known among hunters, as the following extracts show:

"Why [says L. Allen [14]] do Bears leave their teeth marks across a tree or a sapling, as high as they can reach, standing on hind-legs? The highest marks are always freshest. Is it the same Bear that makes the higher mark, to see how much he has grown, or another Bear who can go him that much better?"

"Bears bite trees all through the summer. I think they do that to see who is the tallest one. Only he Bears bite trees. They bite them along their roads, and the one that makes the tallest marks bosses the road. After you kill the big one you don't see another he Bear for a long time on that road. She Bears pass any time."[15]

"In the running season, which is at its height in June, the Bears blaze it [the bear-trail] by biting trees, each leaving his mark as high up as he can reach."[16]

Other Bear-trees are described in the "Biological Survey of Texas," by Vernon Bailey, who writes:[17] "Near one of the trails in the head of Dog Canyon in the Guadalupe Mountains a Douglas spruce a foot in diameter had served for many years as a gnawing tree, while further up the gulch a larger yellow pine was well blazed and deeply scarred by many old and a

[13] Mam. Adir., 1884, pp. 101–2.

[14] L. Allen, East Wareham, Mass., Recreation Magazine, April, 1900, p. 305.

[15] J. B. Burnham, Forest and Stream, March 18, 1899, p. 208.

[16] J. B. Burnham, Forest and Stream, January 7, 1899, p. 3.

[17] N. A. Fauna, No. 25, 1905, p. 188.

few new gashes of powerful teeth. In the Davis Mountains, on the ridge just north of Livermore, a yellow pine a foot and a half through has served as a bear-register for apparently ten or twenty years. It was deeply scarred on all sides, from 4 to 6 feet from the ground, but on one side, from 5 to 6 feet up, the bark had long been cut away, and the dry weathered wood was splintered and gashed with deep grooves of various ages. Two fresh sets of tooth-prints showed on opposite sides of the tree near the top of the ring, and one little Bear had lately tried his teeth in the green bark about 4 feet from the ground. At the head of a gulch on the east side of Limpia Creek stood another big yellow pine that had been similarly treated, and on it, as on the others, the upper limit of reach was about 6 feet from the ground. Apparently the Bear at each visit to one of these register trees had given but a single bite, leaving the marks of an opposing pair of canines."

Finally, I can add my own testimony. I have seen many, yes hundreds, of these bear-trees, chiefly in the Rocky Mountains. They are always by some well-worn pathway or trail of the Bears, and are made and used by Bears of all species.

What is their meaning and purpose? For it is very certain that such a remarkable and universal Bear habit must have some good object.

I think there can be no doubt, as I have elsewhere and years ago said in print, that these are Bear 'sign-boards,' methods of communicating certain information to the Bears. They answer, I believe firmly, the same purpose as the urinary signal posts of dogs, Wolves, and Foxes.

A creature with such exquisite power of smell as a Bear has no difficulty in reading at once, by touch and taint on the register trunk, that here there has recently been a Bear of such sex and species, a personal friend or foe, as the case may be—and the trail shows that he went in such a direction; by following that trail he can overtake that Bear, or, if he prefers it, he can expend his sudden outburst of feeling on the offending but defenceless trunk.

PLATE XCIII.—ASPEN TREE WITH MARKS OF BLACKBEAR CLIMBING.
Colorado, 1901.

PLATE XCIV.—ASPEN WITH GRIZZLY CLAW-MARKS.
From photograph by Mrs. Grace G. Seton.

PLATE XCV.—BEAR'S SIGN-POST, MUCH MARKED.
Wilson Flattops, Colorado, 1901.

PLATE XCVI.—ASPEN ONCE CLIMBED BY BLACK-BEAR. EACH CLAW-MARK IS NOW A BUMP.
Colorado, 1901.

The claw-marks made by a Grizzly differ from those of a Blackbear, first by their size and second by being clearly 5 in number, while the latter often leave but 4. This is due to the shortness of the Blackbear's thumb and claw. In the Rockies the aspen is most frequently used as a bear-register, and it is singularly well adapted for records, as its smooth bark never loses its scars. The claw-marks of the Bear may grow out of pine or cotton-wood, but once in the aspen bark they stay there for life. Thus the bark of a growing aspen carries a record of all that tree's vicissitudes for those who can read. Bear claw-marks, frost-crack, woodpecker borings, insect ravages, horn thrusts from Wapiti, Squirrel gnawings on the mere expansions and sutures of growth, are all there, in plain and legible sight (Plates XCIII-XCVI).

FIG. 243—Quaking aspen, with Bear claw-scars grown out into bumps 1¼ inches long. Colorado, Sept. 25, 1901.

Deep marks such as claw-wounds may even get stronger as years go by. I know of a singular case—a Blackbear climbed an aspen some twenty years before I saw it—there was the record plainly to be seen, but the claw-marks, at first deep pits, had filled up as level black scars, and at length became ever-lengthening bumps, till now each is prolonged into black claw-like warts 1¼ inches long (see Fig. 243).

In addition to the claw- and teeth-marks, it is common to see the bear-tree more or less plastered with mud in which is Bear hair. This was left by some Bear rubbing his back and marking his height after he had been wallowing in the mud. Some observers think that the registers are used only in the running season, but I have reason to believe that in a less degree they serve the year round.

MATING

The sum of evidence shows that in the latitude of the Northern States and southern Canada the mating season of the Blackbear is early June; though in some cases it may be deferred as late as the first week of July.

But little is known of the details. According to some hunters, the males rush along the bear-trails, stopping at every bear-tree to get the latest news—there to learn what Bear has gone by, whether fighting males or unmated females—add their own record, then hurry along the most promising line, to meet, perhaps, a dozen other Bears assembled in some well-known, quiet haunt, where, with much threatening and occasional combat, the sexes are paired—one male to one female. The 'happy couple' set off alone on the 'honeymoon' of a week or more, the female leading the way, the male following, but pausing at times to glance back and hurl, if need be, a rumbling defiance to any other of his sex that seems disposed to follow or dispute his claims.

But the love-bond weakens with the love-moon's waning. So far as known, they part in July, to go their separate ways, and if by chance they meet again that year, it is as strangers or possible enemies.

BEAR STORAGE

Throughout the rest of the summer and autumn their energies are devoted to getting fat. The Woodmouse and the Squirrel store up hordes of choicest food in holes and caves, the Bear and the Woodchuck store it up in their own skins. Autumn with its plentiful nuts, in addition to many other foods, affords the opportunity—and the Bears lose not a minute. Night and day they work, their stomachs are distended to the full, but amazingly good digestion waits on boundless, unbridled appetite, and the Bear grows daily rounder.

Fur is a wonderful protection from the cold, fat is as good. The Bear has four inches of each when winter comes; little it heeds the storm. And yet this, of all the big forest beasts, the best protected, is the only one to shun the battle with winter and seek a safe and sheltered den, there to sink into the sleep that lasts till spring.

WINTER-DENNING

The winter-denning of the Adirondack Bears was studied by Dr. C. Hart Merriam, who says:[18]

"The exact period when the event takes place is determined by the food-supply and the severity of the season. If the beechnut crop has been a failure and deep snows come

[18] Mam. Adir., 1884, p. 97.

early, they generally den near the commencement of winter. If, on the contrary, there has been a good yield of mast and the winter is a mild one (and it is a fact that, with us, good beech-nut years are commonly followed by open winters), the males prowl about nearly, or quite, all winter, and the females only den a short time before the period of bringing forth their young. Indeed, it can be set down as a rule that *so long as a male Bear can find enough to eat he will not den, be the weather never so severe;*[19] for it is evident that he does not den to escape either the low temperature or the deep snows, but to thus bridge over a period when, if active, he would be unable to procure sufficient food. And the female, under similar circumstances, remains out till the maternal impulse prompts her to seek a shelter for her prospective offspring; and in this wilderness they have been found travelling as late as the middle of January."

Quite in accordance with this is the fact that menagerie Bears, sure of their food, almost never go into a true winter sleep.

A tame but free Blackbear, kept on the Red River by Alexander Henry in 1804, began to prepare its winter den as early as November 13.[20]

I have seen one or two Bear dens and have heard many described by the hunters. They vary from a deep, snug, sheltered natural cave in the rocks, to a hollow tree or a hole under an upturned root. Sometimes the Bear digs a den in the level ground, as I once saw in Manitoba, and sometimes it makes a bed under a windfall of logs and brush, or in a dense thicket. But wherever chosen, it is sure to be a dry place where the snow will gather and lie deep all winter.

Great variety in the amount of lining is observable. According to Merriam:[21] "The amount of labour bestowed upon it depends upon the length of time the Bear expects to hibernate. If the prospects point towards a severe winter, and there is a scarcity of food, they den early and take pains to make a

[19] Italics. mine.—E. T. S.

[20] Journal, 1897, p. 253.

[21] Mam. Adir., 1884, pp. 97–8.

comfortable nest; but when they stay out late and then den in a hurry, they do not take the trouble to fix up their nests at all. At such times they simply crawl into any convenient shelter without gathering so much as a bunch of moss to soften their bed. Snow completes the covering, and as their breath condenses and freezes into it, an icy wall begins to form, and increases in thickness and extent, day by day, till they are soon unable to escape, even if they would, and are obliged to wait in this icy cell till liberated by the sun in April or May."

Nevertheless, it seems that the Bear does not truly torpify in hibernation.

It is remarkable that no one yet has found two adult Blackbears in one den. Mother and half-grown cubs have been taken in the same winter quarters, and, of course, the mother with the new-born cubs is the regular thing, but never two old ones together, a fact that speaks for the unsociability of the species.

BREEDING

The breeding of Blackbears was for long shrouded in mystery.

It was formerly believed that Bears would not breed in captivity, but modern methods and care have disproved this. We now have very full data on the breeding of captive Blackbears, and many of the mysteries have been dispelled.

The fullest history of a breeding pair, so far as I know, is that by Arthur B. Baker. It is a complete record[22] of a pair of Blackbears from their first to their fifteenth year. The male was captured as a cub in Central Michigan, July, 1888, and the female, of the same age, was taken about the same time on the south shore of Lake Superior. They were kept captive at R. H. Lodge's Park, Cuyahoga Falls, near Akron, Ohio.

The first litter was born when the parents were 4 years old, that is, the union took place when the old ones were 3½, and this is probably the age at which normally they begin

[22] A notable success in the breeding of Blackbears by Arthur B. Baker, Smithsonian Misc. Coll., Vol. 45, No. 1434, pp. 175–9, January 7, 1904, Washington, D. C.

to reproduce; though it is possible that the female is ready a year earlier.

The period of gestation is about seven months. GESTATION

The young are born in the latter half of January, while the mother is sealed up in her winter den, and—being far from torpid—very well able to devote all her time to the offspring, the case recalling that of the great Indian hornbill, in which the male bird seals up the female in the nest with the eggs, to insure her unremitting attention; only in the case of the she Bear the relinquishment of the world is voluntary and complete, and lasts for several months, during which she neither eats nor drinks, and is yet in full possession of all her appetites, powers, and faculties. YOUNG

FIG. 244—Mastology of Blackbear ♀.
Great Slave River, June 15, 1907.

It was long stated and believed among trappers and hunters that no man ever yet killed a pregnant Bear. The fact is that the embryos are so small that no one, but an expert anatomist, could find them; even up to the time when they are born they are surprisingly small, as well as naked, and rather shapeless. They are much less in proportion than the young of any other mammal outside of the marsupials.

At birth they are about 8 inches long and weigh from 9 to 12 ounces, that is about 1-200 to 1-250 of the mother's weight, while a young Deer is 1-30, a young dog 1-25, the human being 1-20, and the young Porcupine 1-15 of the weight of the mother. A new-born Porcupine is, in fact, as Merriam points out, actually larger and heavier than a new-born Blackbear. SIZE

NUMBER The cubs are usually 2 in number; occasionally there is but 1, especially if it be the mother's first litter; 3 are common, and 4 have been recorded several times.

The Lodge record above referred to runs thus:

1892, Jan. 23.	One male cub, found dead.
1893, Jan. 24.	Two males and one female.
1894, ———	No cubs born, owing to young of previous year having run with mother throughout the summer.
1895, Jan. 23.	One male and one female.
1896, Jan. 24.	Two males and one female.
1897, ———	One male (exact date of birth not noticed, but between January 21 and 27).
1898, Jan. 24.	One male and one female.
1899, Jan. 27.	Three males.
1900, ———	No cubs born, as young of previous year had run with the mother during the summer.
1901, Jan. 26.	Two males and one female.
1902, ———	No cubs born.
1903, Jan. 21.	Two males and one female.

This valuable record proves, among other interesting things, that the wild mother Bear breeds only every other year, unless she has the misfortune to lose her family early in the season.

At birth the cub is blind and is covered with a fine, close dark hair, so thin that it is practically naked.

In all cases observed the mother has hovered and brooded over the young for six or eight weeks, covering them as anxiously as though a single breath of cold air would be their certain and sudden death, as doubtless it might under their natural conditions in certain parts of the range.

A litter of young Blackbears was born in the Brooklyn Zoo in 1899, and the keeper, Edward Walsh, wrote:[23] "The

[23] Forest and Stream, February 4, 1899, p. 84.

PLATE XCVII.—ONTARIO BLACKBEAR.
From life study by E. T. Seton, 1885.
Original owned by H. W. Mickle of Toronto, Canada.

mother seems to spend most of her time suckling them, and is very solicitous for their comfort. The cubs have a habit of whining like puppy dogs, especially when by any chance they are crowded away from their dinner. Their mother licks them and fondles them with her paws and is as proud of them and jealous of interference as any human mother."

Frank J. Thompson, whose account of the Blackbears bred in the Cincinnati Zoo is the earliest that I knew of, gives the following interesting details of their development:[24]

"In about ten days their coats began to show and were of a grayish tint, which gradually passed through the various shades until they became a brownish-black. It was just 40 days before the first one's eyes opened, and 2 days later the second followed suit. From that time forward I watched very closely to ascertain the exact time that would elapse before the young ones would leave the nest, and on the seventy-first day after birth, when the mother, as was her habit, came to the grating to be fed, one of the youngsters left the nest and followed her. So soon as she found it out she immediately drew it gently back, and, on its second attempt, *she cuffed it soundly*, which put a stop to its wandering propensity.

"After a few days she allowed them to wander about at will, provided no one was immediately in front of the den, but so soon as a visitor put in an appearance, they were driven back into the nest, and not allowed to emerge until the strangers were out of sight. For some time she always suckled them in one position, lying over and completely covering them by stretching flat on her belly with her legs drawn up under her and her head tucked down between her front paws. As they grew older and began to run about she would sit on her haunches, lazily lean back against the wall, take a cub on each fore-arm, and hold them up to her breast until they were satisfied. They soon became expert climbers, taking advantage of the slightest inequalities of the stone walls and the cracks between the heavy oaken planks to reach the ceiling of the den on three sides, while the grating in front served capitally for their sky-

[24] Forest and Stream, September 4, 1879, p. 605.

larking. Occasionally they would have a regular sparring bout, standing erect, feinting, countering, and making use of many of the tricks of old votaries of the prize-ring. These frolics would generally end in a clinch fall, and a regular rough-and-tumble fight, when the mother would abruptly put a stop to it by suddenly knocking both of the contestants completely out of time. In fact, as they grew apace, the parental visitations increased so rapidly I began to fear she would put an end to my Bear investigations by chastising the lives out of them, but of late she has slackened in her attentions, and I am in hopes of following the growth of *Ursus americanus* from babyhood to adolescence."

This determination to bring the young up right, no matter how much spanking is needed, is common to most mother Bears, but is very variable individually. I have known an old Bear to punish her young one severely merely because she, herself, had at that time lost her head in a sudden alarm and behaved foolishly. We look not in vain among our own kind for parallel cases.

SPRING

When spring comes with force, in the woods it melts the ice and frees the icebound mother Bear. If the weather is at all settled, and the ground partly clear of snow, she sets forth on her travels in search of food, the little ones stringing behind her like a lot of little pigs. And now they say good-by to the old den. Thenceforth the mother sleeps where and whensoever she is sleepy—and the little ones slumber cuddled in her arms, and more or less beneath her body.

The old Bear usually comes out fat in the spring, but the scarcity of food and the drain of the thriving young family soon reduce her stored-up supplies. And a May Bear is always a lean Bear.

The cubs learn to eat solid food as soon as the bare ground makes it obtainable.

The mother's care of them and their faith in her at this time are ideal, and all pictures of it that have been put on record have a human character that is sometimes exceedingly touching.

A Chicago traveller, whose identity I cannot learn, related the following to a reporter for the *Record-Herald*, August, 1901:

"When I was in Michigan a few weeks ago I had just this experience. I was passing through Harmon City, which is a pretty wild sort of country. A couple of men from the village were doing some work on the outskirts when they caught sight of Bear tracks. They followed for a while and then set a heavy trap. Later they returned, and they had a Bear, sure enough. She was a large brute with dumb, beseeching eyes, from which the tears rolled as they might have rolled from a human being. I went with others and was a witness of the tragedy. The men simply shot her to death as she lay there with her right fore-paw held in that awful grip of steel.

"Then the men waited around until the old Bear, her husband, came in sight. He wasn't trapped, but he was killed just as expeditiously. The poor beasts had no show. But the most pathetic sight, to me, was the three little cubs which had followed their mother to the scene of her death, and which whimpered like sorrowful babies over the killing of their parents.

"When the big Bears were killed one of the little chaps, about the size of a small shepherd dog, climbed to the branch of a tree on which their bodies were suspended and looked down in wonder at the still, dead faces. Another little Bear sniffed feebly at the swaying body of his mother, while the third put his paws, trustingly and pathetically, upon the knees of one of the men whose rifles had done the work. I'm not much of a sentimentalist, but those three little orphan Bears kept me from talking out loud for half an hour."

Notwithstanding her courage and strength in their defence, and her cleverness on their behalf, the mother Bear is sometimes separated from one or more of her cubs; the young ones are lost in the woods. A case of the kind is thus recorded by Dr. Merriam:[25]

[25] Mam. Adir., 1884, p. 101.

"While hunting, June 10, 1878, Dr. C. L. Bagg and the writer followed the old trail from Fourth Lake across Eagle Creek in the direction of John's Lake. In exploring a hardwood ridge a little to the north of the regular course, we were suddenly surprised by a loud and peculiar cry with which we were both unacquainted. It came from the direction of a dense balsam swamp below, and somewhat resembled the squealing of a pig, while at the same time it suggested the noise made by the Great Blue Heron when on its nesting grounds. As the cry was repeated, Dr. Bagg imitated it, and succeeded so well that we soon perceived it to be coming nearer. Fearing that it might change its course, I ran down the hill, and soon saw a dark-coloured animal, about the size of a Raccoon, emerge from the swamp and jump upon a log, rushing headlong in the direction towards Dr. Bagg, and squealing at brief intervals as if in great distress. Bringing my gun (loaded only with No. 4 shot) hastily to my shoulder, I fired, and the report was followed by a shriek of pain and plaintive baby-like sobbing cry that lasted for nearly a minute. On reaching the spot the animal was found to be a cub Bear, and was then quite dead, one of the shot having passed through both ventricles of the heart. It was very thin, weighed but 10 pounds (4,536 grams), and had evidently been lost from its mother for some time. Its stomach contained nothing but beechnuts, and beechnuts that have lain on the ground all winter, and are still fit to eat in June, are certainly few and far between."

Another peep into the pathetic side of the Bear's life is afforded by a letter that I recently received from a little girl in Salmon, Idaho:[26]

"Jim Winn, an old trapper and hunter here in the valley, said that one time he went out hunting, and when he was eating his lunch he heard an awful running and snorting down the hill where his horse was, so when he looked around he saw a little tiny baby Bear trying to catch the horse. He said that he shot at it and hit it in the neck, but did not kill it. He said it cried so pitifully that he was sorry he shot it. Pretty soon the

[26] Personal letter.

mother Bear came and saw the little Bear crying, so she picked it up and spanked it very hard, for she did not know what was the matter with it. Presently she smelled the blood on the little one's neck, and that set her wild. She ran up and down the canyon and cried as if her heart would break. Jim said he had never seen a Bear cry so much like a human being before. Then she came back to the little baby Bear and picked it up (it was still crying) and brought it into the thick woods. Jim said he thought she was going to bury it, for it was nearly dead."

(Signed) Beth Yearian (age 12). Salmon, Idaho, November 23, 1902.

The gambols of a family of little Bears are exceedingly boylike and amusing. They wrestle and box and pretend to fight with all the vigour of gamins at play. Usually they are careful to keep the rules of the game and avoid hurting each other, but ill-tempered Bears are as frequent as ill-tempered boys, and savage quarrels have thus arisen in the family.

A. B. Baker tells[27] of a little reprobate which, when only three and one-half months old, killed his brother in a fight over a pan of milk. This same authority has further given us in context the seamy side of the mother's character:

"The old Bear is a model mother to the cubs as long as they remain under her care, even refusing on their account the attentions of her mate, but when they are taken away, her affection for them seems soon to end. The two cubs of 1898 were removed in May and returned to the mother early in October, after first being kept for two weeks with only a grating between. She had seemed to recognize them, but when they were put together she at once caught the little male by the head and killed him, and only forcible measures prevented her from climbing the tree and repeating the operation on the other cub, which had taken refuge there."

Throughout the summer the old Bear wanders about the home-region that she knows—probably less than a 10-mile

[27] *Loc. cit.*, p. 177.

radius—and the little ones grow in size, but usually dwindle in numbers. Accidents will happen, and little Bears get coughs and colds, or disobey their mothers and come to grief. Consequently, while 3 little Bears are often found in the mother's den, rarely more than 2 are seen roaming by her side in summer, and autumn, in many cases, sees the number reduced to 1.

The young, whatever the number, are believed to den up with the mother the following winter, and probably remain with her until the mating season comes with June, and finally the family feeling is dispelled by the newer thought. The young ones scatter, and thenceforth when they meet the mother they are little better than any other stranger.

THE FATHER

So far as I can determine, the father Bear takes no interest in the young ones. Menagerie keepers have learned, through many disastrous experiences, that the less he is with the cubs the better for them. In the vast majority of cases the mother is the only adult seen abroad with the young.

On the other hand, there are on record a few cases in which an adult male has been found associated with the mother and little ones, and Miles Spencer, of Fort George, Hudson Bay, after years of experience in that region, backed by the opinions of the native hunters, puts his view on record in this brief sentence: "The male assists in rearing the young."[28]

DISPOSITION

Notwithstanding a widespread idea, the Blackbear cannot be called a fierce or dangerous animal. On the contrary, it is one of the shyest and most timid of wild creatures. I have met scores of them in the woods, and, almost without exception, they fled like Rabbits as soon as they knew I was near.

The exceptions to this rule are: A captive Bear that has, through daily association, lost its fear of man, and, through cruelty, begotten a hate of him; a female whose young are threatened; and finally, a wounded, cornered, crippled, or

[28] Low, Expl. James Bay, Can. Geol. Surv., 1888, Pt. J, App. III, p. 78.

PLATE XCVIII.—HOME LIFE OF A BEAR FAMILY.

otherwise disabled Bear, which will fight just as surely as a Rat or a Chipmunk will in like conditions.

Outside these special circumstances, there is little need to fear the wild Blackbear.

When cornered or forced to fight, it is a dangerous enemy. It can disable a dog or a man with a blow of its paw. With its jaws it can crush ribs and limbs. But its claws, sharp and driven by muscles of far greater power than those of the strongest man, are its truly terrible weapons, and, in spite of current legends, we may rest satisfied that no man, however powerful, if armed only with natural weapons, would have the slightest chance in combat with a full-grown Blackbear.

INTELLIGENCE

Among hunters I find the greatest difference of opinion regarding the Bear's intelligence. Some will tell you that any one, with any kind of a trap, can catch a Bear. Others maintain that the smartest Fox that ever lived is a fool to an old Blackbear, and stoutly contend that a successful Bear hunter and trapper has attained the acme of woodcraft.

There are doubtless exceptional Bears, whose performances have raised the trappers' opinion of the whole race. It is well known also that each specialist is apt to give first place to his own craft, and ascribe to his prey an intelligence that he denies to creatures which he knows less about.

There is little doubt that the Bear is high in the scale of intelligence, though it cannot compare with a dog, Fox, or Wolf. It is gifted with marvellous powers of smell and hearing, and has a deep-rooted shyness about all things strange, or doubtful, which saves it again and again from traps of various kinds. Its fixed, safe, and saving motto is: 'In case of doubt, *run*,' and it is nearly always in doubt.

I have frequently seen a Bear at a distance of half a mile cease feeding, throw up his nose as the tell-tale wind brought tidings of my presence, then fly to the woods, to journey miles before again assured of peace.

On the other hand, the Blackbear's eyes are not very good, and twice on the open plains in Manitoba I have, at broad daylight, walked right into a Blackbear that did not know I was near, simply because I had the wind.

A MISCHIEF MAKER

Many tales are told of Bear depredations in western camps during the absence of the owners—depredations in which hunger had no incentive part. Merriam gives[29] one that occurred in the Adirondacks at the Government Survey Camp in 1882:

"Returning one day, after a temporary absence, the members of the party were astonished to find their tent torn down, and blankets, books, and instruments strewn about upon the ground. The foot-prints of a Bear revealed the identity of the marauder, and Mr. Colvin, Superintendent of the Survey, afterwards fired at and wounded the beast, but did not succeed in capturing him."

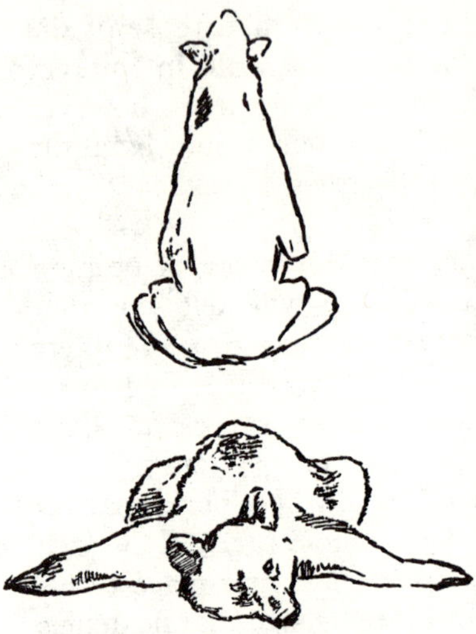

FIG. 245—Bear poses. (From life.)

J. Blackwell, of Tacoma Hotel, Seattle, told me that for long they kept a pet Bear cub that developed an extraordinary love of mischief. Whenever he could break away he left a trail of destruction behind. His wickedest exploit, the one which finally turned the women against him, and of course ended in his ruin, was entering a house while the family was away and deliberately seeking out and tearing to shreds all the bonnets in the wardrobe. He would face and fight anything but the *unknown*. And the only thing that seemed unknown to him was a wheelbarrow. A small boy could drive him anywhere with a wheelbarrow.

Another old Bear that he had was very savage; no one, not even the keeper, dared venture within the radius of the chain.

[29] Mam. Adir., 1884, p. 103.

But one day a cripple who came to the hotel got very drunk and disappeared. Next morning they heard a voice in the Bear's den, "Lay over—who are ye shovin'"; and there was the missing cripple sleeping with this ferocious Bear. They had much difficulty to get him out, partly because he didn't want to come, and partly because the Bear fought them for possession of his guest.

A tame Bear was kept at Park River House on the Red River (Minn.) by Alexander Henry in 1804. The old chronicler says:[30]

"He is so tame as to require no care or confinement, but associates with the dogs, and even follows them and the men into the plains and woods."

The same old scribbling pioneer gives us another glimpse of Blackbear character in the following:[31]

"While we were arranging camp I saw a Bear on the east side of the river, a little above us, coming down to drink. I crossed over and followed him; he soon stopped within a few paces and ran up a large oak. I shot him between the shoulders and he fell to the ground like a log, but in a moment was scampering away as fast as he could. I traced him by the blood, and soon found him sitting under a brush-heap grumbling and licking his wounds. A second shot dispatched him. By the hideous scream he uttered when he fell from the tree I imagined he was coming at me, and was waiting for him with my second barrel cocked when he ran off. I went for my two men, and it was hard work for us three to draw him to the canoe; he was very fat. I found that my first ball had gone through his heart. I was surprised that he should have been so active after a wound of that kind."

Of course, it is well known that a Blackbear is a good climber, but I shall never forget the surprise I got when first I saw a wild one climb a tree. I had pictured to myself something like the slow moving up of a man or a sloth, or at best the action of an expert sailor going up the shrouds. But what I CLIMBERS

[30] Journal, 1897, p. 253.

[31] *Idem.*, p. 87.

really saw was more like the action of a cat. It had not the spring and agility of a Squirrel or Marten, but this Bear went up three times as fast as any man could, and quite as well as any monkey that ever I saw. In coming down he travels tail first. It is quite a common thing for a Bear up a tree, when fired at, to throw himself to the ground from a considerable height. Those hunters who do not know this trick are apt to think the game is killed, and are generally surprised to see him bound off, as though quite unhurt.

A friend of mine had a Blackbear cub that used to play some very curious tricks on a dog that he alternately played and quarrelled with. Hiding in a tree, under which the dog sometimes slept, he would await a good chance to leap from a height of fifteen feet and land with crushing force on his enemy's body, not breaking any bones, but knocking his breath out, and driving him away in ignominious rout.

SWIMMING

Bears are good swimmers. It is quite a common thing to see Bears in Muskoka and in northern Manitoba make voluntarily long swims across lakes and rivers. In the August of 1906, Dr. Gordon Bell, with the other Water Commissioners of Winnipeg, secured a large Bear swimming in Shoal Lake.

It is easy, if one have boat and rifle, to overtake and kill the swimming Bear, but without the latter it is a risky business, for the Bear, on seeing the boat come near, may turn and climb into it. He does not usually attack the other occupant, under these circumstances, but his ideas of 'trim' are so inadequate that it usually ends in the hunter having to swim for it.

WALLOWS

Old hunters who have lived their lives among the one-time swarming Bears of the Rockies, tell you that a Bear is a kind of a pig. What a pig will eat a Bear will eat; what a pig will do a Bear will do; only a Bear is smarter and he can climb. Many of them apply pig nomenclature to Bears, speaking of them as 'boars,' 'sows,' 'droves,' etc.

In the Colorado Mountains I once saw a black muck wallow much like that of a Wapiti, but all about were evidences

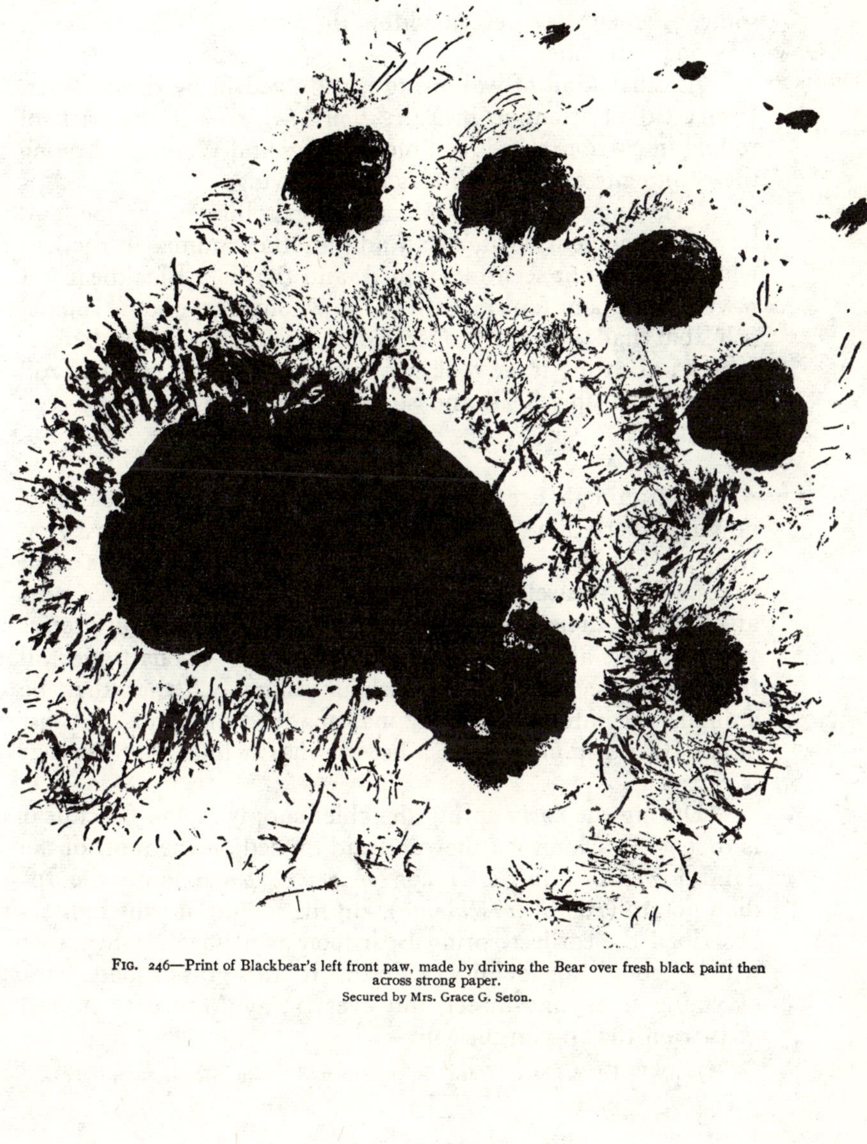

FIG. 246—Print of Blackbear's left front paw, made by driving the Bear over fresh black paint then across strong paper.
Secured by Mrs. Grace G. Seton.

that a mother Blackbear and her young had been using it in pig fashion. The hunters there said wallowing was a constant practice of Blackbears in hot weather when flies were troublesome, or when they were shedding their coats. This, however, was on September 21.

DOPING Another kind of wallowing is indulged in by these Bears. I once called attention in *Recreation Magazine* to the trick of rolling in carrion, that is common to dogs and Wolves. Among the comments called forth was the following:

"Having read in *Recreation* of dogs rolling in carrion, and having seen them do it, I can add another animal to the list, which I have not seen mentioned, and that is a Blackbear. I have a cub about four months old to whom I gave some cooked fish that had been left over from supper the night before. Instead of eating it, he took it out of the pan and began to roll in it, rubbing his head and shoulder the same as I have seen dogs do."[32]

FOOD 'All animals are omnivorous, especially the Blackbear,' might properly have appeared in a certain celebrated essay on Beasts. The Bears, like the Coons, are quite omnivorous at all times. The Weasels will eat fruit, if hard put, but prefer meat at all times. The Muskrat will eat fruit if starving, but prefers vegetables at all times. But the Bears and Coons prefer all things eatable at all times without asking whether they be animal, vegetable, or unholy man-made compounds. A list of the Bear's staples is not a list of what it likes, but of what it can get.

During the early spring the chief supply of the Blackbear is roots. In Manitoba they are said to feed on the roots of the Sand-flower or Prairie Crocus (*Anemone patens*) and the Indian potato (*Psoralea esculenta*). In the mountains the hunters described the earliest spring Bear food as a fibrous white root which I could not identify. To this it adds grass shoots, bark of young trees, any insect, and every stray Mouse or morsel of carrion that it can pick up.

[32] James W. Nicol, Moore, Wash., in Recreation Magazine, March, 1900, p. 223.

In many of the northern lakes a new food supply is added in the myriads of Mayflies that are drowned and washed up on the beach. About Shoal Lake, Man., the residents assured me that in some seasons the shores of the lake are covered with a pile of dead Mayflies 6 feet wide, 6 inches high at the highest point, and about 20 miles long.

E. A. Preble, in his notes on the Blackbear in Keewatin, writes:[33]

"One was seen near Robinson Portage by Mr. W. C. King, who passed this point a day or two ahead of us on his way towards York Factory. This Bear was feeding on the piles of Mayflies (*Ephemeridæ*) which perish in myriads and are washed up on the shores in long windrows. These are said to constitute a favourite food of the Bear."

Hearne says[34] of the Blackbears he killed between York Fort and Cumberland House: "Their flesh was abominable. This was in the month of June, long before any fruit was ripe, for the want of which they then fed entirely on water-insects, which in some of the lakes we crossed that day were in astonishing multitudes. [Foot-note says, 'lying in putrid masses to the depth of 2 or 3 feet.']

"The method by which the Bears catch these insects is by swimming with their mouths open, in the same manner as the whales do when feeding on the sea-spider. There was not one of the Bears killed that day which had not its stomach as full of these insects (only) as ever a hog's was with grains, and when cut open the stench from them was intolerable."

An abundant spring food-root in much of the Bear's range is the arum, and, so far as other vegetarians are concerned, it is probable that the Bear is welcome to every root of the kind it can find in the woods. I know of no other creature that can stand its pungent terrors.

Audubon and Bachman comment thus on the habit:[35]

"Perhaps the most acrid vegetable eaten by the Bear is the Indian turnip (*Arum triphyllum*), which is so pungent that

[33] N. A. Fauna, No. 22, 1902, p. 64.
[34] Journey, 1795, p. 370.
[35] Quad. N. A., 1849, Vol. III, p. 190.

we have seen people almost distracted by it when they had inadvertently put a piece in their mouth."

Richardson, when on the Churchill River in 1848, was shown a root that evidently supplied the Bears with food. He says of it:[36]

"The *Actaea alba* grows abundantly here. It is called by the Canadians *le racine d'ours*, and by the Crees *musqua-mitsu-in* (Bear's food). A decoction of its roots and of the top of the spruce fir is used as a drink in stomach complaints."

Throughout the summer all kinds of insects, and especially ants, are important Bear food.

In the sandhills about Carberry, in the woods about Lake Winnipegosis, throughout the Bitterroot Mountains of Idaho, and on the ranges of the upper Yellowstone, as well as in the Rockies of Colorado and the Low Laurentians of the upper Ottawa, I have found that ants' nests furnished the Bear with an important article of food. Following the trail of one, I have found that it invariably turned over every log and flat stone that it came to, and ripped open every rotten log and stump in its search for insects, the greater part of which must have been ants. Among the Bitterroot Mountains I have, in a single day, passed hundreds of these demolished logs and stumps.

In the Adirondacks, according to Merriam,[37] the Blackbear "delights in tearing open old stumps and logs in search of the ants that make their homes in such situations. * * * "

"While fishing in the North Bay of Big Moose Lake, during the summer of 1881, Mr. Harry Burell Miller, of New York City, heard a Bear tearing down an old stump that stood on a point in the bay. His guide, Richard Crego, noiselessly paddled him to the spot, and he killed the Bear with one ball from his rifle. Its stomach contained about a quart of ants and their eggs."

As summer wears on, the Blackbears of the Pacific watershed have a new supply in the myriads of salmon with which

[36] Arc. Search Exp. of 1848 (1851), Vol. I, p. 82.
[37] Mam. Adir., 1884, p. 95.

every clear stream is teeming. And those of the east find a corresponding, though smaller, supply in the suckers and other swimming spawners.

Berries, now, begin to ripen, and furnish another bountiful resource. Strawberries, blackberries, and raspberries allure the Bear from other foods, and strike an equitable bargain for their gifts. In return for their delicious pulp, the Bears scatter excrementally the unimpaired seeds of these fruits, and thus sow the waste places near and far. Consequently, when the conditions become at all favourable, there is an abundant supply of berry seed to take advantage of the opportunity. Thus a region devastated by fire is immediately and unwittingly sown with berry seeds by the Bears, who, however unconscious they may be of their share in the planting, are not slow to come and glean their own harvest.

A pleasing variation of late summer foods is found in the nests of several species of wasps, as well as of wild bees.

According to Merriam,[38] the Bear "digs out the nests of the 'yellow-jackets,' devouring both the wasps themselves and the comb containing their honey and grubs. So fond is he of honey that he never misses an opportunity to rob a 'bee tree,' manifesting no fear of the bees that angrily swarm about him, his thick hair and tough hide protecting him from their stings. When plundering the apple orchard he is said to touch only the sweetest fruit."

Similar testimony is given by Captain J. P. McCown,[39] who observed the Blackbear in Tennessee.

On September 21, 1905, at Lake Caughnawana, 40 miles east of Kippewa, Quebec, I found that a Bear I was tracking had stopped at a rotten log to dig out a wasp's nest a few minutes before; the comb, in pieces, was scattered about, and a number of the yellow-jackets hovered angrily over the ruins of their home.

On the upper Red River, in 1800, when Blackbears abounded, Henry writes,[40] September 22:

[38] *Ibid.*, pp. 95–6.
[39] Aud. & Bach., 1849, Vol. III, p. 195.
[40] Journal, 1897, p. 101.

"Bears make prodigious ravages in the brush and willows. The plum-trees are torn to pieces, and every tree that bears fruit has shared the same fate; the tops of the oaks are also very roughly handled, broken, and torn down to get the acorns. The havoc they commit is astounding; their dung lies about in the woods as plentiful as that of the Buffalo in the meadow."

Autumn arriving with its showers of beechnuts, acorns, and other such food, gives all wild creatures a notable chance to gather for the coming cold. All work hard to profit by the opportunity, storing inside their skins or dens as their custom may be, and none work harder or more successfully than the Bear.

It labours without ceasing till nature comes with the snow-clouds and abnormally ends the feast by spreading the white cloth.

But the Bear has no notion of quitting the delights of active life for the dull monotony of a winter's sleep any sooner than possible, and, roaming still in search of food, is often led into mischief, killing on such occasions calves and sheep in the field or even pigs in the sty.

At such times it has even been guilty of cannibalism. George Crawford (Mittigwab), an Indian of Kippewa, told me of a case that he observed on the Upper Ottawa, about 1890. He tracked a very big Bear in the snow to where it had hunted out another that was already denned up, and had devoured it all but the jaws. The big fellow had feasted for two days, sleeping in the den of the victim, and coming out only to eat or else to drop dung, of which there was an immense quantity outside.

Colder weather, deeper snow, and scarcer food at length unite in peremptory order to the Bears—'Go now and den for a time.' First to obey are the females, that must have a good abode suitable for nursery, rather than a mere storm-break; last the males, that are content to rough it, and by being late to bed and early to rise, get that much more out of life. When the Winter King is reigning supreme, there is no evidence of such a thing as a Blackbear in the woods.

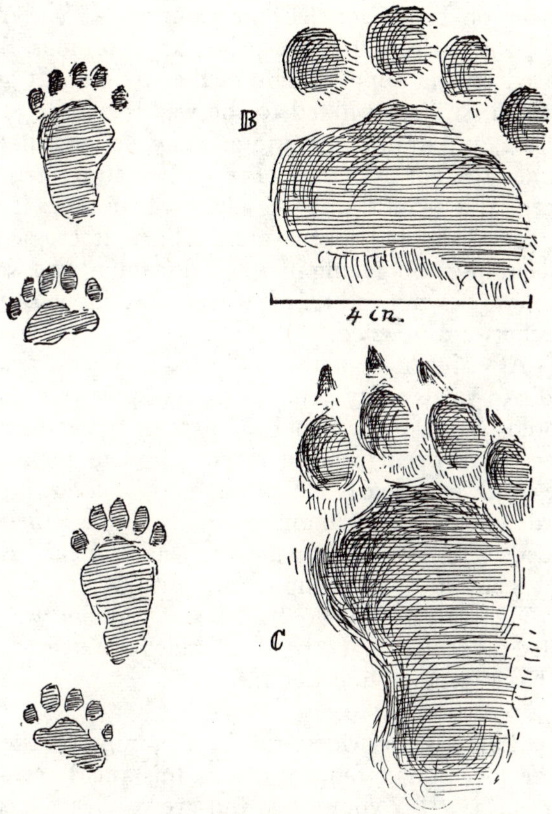

FIG. 247—Bear-tracks

The series on the left side are tracks of a large Kadiak Grizzly, sketched in Washington Zoo, after he had walked wet-footed across the dry asphalt. They illustrate the juxtaposition of all Bears' feet in walking.

The two large ones on the right are tracks of a Blackbear observed in the mountains of Colorado. *B*, the right front; *C*, the right hind-foot. These showed but 4 toes each and rarely any claws, although in all respects the individual was normal.

SPRING ADVENT

Countless Bears have been seen and slain in their winter dens, but I know of only one man who had the good luck to see an old Bear at the time when first it emerged in the spring. That man was Bert A. Dobson, my Adirondack guide.

It was in mid-April of 1906; he and his partner were log driving on Fish Creek, Lewis County, N. Y. As they sat on the bank waiting for the drive, an enormous Blackbear appeared on the opposite side of the stream. It gazed stupidly at them, sniffed, walked to the water, within 30 yards of the loggers, and drank long and greedily, so long that its front legs seemed tired of stooping. Presently it straightened up, sniffed, gazed, and again drank heavily. Three times it did this, till it seemed distended with water. Then it crossed the stream, still sniffing and gazing in a dazed fashion, and walked past the men. They noticed that it was an enormous Bear and that its cheeks were grizzled.

After it was gone they followed its back track in the snow, and 250 yards away came to a small pine tree that was deeply scored by the teeth of a Bear at 5 or 6 feet from the ground. Pieces had been torn out recently and to such an extent that the trunk was nearly cut through. On following the trail 100 yards farther, it led to an enormous hollow pine tree and there ended. This proved to be the winter den. Evidently they had seen the Bear taking its first drink.

This Bear, known by its size and its gray cheeks, was killed a year later at a place 2½ miles from the den, and is said to have weighed 400 pounds.

MEAT

There is the widest range of opinion on the fitness of the Bear for human food, and circumstances seem to justify all extremes. If a young but full-grown Bear, recently fattened on grains, roots, and berries, be quickly and mercifully killed and the carcass properly cleaned, skinned and prepared, the meat is excellent. But if an old Bear, tough, lean, or carrion-fed, or finally killed after a long pursuit in warm weather, the meat is little better than poison. The very dogs will pass it by in disgust; it has attractions for nothing but flies.

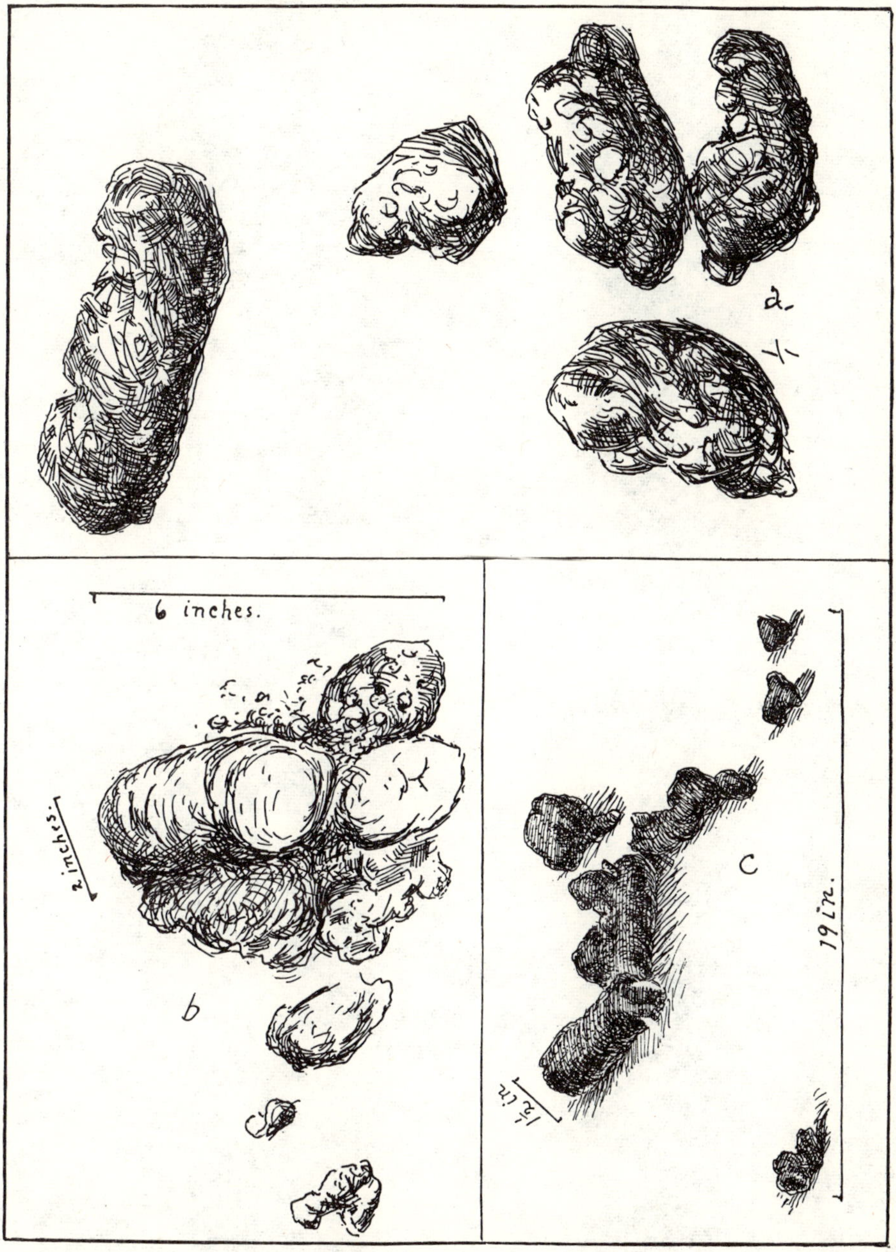

PLATE XCIX.—SCATOLOGY OF CERTAIN BEARS.

a. Blackbear. Yellowstone Park, August 7, 1897 (natural size). Chiefly grass with fruit and seeds.
b. Blackbear. Athabaska River, June 15, 1907 (one-third natural size). Chiefly fruit and leaves of arctostaphylos and poplar.
c. Grizzly Bear. Routt County, Colorado, September 14, 1901 (one-fifth natural size). Jet black, chiefly vegetable remains, but also much fine hair that might have been from a ground-squirrel or from the legs of a fawn. Each piece was crawling with maggots inside. The pile was many days, maybe two or three weeks old.

This Bear is a shy, inoffensive animal. A dangerous Blackbear is much less frequent than a dangerous dog or bull, and I am in favour of not only repealing all laws granting bounties for killing Bears, but of putting the Blackbear on the protected list, as a high-class game animal.

BUG-EATERS
ORDER INSECTIVORA

XLVIII.

Common Shrew, Cooper Shrew, or Masked Shrew.

Sorex personatus I. G. St. Hilaire.

(L. *Sorex*, a shrew; *personatus*, masked, probably because its eyes and ears are hidden.)

Sorex personatus I. GEOFF. ST. HILAIRE, 1827, Mém. Mus. His. Nat., Paris XV, p. 122.

TYPE LOCALITY.—Eastern United States, probably New York.

FRENCH CANADIAN, *la Musaraigne.*

The Family, *Soricidæ* or Shrews, comprises small *mouse-like creatures*, but most of them are smaller than any Mouse, and in anatomy as different from the Mice as a small Badger is from a big gray Rabbit. This we should realize if we could set together a Mouse and a Shrew, each magnified to the size of a sheep. Their most striking peculiarity is the absence, or apparent absence, of eyes and ears; next, their long sharp nose, and last, but of most importance, their teeth. On comparing the skulls of Mouse and Shrew we shall see more clearly the distinctive peculiarities of these. Even in color they differ, those of Mice being clear yellow or whitish, while those of Shrews are usually more or less stained with chestnut at the tips.

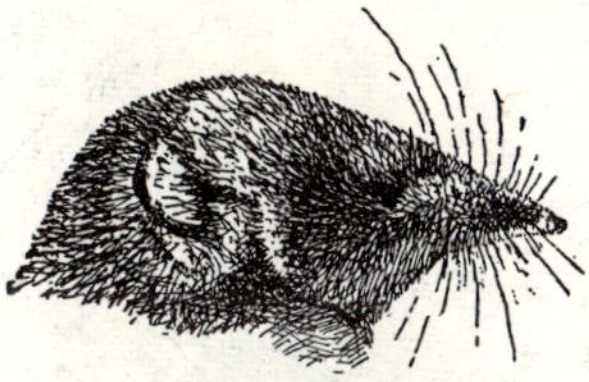

FIG. 248—Head of *S. personatus* to illustrate the mask. (Twice life size ad. nat.)

Side by side on a large scale (Figs. 249 and 250) the great divergencies of their skulls appear. They suggest the hippopotamus and crocodile. And every little bump and line has a meaning—is evidence and record of the old old fight. There are indeed wonderful histories written in these small teeth and

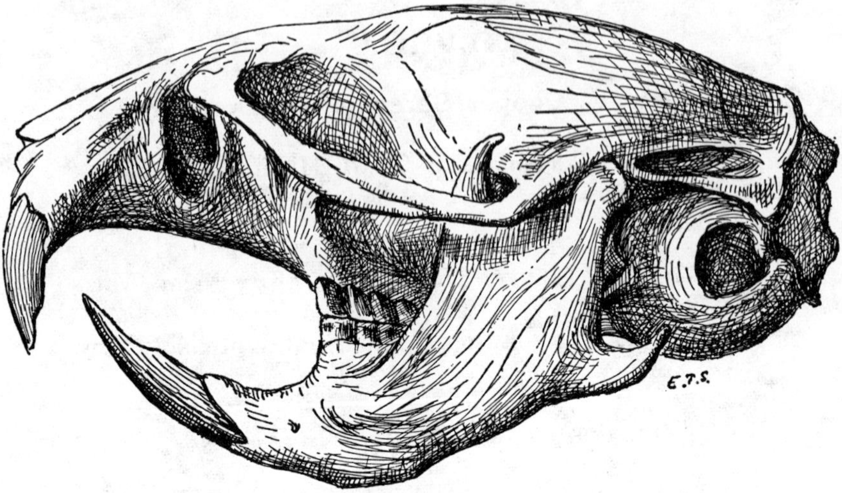

FIG. 249—Skull of Meadow-mouse (*M. pennsylvanicus*). (Five times natural size.)

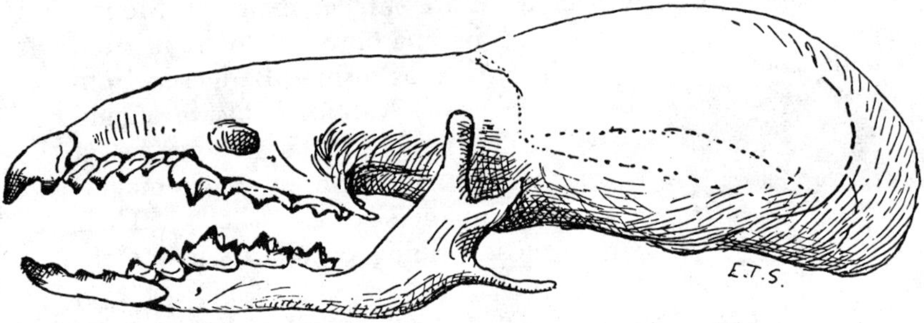

FIG. 250—Skull of Cooper Shrew (*S. personatus*). (Eight times natural size.)

skulls. Not in greater degree, perhaps, than in every tissue and fibre, but these are more lasting and preservable, and though the clumsy hand may destroy them by violence, it cannot tamper with them as records; there they are in their superlative delicacy and meaning for those who can read them.

And in so doing it is difficult to give them overvalue. The heedless crushing of one of these wonderful little caskets of information shocks the naturalist with much the same feeling as that an art connoisseur might experience on seeing some imbecile destroy a Rembrandt etching or a Tanagra figurine.

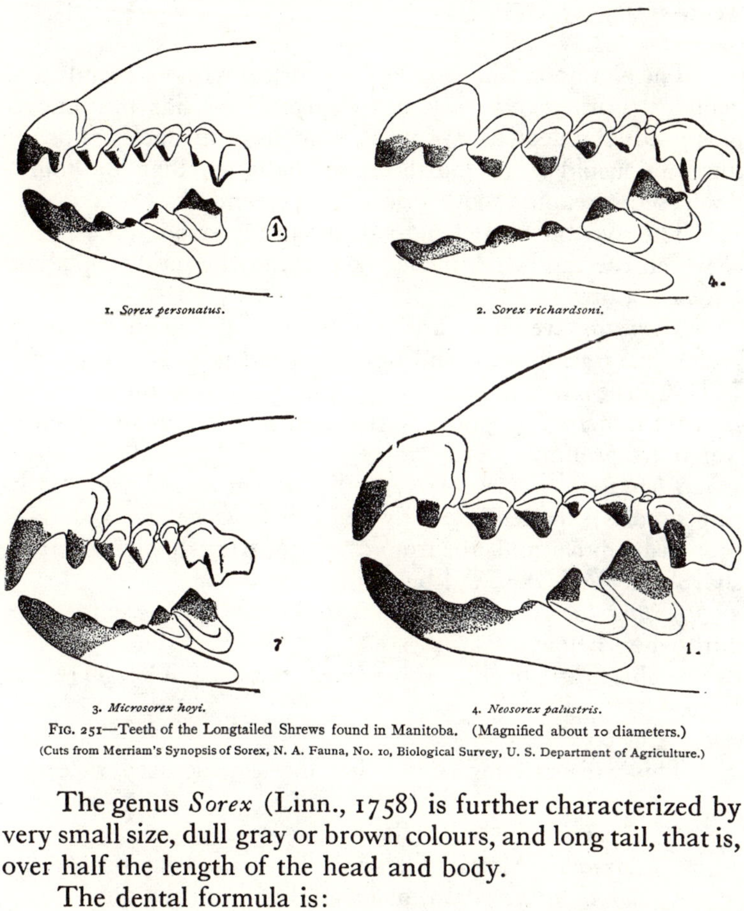

1. *Sorex personatus.* 2. *Sorex richardsoni.*

3. *Microsorex hoyi.* 4. *Neosorex palustris.*

FIG. 251—Teeth of the Longtailed Shrews found in Manitoba. (Magnified about 10 diameters.)

(Cuts from Merriam's Synopsis of Sorex, N. A. Fauna, No. 10, Biological Survey, U. S. Department of Agriculture.)

The genus *Sorex* (Linn., 1758) is further characterized by very small size, dull gray or brown colours, and long tail, that is, over half the length of the head and body.

The dental formula is:

$$\text{Inc. } \frac{4-4}{2-2}; \text{ can. } \frac{1-1}{0-0}; \text{ prem. } \frac{2-2}{1-1}; \text{ mol. } \frac{3-3}{3-3} = 32$$

But the teeth are so unlike those usually called by these names, that it is easy to wrongly identify them.

It is common to call the upper incisors (exclusive of the first) and the canines, *unicuspids*, because of their single point or cusp. The form of the teeth is greatly changed by wear.

The Common Shrew (Plate C) agrees with its Family and genus; it is further a very small species and has the unicuspids gradually decreasing in size from front to back; thus the third unicuspid (*i. e.*, fourth incisor) is larger than the fourth unicuspid or canine tooth (see Figs. 250 and 251—1).

SIZE The average total length is about $3\frac{15}{16}$ inches (100 mm.), of which the tail is $1\frac{9}{16}$ inches (40 mm.); the hind-foot about $\frac{1}{2}$ inch (13 mm.).

WEIGHT Merriam weighed an Adirondack specimen at 43.95 grains (2.85 grammes), and he considered it a very large one.[1] J. H. Linsley weighed one at 47 grains,[2] and Professor S. F. Baird another at 37 grains.[3] I found a Cos Cob specimen to weigh 3.1 grammes (47.8 grains).

COLOUR *The summer coat*, above, sepia brown mixed with dark-tipped hairs; shaded into ashy white or fawn below; tail, dark brownish above and all around at tip, whitish below, rather sharply defined, *i. e.*, tail bicolored.

Winter coat, more grayish above and lighter, sometimes pure white below.

A chestnut phase is sometimes found. The sexes are alike.

This wide ranging species has but 4 recognized races:

personatus G. St. Hilaire, the typical form.
streatori Merriam, slightly larger and darker.
arcticus Merriam, paler, and with longer tail.
miscix Bangs, larger and paler than true *personatus*.

[1] Mam. Adir., 1884, p. 174.
[2] Zoöl. N. Y., 1842, Pt. I, p. 23.
[3] Pac. R. Rep., 1857, Vol. VIII, p. 26.

MAP 57—RANGE OF THE COMMON SHREW AND ITS FOUR RACES.

Sorex personatus **I. G. St. Hilaire.**

The outlines show the theoretical range. The spots are actual records: they are chiefly from C. Hart Merriam's Synopsis, and G. S. Miller's Long-tailed Shrews, with others by Messrs. S. N. Rhoads, W. H. Osgood, E. A. Preble, A. P. Low, J. A. Allen, O Bangs, L. M. Turner, S. F. Baird, J. D. Figgins, J. Macoun, and E. T. Seton.

Life-history.

RANGE

This is the widest ranging of our Shrews, as will be seen by Map No. 57. It has been taken at Norway House and Turtle Mountain. I found it at Kenora, Carberry, Winnipeg, Pilot Mound, and Duck Mountain, so doubtless it is generally distributed throughout Manitoba.

INDIVIDUAL RANGE

We have but little evidence on the individual range. Analogy would lead one to believe that an acre was an ample kingdom for such a pigmy, but the light afforded by Nelson's notes on the Yukon, cited later, shows that it may travel a mile or two from home at certain seasons.

MIGRATION

These Alaska observations seem to point also to an extensive migration at the beginning of winter. It may be a regular migration or it may be a mere land rush for good claims whereon to settle for winter.

ENVIRONMENT

Although considered a woodland species, I found it abundant on the grassy prairies, near scrubby hollows and sloughs. It is but slightly subterranean and is incapable of climbing. Its favourite surroundings are in grassy tangles and brushy labyrinths of roots and fallen branches by the side of streams or ponds; it is never found far from water and yet it is not in the least aquatic.

Its perfect fitting into its allotted surroundings is thus graphically pictured by Merriam:[4]

"The naturalist well knows that, however cautiously he may walk, the stir of his footstep puts to flight many forms of life that will reappear as soon as quiet is restored; therefore, in his excursions through the woods, he waits and watches, frequently stopping to listen and observe. While thus occupied it sometimes happens that a slight rustling reaches his ear. There is no wind, but the eye rests upon a fallen leaf that seems

[4] Mam. Adir., 1884, p. 173.

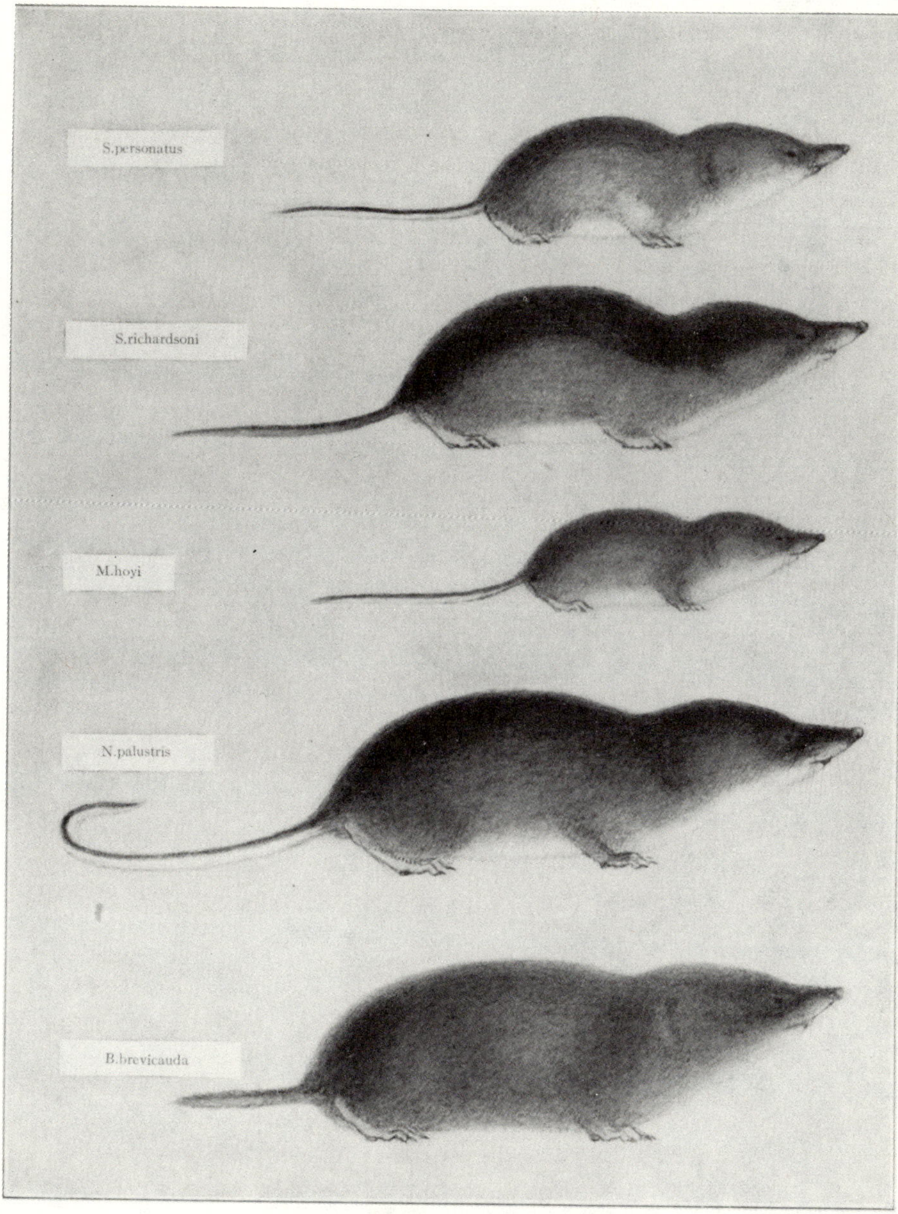

PLATE C.—THE SHREWS FOUND IN MANITOBA.
Drawn by E. T. Seton.
From specimens supplied by the Biological Survey, U. S. Dep. Agr.

to move. Presently another stirs, and perhaps a third turns completely over. Then something evanescent, like the shadow of an embryonic Mouse, appears and vanishes before the retina can catch its perfect image. Anon, the restless phantom flits across an open space, leaving no trace behind. But a charge of fine shot, dropped with quick aim upon the next leaf that moves, will usually solve the mystery. The author of the perplexing commotion is found to be a curious, sharp-nosed creature, no bigger than one's little finger, and weighing hardly more than half a dram. Its ceaseless activity and the rapidity with which it darts from place to place is truly astonishing, and rarely permits the observer a correct impression of its form."

At Carberry it was the most abundant of the Shrews. I captured a dozen of them in 6 traps during August, 1884, the traps being kept in the same places along the edge of a slough half a mile in extent. NUMBERS

In the Adirondacks they abound, and Merriam speaks of killing 11 in one day under haycocks that had been standing a few days in the rain.

Like most of our small mammals, it is subject to periodic fluctuation of numbers without evident cause. In 1882 and 1884, it was unusually numerous in Manitoba.

The human ear can hardly hear the high-pitched squeaking that is the only known sound of this small beast, and there is every reason to believe that this is varied to express their simple emotions, and even to serve as a song. VOICE

Many times in the quiet summer and early autumn evenings I have heard in the low thickets about the sloughs of Manitoba, a faint small voice, a twittering, so fine and high that it was not easily traced. It could not have been far away, and no bird was seen to suggest the singer. I think it probable that this was the song of a *Sorex*.

Most, perhaps all, our Shrews have smell-glands on their sides or lower parts; the extent and purpose of these have not yet been worked out, and are not likely to be very soon, for they are at present beyond our sense power to gauge or analyze. The genius who invents an odorometer will open a new and wonderful world to the ken of man.

SOCIABILITY

Although so numerous, the Masked Shrew is neither gregarious nor sociable. Indeed, it is unsociable to the point of ferocity. Whenever two of them fell into one of the pitfalls that I used for their capture, the stronger one invariably attacked and devoured his weaker brother.

Merriam also says:[5] "I once confined three of them under an ordinary tumbler. Almost immediately they commenced fighting, and in a few minutes one was slaughtered and eaten by the other two. Before night, one of these killed and ate its only surviving companion, and its abdomen was much distended by the meal."

The only record I can find of two being together without a murderous fight is that given by Herrick (quoted later), and these I presume were a pair.

DEN

This species is not known to tunnel or burrow. Its life is spent above ground, and its home-nest is in any sheltered spot, not too far from the level of the earth. Merriam says:[6]

"Whenever a tree or a large limb falls to the ground, these Shrews soon find it, examining every part with great care, and if a knot-hole or crevice is detected, leading to a cavity within, they are pretty sure to enter, carry in materials for a nest, and take formal possession. Hence their homes are not infrequently discovered and destroyed by the wood-chopper."

MATING

We have very little light on the mating of the Shrews. All the evidence goes to show that the species may mate and multiply with little regard to season; that is, young may be found at any time, except in winter.

Whether they pair or are polygamous or promiscuous is not known. But it is a general rule that animals which are

[5] Mam. Adir., 1884, p. 174.

[6] *Ibid.*, p. 174.

so ferociously quarrelsome and cannibal as these, do indeed pair, though the two continue together for a brief season only, perhaps an hour or two during the season of meridian ardour, then part to meet no more as friends.

The only facts bearing on the case I find in Herrick's "Mammals of Minnesota." His account is so charming, and new, that I give it in full.[7] Not the least important feature is the date—November.

"In November, 1883, the writer lay encamped under the canopy of the sky in Pine County, Minn., endeavouring to escape the chill of the frosty air by drawing the blanket close and hovering nearer the camp-fire. To a person alone in the woods for the first time after a long interval, every sound is novel and more or less charged with mystery. The wind stirred the tree tops, and impinging boughs clattered, and the trunks groaned under the tortion, each tree with its own doleful note. The few remaining pines added their sighing to the many melancholy sounds belonging to an autumn forest at night. But amid all the sounds nothing could be identified as coming from anything living, even the distant howling of Wolves was silenced, and I began to feel that the attempt to gain personal knowledge of the ways of the woodsy mammals by night study would prove futile, and composed myself to sleep. The half-somnolent reverie which forms the prelude to slumber was broken by faint melodious sounds on an excessively high key—so high that it seemed that I might be simply hearing the lower notes of an elfin symphony, the upper registers in which were beyond the powers of human ears to distinguish. The sounds were distinctly musical, and reminded me of the contented twitter of birds finding resting places among the boughs at night. Without moving, I turned my eyes upon the fire-lit circle, about which the darkness formed an apparently impenetrable wall. Only the most careful scrutiny enabled me to discover the tiny musicians. Within a few feet of my head, upon a decayed log, raced a pair of Shrews (*S. cooperi*),[8] so minute as to escape my observation at first. Up

[7] Mam. Minn., 1892, pp. 41–2.

[8] *S. cooperi* = *S. personatus*.

and down, with the most sprightly motions imaginable, they ran, twittering incessantly. Hither and thither they scampered, over my clothing and almost into my pockets, like veritable Lilliputians, seizing now a crumb of cheese, with which my traps were baited, and now a bit of fish fallen from my improvised supper table. During the eating the conversation was not interrupted. The little visitors were not bashful about criticising the housekeeping of their host, if their apparent amusement can thus be interpreted, but it was a most good-humoured little party, nevertheless, which thus unceremoniously ransacked my larder. The party increased in numbers and merriment, until I was almost forced to believe myself an uninvited guest within the magic circle of Queen Mab's domain. I watched with interest the result of their intrusion into the traps which stood about for the capture of any Red-backed Mouse that might invade my camp, but Sorex passed entirely within, and, daintily arching his back, contentedly nibbled the cheese; when the spring rose it usually suffered but a short fright, and returned to finish the interrupted meal. Canned fish seemed to be more acceptable than any other food I had to offer. Tiring of the watching, I again lay down to sleep, during which time elfin voices sounded in my dreams. About midnight one of the little imps sprang across my face in so violent a way as to partially awaken me, and thus, as good fortune had it, I was awake sufficiently to recognize the meaning of a sharp crack overhead, and sprang out of my bed in time to see it occupied by a massive tree trunk which the fire had burned off not far from the ground."

HABITS

These tiny creatures run by day as well as by night, and the only apparent use of their rudimentary eyes is to tell them when they are emerging from safe shadow into the open light with its great increase of danger.

I captured a number of them by making little pitfalls of pickle-jars, sunken level in runways, made by lifting a pole that had long lain in the rank grass, and found that the efficacy of this trap was greatly increased if a raised cover were put over each, so the jar was a shady place in the middle of an open run. The Shrew seemed to rush along the tunnel with more haste than discretion, and so was made prisoner.

When trapping for larger creatures and using a meat bait, one often finds the trap sprung without catching the meddler. In many cases I have traced the matter home to this little Shrew.

In the early spring, while yet the snow is deep everywhere, the sun-heat gathered by the long projecting grass, melts holes through the drift to the ground below. These shafts, with a strong stalk up the middle of each, afford a tempting opportunity to scramble into the big world. The little Shrews often yield to the temptation. They clamber up onto the vast expanse of snow and, setting out to explore, they forget the way back to the 'elevator-shaft,' and get lost. The snow is commonly crusted at this season, so they cannot burrow, and usually they die in a few hours, not from cold, but from hunger.

A number of curious notes on the subterranean migration of this Shrew, when the ground is covered with snow, are given by Nelson in his "Natural History of Alaska."[9]

"In fall [he says] the first severe weather brings them about the trading stations and native villages, and there they forage and penetrate every corner of the houses with all the persistence of the domestic Mouse. Scores of them were killed about our houses at Saint Michael every winter, and they were equally numerous at the other stations throughout the interior.

* * * * * * *

"After snowfalls they travel from place to place by forcing a passage under the snow, and frequently keep so near the surface that a slight ridge is left to mark their passage. On the ice of the Yukon I have traced a ridge of this kind over a mile, and was repeatedly surprised to see what a direct course the Shrews could make for long distances under the surface. These minute tunnels were noted again and again crossing the Yukon from bank to bank.

[9] Nat. Hist. Alaska, 1887, pp. 270–1.

"These little adventurers sometimes tunnel far out on the sea-ice, and the Norton Sound Eskimo have a curious superstition connected with such stray individuals. They claim that there is a kind of a water Shrew living on the ice at sea which is exactly like the common land Shrew in appearance, but which is endowed with demoniac quickness and power to work harm. If one of them is disturbed by a person it darts at the intruder, and burrowing under the skin, works about inside at random, and finally enters the heart and kills him. As a consequence of this belief, the hunters are in mortal terror if they chance to meet a Shrew on the ice at sea, and in one case that I knew of, a hunter stood immovable on the ice for several hours until a Shrew he happened to meet disappeared from sight, whereupon he hurried home, and his friends all agreed that he had had a very narrow escape."

ENEMIES Many hawks and owls are known to destroy this Shrew. At Carberry, September 29, 1884, I collected a great gray-owl whose capacious maw contained nothing but one of these mites. I recently received another specimen taken from the stomach of a rough-legged hawk at Winnipeg, October 21, 1907. Many beasts of prey will prey on this least of beasts; that is, they will kill it, but it has a defence somewhat like that of the Skunk—a most obnoxious smell, which usually makes them think twice before swallowing it. I have often found Shrews dead on the path with skulls crushed, but otherwise uninjured. These I believe were killed by Weasels or cats that chanced to catch them scampering by, but which, on second thought, could not stomach the rank, protective smell with which the Shrew had enveloped itself—a little too late.

Another enemy that will slay but hardly eat the *Sorex* is the Mink. Miller, in recording the capture of a remarkably large Mink at Peninsula Harbor, Ont., says:[10] "He had followed the water's edge closely most of the way, but occasionally had made short excursions up the beach in search of prey.

[10]Mam. Ont., Proc. Bost. Soc. Nat. Hist., 1897, p. 43.

On one of these side expeditions he had captured a Shrew (*Sorex personatus*) which had also left its track in the sand. The tracks of the two animals showed that the Shrew was not killed until he had led his enemy a sharply zigzag chase. The Mink left the tail and hind-quarters of the Shrew lying on the sand, and continued his way directly to my trap."

At Duck Mountain, Man., June 15, 1884, I found a dead Masked Shrew in a catbird's nest about six feet from the ground. As the Shrew could not have climbed there itself, I suspect that a jay or shrike had killed it for food, but changed his mind about eating it, on fully realizing the foul deed he had done.

In my Journal for 1882 I find a note that may record a similar incident:

October 27, 1882, while examining an old stump in the woods to the north of Carberry, I met with an excellent illustration of the aptitude of the Spanish name for the woodpecker, 'Il Carpentero,' as applied to our flicker. I mean in the sense of its being a worker in wood and house-provider for others. The history of the case was briefly this, as far as the circumstantial evidence revealed it:

First came the hard-working flicker and excavated the hole, perhaps while yet the stump was sound, and in the years that followed we know not how many young flickers cracked their glass-like shells in this narrow chamber; and after the flickers came no more it was taken by some bird, a grakle perhaps, that, like the 'foolish man, founded his nest on sand,' finishing its superstructure with mud, sticks, and straw. Next came a new possessor, who built a strong shapely nest of moss and mud; but for the situation it might have been the work of a robin. Lastly, this many-storied tenement house became the eyrie of a sparrowhawk, whose household furniture of straw and moss reached half-way up to the doorway. A strange tale of a hole, surely; but there was more yet to be learned from the old stub, and, allowing fullest weight to circumstantial evidence and accepting the supposititious as fact, I may be allowed to relate, as a matter of established history, that on a certain day

Sir *Falco sparverius* brought home to his brood a tiny Shrew, of the species yclept by scientists the *Sorex personatus*. Now, it chanced that the young hopefuls of the robber baron were not just then very hungry—oh! marvellous chance—so that the *Sorex personatus*, being left to his own devices, set about to escape, and so far succeeded that he burrowed down through the home effects of the kestrel and the moss-builder, but here the hard mud floor barred further progress, and the poor little captive, weary and wounded, soon died in the buried nest; and there I found him, like Ginevra in the oaken chest, when long afterwards I broke open the rotten timber and made it disclose a tragic tale that, maybe, never happened at all.

FOOD

The food of this Shrew is chiefly insects and worms, with a welcome variation when flesh meat of any kind happens to fall in its way. There is reason to believe that it will occasionally kill and devour Field-mice, and, as we have seen, it is very ready when in captivity to destroy for food the weaker ones of its own species. As related, Merriam put three of them together. One was killed and eaten by his companions; then the stronger of the survivors served the other in the same way. "Hence," he says, "in less than eight hours one of these tiny wild beasts had attacked, overcome, and ravenously consumed two of its own species, each as large and heavy as itself. The functions of digestion, assimilation, and the elimination of waste are performed with wonderful rapidity, and it seems incomprehensible that they should be able to procure sufficient animal food to sustain them during our long and severe winters; indeed, I incline to believe that their diet is more comprehensive than most writers suppose, and that they feed upon beechnuts and a variety of seeds, and possibly roots as well, though I confess that I have no direct evidence to adduce in support of this supposition."

There is, however, much indirect evidence. Some allied species and many carnivores are known to do this very thing when they fail to find the necessary rations of flesh meat.

The appetites of Shrews are enormous and their digestion is rapid. No doubt they eat their own weight of food every twenty-four hours; and they die of starvation in half a day.

They do not hibernate, neither are they known to lay up a store of food, so the problem of food-supply must keep them eternally vigilant under the snow.

XLIX.

Richardson Shrew, Black-backed or Saddle-backed Shrew.

Sorex richardsoni Bachman.

Sorex richardsonii BACH., 1837. Jour. Acad. Nat. Sci., Phila., VII, pt. II, 383, Pl. XXIV, Fig. 5.

TYPE LOCALITY.—Probably plains of Saskatchewan.

FRENCH CANADIAN, *la Musaraigne de Richardson.*

In general, this species (see Plate C) combines large size and a rather short tail, with the family and generic characters. It has the third unicuspid larger than the fourth (see Fig. 251—2), and its fur is of the tricolour style, that is, *in bands, back dark, sides lighter, below lightest*; each colour rather abruptly defined against the next.

SIZE Total length, about $4\frac{1}{2}$ inches (114 mm.); tail, $1\frac{9}{16}$ inches (40 mm.); hind-foot, $\frac{9}{16}$ inch (14 mm.).

COLOUR Above, clear dark brown not sprinkled with hoary; sides, dull yellowish-brown contrasted against back and belly colour; below, pale brownish-white; tail, dark above, also below on tip third, else below, pale brown, that is, bicoloured with the colours of the flanks and back.

LIFE-HISTORY.

RANGE Very little is known about the range of this Shrew. The map shows all the reliable records. I secured 3 specimens in Manitoba, 2 at Carberry, and 1 at Shoal Lake. Preble got 9

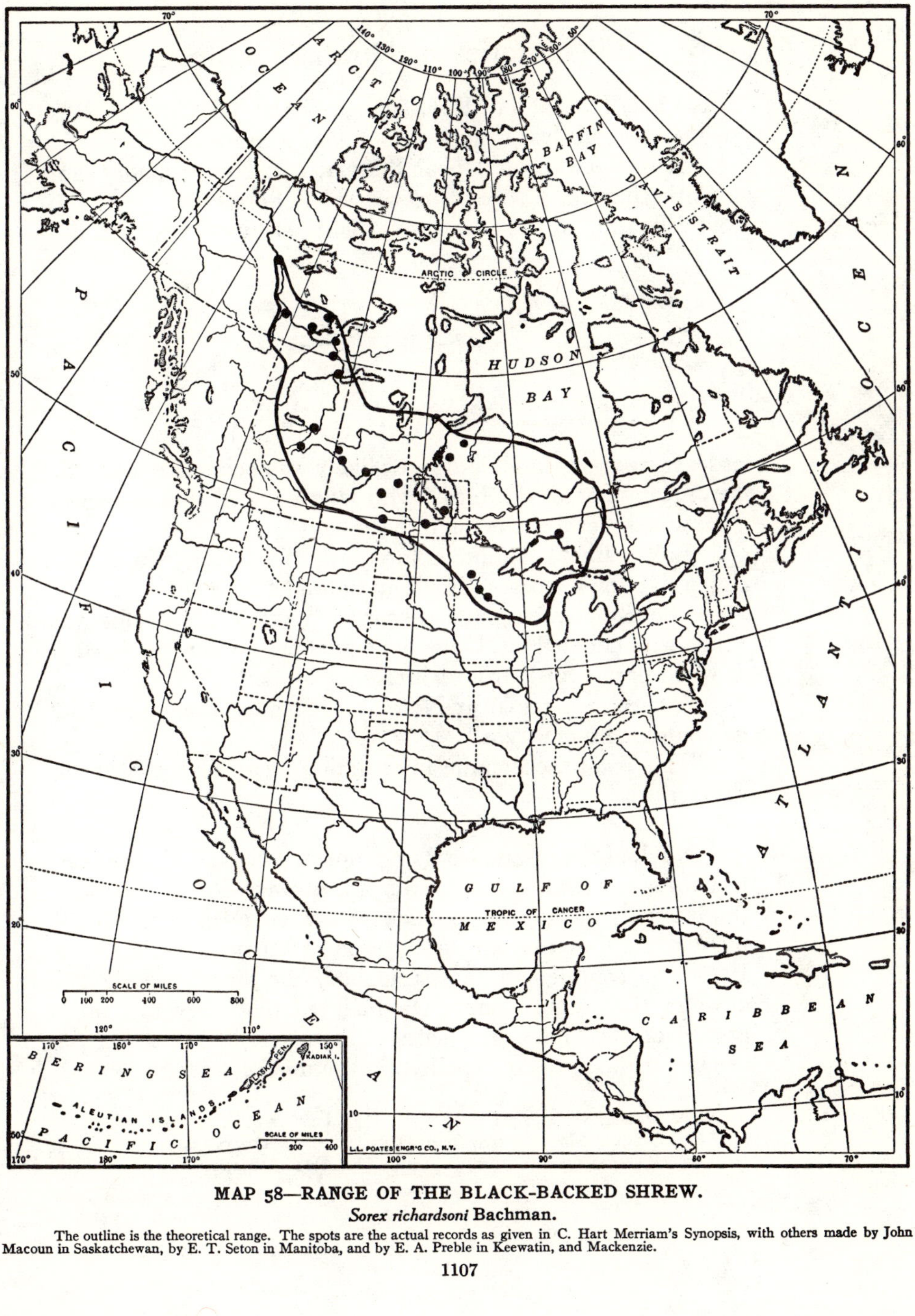

MAP 58—RANGE OF THE BLACK-BACKED SHREW.

Sorex richardsoni **Bachman.**

The outline is the theoretical range. The spots are the actual records as given in C. Hart Merriam's Synopsis, with others made by John Macoun in Saskatchewan, by E. T. Seton in Manitoba, and by E. A. Preble in Keewatin, and Mackenzie.

at Norway House[1] and E. Hollis got one at Touchwood Hills,[2] so that it doubtless ranges throughout Manitoba.

ENVIRONMENT

It is always found near the water and is probably more aquatic than the Cooper Shrew. The Shoal Lake specimen I found floating in the water in a night-heronry. It was a mile

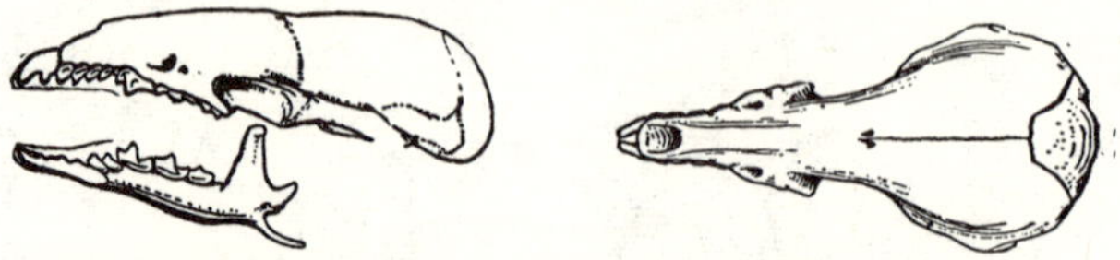

FIG. 252—Skull of Richardson Shrew (*Sorex richardsoni*). (Double natural size.) From Merriam's Synopsis. Plate VI, N. A. Fauna 10, 1895. Biological Survey, U. S. Dept. Agr.

at least from dry land, and most likely was being carried to the heronry by one of the birds, who later decided that the specimen was not savoury enough to eat.

Bailey found it quite common at Elk River, Minn., and much like the Common Shrew in habits.[3]

The above-mentioned specimen, recorded by Hollis, was taken with cheese bait, November 7, after the snow had fallen. It was but ten yards from water.

A specimen taken at Norway House by Preble on June 22 contained 6 embryos.[4]

This is all the definite information I can find concerning the species.

ITS CONGENER

No doubt it resembles its European congener (*S. araneus*) in habits as closely as it does in appearance. Bell describes[5] the latter as feeding on mollusks, worms, and insects, and so pugnacious that two are rarely seen together except fighting. If two be confined in the same box, the stronger kills and devours the weaker one. He believes that they are much preyed on by Moles and owls. Cats also will kill them, but rarely eat them, being repelled by the musky smell.

[1] N. A. Fauna, No. 22, 1902, p. 73.
[2] Zoölogist, August 15, 1902, p. 297.
[3] Rep. Orn. and Mam. for 1887 (pub. 1888), U. S. Dep. Agr., p. 435.
[4] *Loc. cit.*, see Note 1.
[5] Br. Quad., 1874, pp. 143–4.

L.

Pigmy Shrew, or Hoy Shrew.

Microsorex hoyi (Baird).

(Gr. *micros*, small; L. *sorex*, a shrew; *hoyi* of Dr. P. R. Hoy who discovered it.)

Sorex hoyi BAIRD, 1857, Mam. N. A., p. 32.
Microsorex hoyi ELLIOT, 1901, Syn. Mam. N. A., Field. Mus., Zoöl. Ser., Pub. No. 45, Vol. II, p. 377.

TYPE LOCALITY.—Racine, Wis.

FRENCH CANADIAN, *la Musaraigne de Hoy*.

The genus *Microsorex* (Baird, 1877) may be known by its small size and peculiar teeth which resemble those of *Sorex* in number but differ in their proportionate sizes, etc. (see Fig. 251—3). Its third upper unicuspid is very minute and nearly hidden between the second and fourth. It has on the inner side of the first and second unicuspids a *distinct secondary cusp*. Its mandible is short and heavy; the feet are *not* fringed with bristles, that is, they resemble those of the Common Shrew.

At first glance the Hoy Shrew (see Plate C) looks like a very small Masked Shrew, but the generic characters distinguish it.

Total length, about 3⅜ inches (85 mm.); tail, 1⅛ inches (29 mm.); hind-foot, ⅜ inch (10 mm.). SIZE

Above, dull brown shaded gradually into pale gray below; chest, usually tinged with rusty; tail, faintly bicoloured. Every word of which will apply equally well to COLOUR

MAP 59—RANGE OF HOY SHREW AND ITS THREE RACES.

Sorex hoyi **Baird.**

The outline shows the theoretical range. All the records I can find are spotted on the map. They are from C. Hart Merriam's Synopsis, G. S. Miller's Mam. Ont., S. F. Baird's Mammals, E. A. Preble's Keewatin, etc., W. H. Osgood's Cook's Inlet, C. C. Adams, Northern Michigan; specimens in Field Museum.

personatus, so that we must rely on size and dentition for identification.

Three races are recognized:

hoyi Baird, the typical form.
eximius Osgood, larger and paler.
alnorum Preble, like *hoyi*, but larger, and lower parts not tinged with buffy.

This, the least of our beasts, was first discovered at Racine, Wis., but has since been found in various localities, from Nova Scotia to British Columbia, as shown on the map, No. 59.

Dr. C. Hart Merriam records[1] a specimen from Red River Settlement. On this rests its claim to being Manitoban.

Little is known of its habits, but probably they differ little from those of *Sorex personatus*.

The following brief account is the longest I can find:[2]

"Hoy's Shrew avoids bogs and heavy woods. At North Bay I invariably found it in dry clearings and gardens. Several fell into pitfalls dug in a garden and others entered traps set beneath stumps in a meadow. The one taken at Peninsula Harbor was found by a dog under the rotting trunk of a small tree in an open upland prairie.

"A female taken at North Bay on August 22 has only 4 mammæ; all inguinal. In this character it differs from *Sorex personatus* and agrees with *Blarina brevicauda*. A reduction in the number of mammæ is probably characteristic of the subgenus *Microsorex*." (*Miller.*)

[1] N. A. Fauna, No. 10, p. 90, 1895.

[2] Mam. Ont., 1897, p. 37.

LI.

Marsh-shrew, Water-shrew, or Black-and-white Shrew.

Neosorex palustris (Richardson).

(Gr. *neos*, new; L. *sorex*, a shrew; L. *palustris*, of marshes.)

Sorex palustris RICH., 1828, Zoöl. Journ., III, No. 12, p. 517.
Neosorex palustris ELLIOT, 1901, Syn. Mam. N. A., Field Mus., Zoöl. Ser., Pub. 45, Vol. II, p. 378.
TYPE LOCALITY.—Region between Hudson Bay and Rocky Mountains.

FRENCH CANADIAN, *la Musaraigne de Marais*.

The genus *Neosorex* (Baird, 1857) is much like *Sorex*, but differs in having greater size, different colour pattern, and peculiar feet; these are adapted for swimming, having beautiful white fringes of bristle-like hair; they are, indeed, much like the feet of a Muskrat in miniature.

Its teeth differ somewhat from those of *Sorex* (see Fig. 251—4); but many consider the differences to be only subgeneric. It is easily distinguished from the other Long-tailed Shrews in Manitoba by its much greater size and black-and-white style of colouration.

SIZE Total length, 6⅛ inches (155 mm.); tail, 2 9/16 inches (65 mm.); hind-foot, ¾ inch (19 mm.).

COLOUR All above dusky brown, or very dark gray, sprinkled with hoary; below white, silvery in some lights; dorsal and ventral areas rather sharply defined from snout to tail root; tail, bicoloured, blackish above and all around near tip, white below.

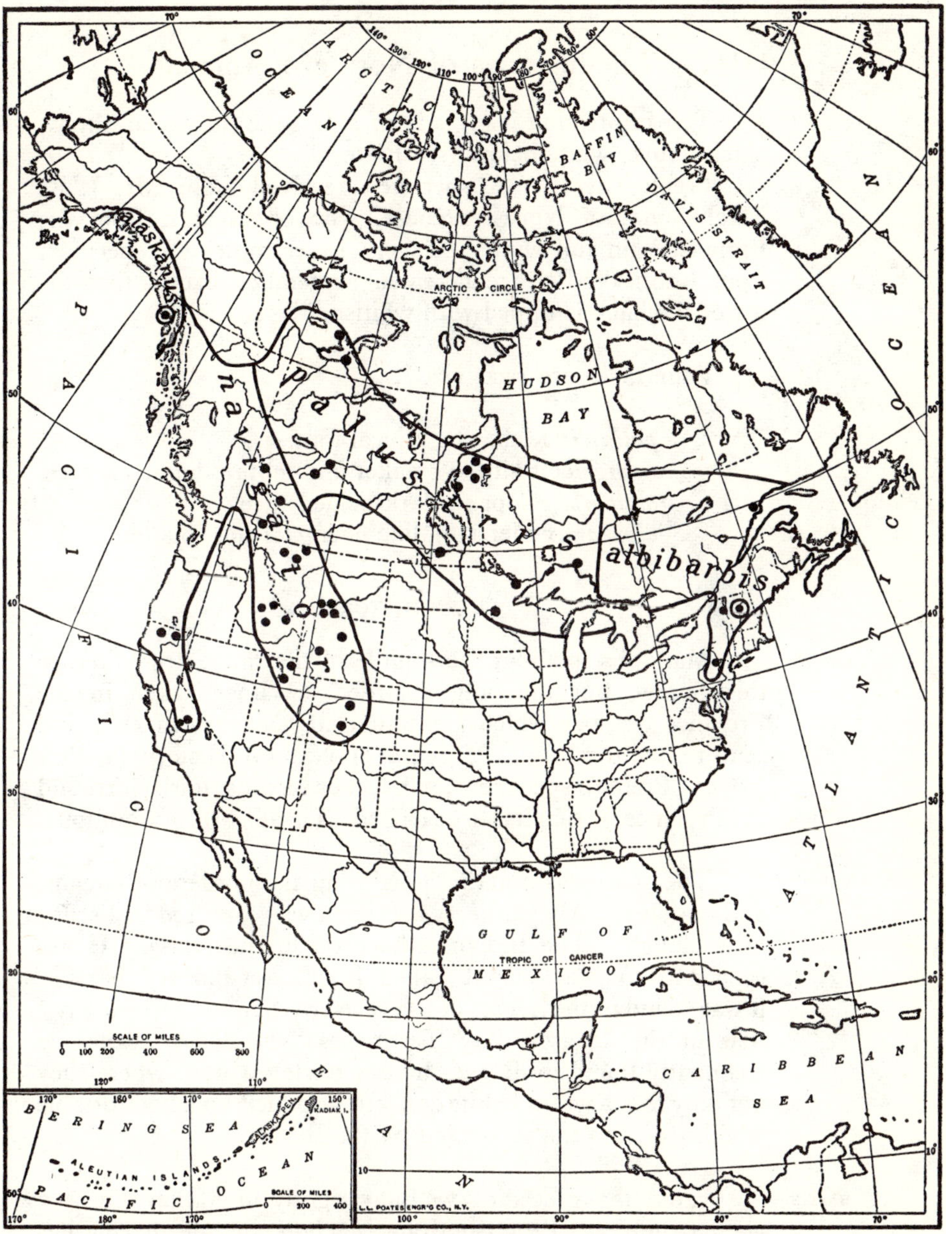

MAP 60—RANGE OF THE WATER-SHREW AND ITS FOUR RACES.

Neosorex palustris **(Richardson).**

The outlines enclose the theoretical range. The spots are the actual records; they are chiefly from C. Hart Merriam's Synopsis, with others by S. N. Rhoads, E. A. Preble, J. Macoun, and E. T. Seton.

The impression it gives is of a large Long-tailed Shrew, black above, white below (see Plate C).

Preble says:[1] "A specimen received from Mr. J. K. MacDonald, of Norway House, who obtained it from the Indians, is in full winter pelage. It resembles the specimen from Robinson Portage taken in September, except that the fur of the back is tipped with whitish."

Four races are recognized:

palustris Rich., the typical form.
navigator Baird, a smaller and more plumbeous race.
albibarbis Cope, a dark-bellied race.
alaskanus Merriam, like *navigator* but smaller.

Life-history.

RANGE Manitoba is about the middle of the range accredited to this Shrew, but the only recorded specimen taken in the Province was one which I captured at Carberry in 1884 and sent to the American Museum of Natural History in New York. Preble's specimens, with other records, fairly surround the Province, so that it is to be looked for in all parts of Manitoba.

ENVIRONMENT The species is aquatic, inhabiting the borders of streams and marshes. My Carberry specimen was captured in a sedge by a slough. The Indians who brought the Norway House specimen to MacDonald called it 'Beaver-mouse,' because it was found in the houses of the Beaver. Samuel Hearne says[2] that in the Hudson Bay Territories "the Shrew Mouse is frequently found in Beaver houses during winter, where they not only find a warm habitation, but also pick up a comfortable livelihood from the scraps left by the Beaver."

HABITS From the preceding we may rightly infer that it was a good swimmer. Professor John Macoun tells me that he has

[1] N. A. Fauna, No. 22, 1902, pp. 71–2. [2] Journey, 1795, p. 386.

seen it in the clear open waters of a mountain brook at Crow's Nest Pass, B. C. It darted about swiftly in the current, without apparent effort, the snout and back only out.

V. Bailey found it rather scarce at Elk River, Minn., and adds,[3] "I have always found them living in holes in creek banks; in the spring of 1886 a neighbour caught and gave me one that he found swimming in a small pond of snow water in a hollow near his house."

My Carberry specimen was captured August 28, 1884, in the runway of a Marsh-mouse, and I doubt not it preys on them regularly. It was a female evidently suckling young.

Here our knowledge of its habits ends. The best we can do is turn to the uncertain light of analogy.

The Crossopus is a European congener that may help in this. It is a species that leads the life of a miniature Otter, swimming and diving with the greatest ease, and taking to the water as an escape from its enemies. It feeds on flesh, insects, and mollusks, and nests in a hole in a bank by the water.

According to Bell[4] it produces 6 young at a brood.

[3] Rep. Orn. Mam. U. S. Dep. Agr. for 1887 (1888), p. 435.
[4] Br. Quad., 1874, p. 154.

LII.

The Short-tailed Shrew or Mole-shrew.

Blarina brevicauda (Say).

(*Blarina*, a name made up by Gray in 1837; *brevicauda*, from L. *brevis*, short, and *cauda*, a tail.)

Sorex brevicaudus SAY, 1823, Long's Exp. Rocky Mts., I, p. 164.
Blarina brevicauda BAIRD, 1857, Mam. N. A., pp. 42–45.
TYPE LOCALITY.—Near Blair, Neb.

FRENCH CANADIAN, *la Taupe Musaraigne*.
OJIB. & CREE, *Kin'-skee-sha-wah-wah-bee-gah-note'-see* (sharp-nosed, short-tailed Field-mouse).

The genus *Blarina* (Gray, 1837) differs from *Sorex* exteriorly in its much shorter tail. The latter is about one-fourth the length of the head and body. The tooth formula is as in *Sorex*, but the relative proportions of the unicuspids are different; the first two are very large, and the next two much smaller. The fifth unicuspid, that is, the canine, or sixth tooth from the front, is so small as to be hard to see. These peculiarities are very marked in *Blarina brevicauda* (see Fig. 253 and Plate C).

SIZE

In total length it is about 5 inches (127 mm.). Tail vertebræ, 1 inch (25 mm.); hind-foot, ⅝ inch (16 mm.). It is largest in the type region, and smaller in the northern, southern, and eastern parts of its range, and especially so on the Atlantic coast.

A large male taken at Cos Cob, October 24, 1908, weighed 22.3 grammes; an adult female taken at same place, September 23, weighed 17.1 grammes. WEIGHT

All above, dark brownish-gray, shading on the under parts into a much paler tint of the same colour; everywhere the coat is silky and glossy. The under-fur is lead colour. In general, summer coat is palest. Sexes alike. COLOUR

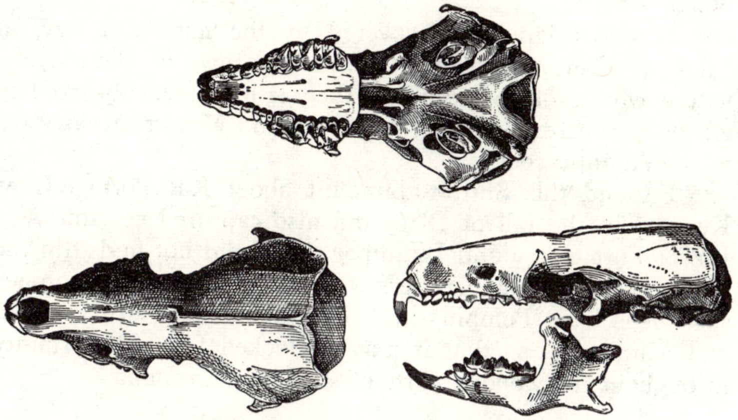

FIG. 253—Skull of *Blarina brevicauda*. (Double natural size.)
(Cuts from Merriam's Revision, N. A. Fauna, No. 10, Plate I. Biol. Surv., U. S. Dept. Agr., 1895.)

When seen alive it looks like a lead-coloured bobtailed Mouse, without eyes or ears. Its superior size and very short tail will easily distinguish it from others of our Shrews.

There are six recognized races:

brevicauda Say, the typical form.
carolinensis Bach., much smaller.
hulophaga Elliot, very small, pale, and short-tailed.
peninsulæ Merriam, like *carolinensis*, but with larger hind-foot and more slaty colour.
aloga Bangs, a small pale brown form from Martha's Vineyard.
compacta Bangs, a small slaty form from Nantucket.

Life-history.

RANGE The map, No. 61, shows the range of the species to be in the forested part of the moist temperate region of eastern North America.

It was originally discovered by the naturalist Say, at Engineer Cantonment, near the present town of Blair, Neb., in the winter of 1819. Since then it has been observed in all the region from western Nebraska and Manitoba, eastward to the Atlantic coast.

I found this Shrew abundant about Rat Portage (now Kenora) in the fall of 1886, and also captured specimens at Lower Fort Garry and Winnipeg, but I did not find it in the prairie region about Carberry. Dr. Merriam has recorded[1] specimens from Pembina and many points in North Dakota and Minnesota, so that it may be looked for with certainty throughout the wooded parts of southern Manitoba.

ENVIRONMENT All the eyeless bug-hunters that form the present group of Insectivora are creatures that prey where eyes are of little use, which means under brush, moss, or ground. If we set them in a scale of subterraneousness, a scale corresponding with their degree of eyelessness, we shall put:

1st. The Long-tailed Shrew, that prowls in labyrinth and thicket, but rarely digs.

2nd. The Blarina, that hunts still lower, a threader of mouse-tunnels, and a digger in moss, fallen leaves, and loamy soil, like an inexperienced Mole.

3rd. The Star-nosed Mole that digs in moss, softer mould, and occasionally in mud.

4th. The true Mole that never ceases to work for a living by tunnelling in the ground, be it never so hard.

The favourite localities of the Blarina are woodlands, under log-piles, and among tangles of brush and sedgy grass,

[1] N. A. Fauna, No. 10, 1895, p. 13.

MAP 61—RANGE OF THE SHORT-TAILED SHREW AND ITS SIX RACES.

Blarina brevicauda (Say)

The outlines enclose the theoretical range, fairly well established in the south and east. The spots are the actual records; they are chiefly from C. Hart Merriam's Revision, with others by O. Bangs, G. S. Miller, S. N. Rhoads, V. Bailey, D. G. Elliot, and E. T. Seton.

along streams and in hardwood bush; here they live somewhat after the manner of a Field-mouse, but also digging and tunnelling in a way that recalls the Moles.

The furrowed—sometimes tunnelled—track that this animal leaves in snow is an exact expression of its methods and of its summer life beneath the leaves and rubbish in the woods.

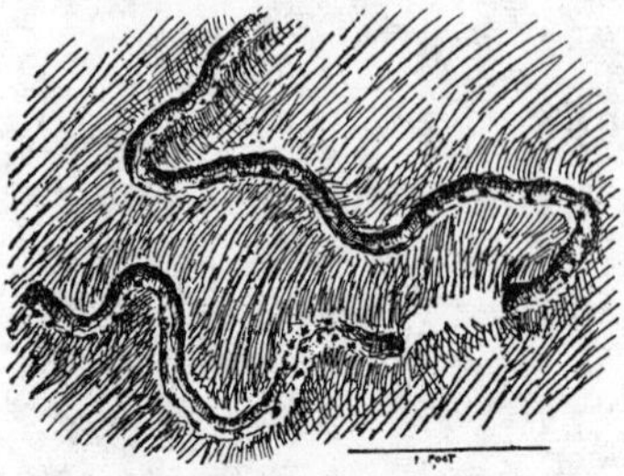
FIG. 254—The furrowed trail of the Mole-shrew or Blarina.

I never realized this fully until I chanced to see one in its daily life at Cos Cob. The note in my journal runs thus:

June 30, 1905.—This morning at nine I sat watching a Hare in the woods; there was a rustling of leaves. Then I saw the dry brown carpet near me lifting along a crooked line. At first I thought it must be a Chipmunk driving a new air-shaft from below, but the rustling continued. At one point a sharp nose appeared and worked about in the air, then speedily was withdrawn. The heaving of the leaf-bed continued at the rate of a very slow walk; then at a bare place the heaver emerged for a moment. It was a Blarina. He disappeared at once under the next leaf-bed, and so went on burrowing his way, not mole-like, in the earth, but in a fashion of his own beneath the leaves. Twenty-five feet farther I lost all trace of the leaf-heaver. I examined his trail, but found no tunnel; all had closed behind him. Free as a Mole in the soil, he drove his sub-leaf gangway where he would, and doubtless lived on the country as he went. This, then, was his way of life—this little inter-world betwixt floor and carpet was for him; and thus I learned why he had bartered his eyesight for keener powers of smell and touch.

Less aquatic probably than either Marsh-shrew or Star-nosed Mole, the Blarina is nevertheless rarely far from water. All that I have seen or taken were within 100 yards of a stream or pond, and most of them on the water's very edge.

NUMBERS

With 6 traps set for two weeks in a sedgy corner of an island at Lake of the Woods, I captured 5 of this species and had evidence of others remaining. As the sedgy tract was

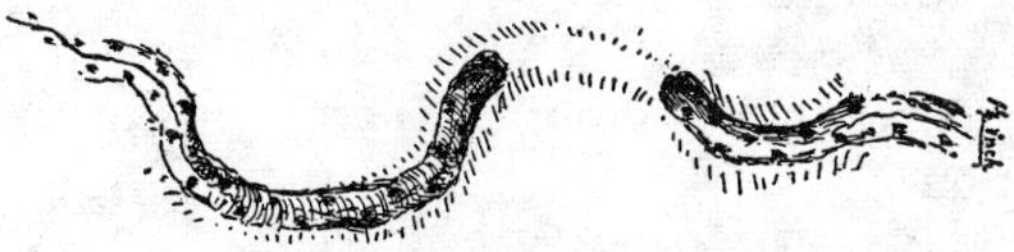

FIG. 255—Portion of Blarina labyrinth on snow.

less than two acres in extent, this furnishes a hint of their possible numbers.

One day in January, 1907, after a fresh fall of snow, I walked for a mile through the woods at Cos Cob, and found labyrinths of fresh Blarina tunnels about every fifty yards.

There was doubtless a Blarina for each system of tunnels, and many of the species were probably not yet represented, as it was but three or four hours since the fresh snow came. This, therefore, affords a minimum gauge of the creature's numbers.

In late September, 1908, I noticed a great many small, round holes opening from underground galleries. They were sometimes 1 inch across, sharp, round, and opened from below; sometimes 1¼ inches with a little loose earth scratched

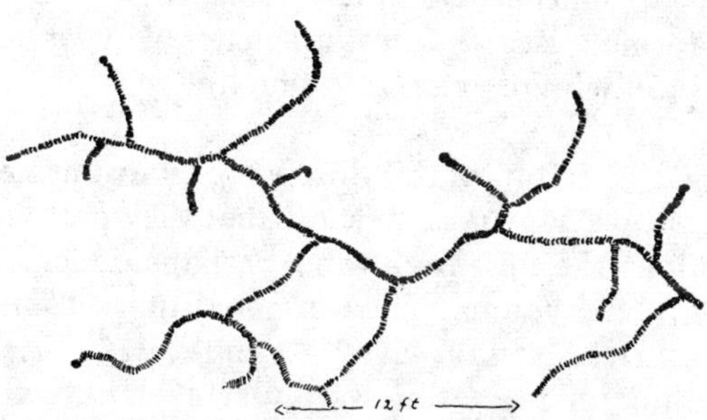

FIG. 256—Blarina labyrinth on snow. March 6, 1907. Cos Cob, Conn.
The black spots are plunge holes straight down.

out. The former went down plumb, the others sloping. They were so numerous that I found 6 along a little woodland pathway 94 yards in length; 6 traps set by these resulted after 3 days in the capture of 8 Blarinas. At another place they were even more plentiful, as I counted 17 holes in a space 8 by 10 feet. In this woods there were certainly 50 Blarinas to the acre.

SOCIABILITY Like the rest of the group, this animal is neither sociable nor gregarious.

TUNNELS The tunnel inhabited by a pair of these animals in October, 1908, is shown in Fig. 257. I found no nest. When I reached the place near the lowest stone on the diagram, and noticed first the *plugging of the gallery*, second the *plunge hole*, I thought that the nest was close at hand, as these are the usual indications. But the plunge ended on a hard rock.

This gallery was everywhere 1 to 2 inches down, and 1¼ to 2 inches wide.

NESTS There is evidence that the species not only makes a warm nest for its young, under a log or stump below ground, but also keeps up a soft and comfortable home for itself the year round. A. F. Shull, whose studies have shed so much light on this Shrew, says[2] that the Blarina nest differs from that of a *Microtus*—thus, the former uses the nesting material as it finds it, the latter tears and shreds it up into fine lint.

BREEDING The mating habits of this Shrew are almost unknown. The little evidence we have goes to show that they pair, but that the male abandons the nest for a time, perhaps under pressure, on the arrival of the young. The first pairing season occurs in early April. Dr. Merriam says:[3] "On the 22d of April, 1878, I found a couple of these Shrews under a plank-walk near my museum. They proved to be male and female, and the latter contained young, which, from their size, would probably have

[2] Habits of the Short-tailed Shrew, *Blarina brevicauda* (Say), Am. Nat., August, 1907, pp. 495–522, 5 illustrations.

[3] Mam. Adir., 1884, p. 172.

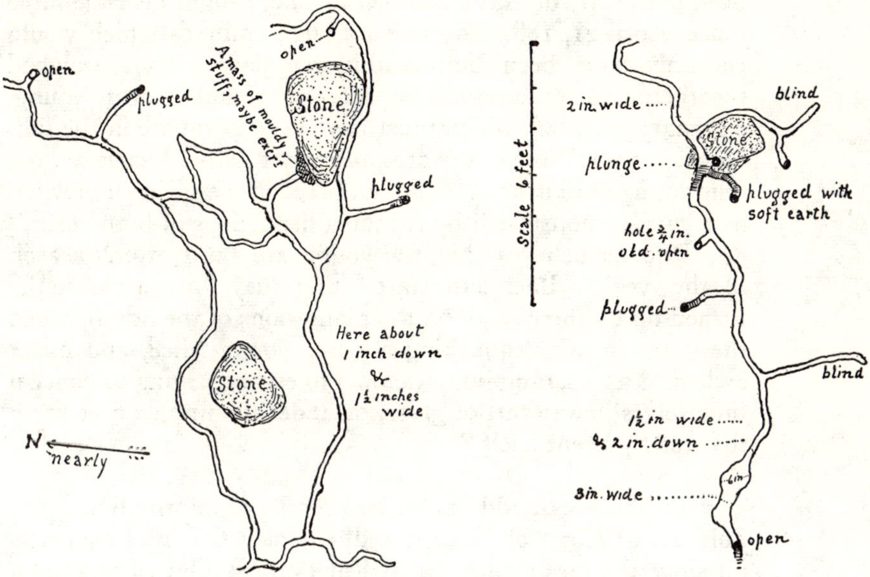

Tunnels of Blarina, Oct. 6, 1908, Cos Cob, Conn.

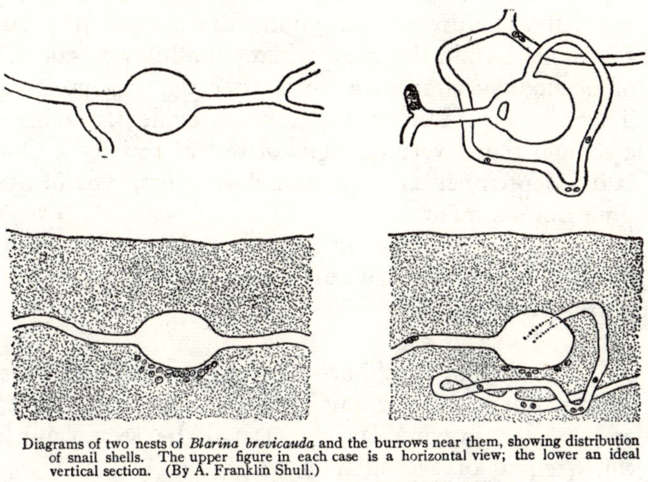

Diagrams of two nests of *Blarina brevicauda* and the burrows near them, showing distribution of snail shells. The upper figure in each case is a horizontal view; the lower an ideal vertical section. (By A. Franklin Shull.)

FIG. 257—Tunnels and nests of Blarina.

been born early in May. Another female, caught near the same place, April 21, 1884, contained 5 large embryos which would certainly have been born within ten days. They weighed together 4.20 grammes. I procured a half-grown young, February 10, 1884, which must have been born late in the fall. Hence, two or three litters are probably produced each season. The young born in autumn do not breed in the spring following, as I have demonstrated by repeated dissections of both sexes."

Rhoads believes that the young are born at any season of the year.[4] Bachman states[5] that they "are occasionally turned up by the plough on the plantations of the South, when they utter a faint squeaking cry like young Mice, and make awkward and scrambling attempts to escape, trying to conceal themselves in any tuft of grass, or under the first clod of earth that may present itself."

To this I can add one or two scraps of information. On July 21, at Cos Cob, I captured a female that was evidently nursing, and on October 20, at Rat Portage, Ont., I secured a large male that was in rut.

In the above-mentioned investigations made September, 1908, the capture of each young Blarina ended the digging at that hole, but, in the cases where an adult was taken, the signs of occupancy continued until after the capture of a second. These two were male and female, evidently living together, although there were no signs of sexual activity. One female, taken September 21, had, as below, 5 embryos of about half-time development.

Shull's paper also implies that the species lives in pairs during winter. Thus we are far on the way to proving life-long partnership.

FOOD The diet of the Short-tailed Shrew is chiefly insects and worms, but it will eat any kind of living food that it can find and master, preying largely, as will be seen, on Field-mice, which equal or exceed it in weight.

[4] Mam. Penna., 1903, p. 195. [5] Quad. N. A., Vol. II, 1849, p. 177

In the following are detailed the stomach contents of 13 Blarinas taken at Cos Cob, Conn., 1908, by E. T. Seton, determined by A. Franklin Shull, of Columbia University, New York:

1. Male.—Without date. Stomach: Earthworms, almost whole; membranous wings of insect (beetle).

2. Female.—Without date. Very badly preserved. Pickled long after death apparently; hair nearly all off; that remaining very loose. Gorged with connective tissue, cartilage, and muscle of——? Intestine filled with same material, digestive organs not properly functioning, or this would not be possible. Or is this material indigestible?

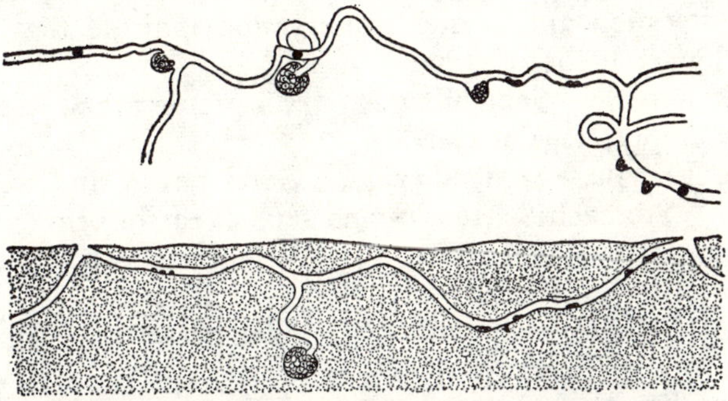

FIG. 258—Diagram of a typical burrow of *Blarina brevicauda*, showing distribution of snail shells, and an underground storage chamber with spiral descent. The upper figure is a horizontal projection; the lower an ideal vertical section. The black circles in the upper diagram are points where the burrow descended abruptly into the ground. (By A. Franklin Shull).

3. Female.—September 21. Good condition; well preserved stomach: Earthworm setæ; parts of insects; some of its own (Blarina's) hair, probably swallowed with food. Uterus contained 5 embryos, about 5 mm. long from caudal end to cervical flexure; probably between one and two weeks old.

4. Male.—September 22. Skull crushed by trap; excellent preservation. Stomach full: Earthworms, almost undigested. Must have been preserved soon after death, the digestion having been largely prevented, and hair being firm.

5. Female.—September 22. Poorly preserved, had been dead some time before pickling. Stomach almost empty; earthworm setæ only thing that could be identified; flesh unrecognizable.

6. Male.—September 22. Good condition. Stomach: Insects wholly; soft parts almost completely digested.

7. Female.—September 22. Stomach: Insects only; small quantity. Must have been taken at same place as No. 6, food being precisely same [they were living together]. More complete digestion and poorer preservation indicate that it was dead longer before pickling.

8. Female.—September 22. Fair preservation. Stomach nearly empty; legs of Isopod.

9. Female.—September 22. Good health and well preserved. Stomach: Muscles and setæ of earthworm only.

10. Male.—September 23. Well preserved. Stomach: Earthworms entirely.

11. Female.—September 23. Well preserved. Stomach: Earthworms; insects probably.

12. Female.—September 23. Liver somewhat inflamed and hardened, may have been due to injury in capture. Stomach: Isopod legs. What appear to be hyphæ and sporangia of some fungus; parts of insects.

13. Female.—September 25. No especial sexual development; in good condition, well preserved. Stomach: Earthworms; some arthropod, probably a sow-bug, recognizable by chitinous pieces and legs; connective tissue and striated muscle probably of a small mammal (rodent?) flesh found in teeth [doubtless the bait of the trap].

Merriam found the Blarina partial to beechnuts, and ready to eat corn and oats at a pinch.[6]

In a feed-box at one corner of an outbuilding I saw by the tracks that a Mouse came daily to steal crushed corn, so set a trap, and was surprised and sorry next day to find that the thief, already dead, was not a Mouse, but a female Blarina,

[6] Mam. Adir., 1884, p. 169.

evidently nursing a brood, although this was July 21. On examination her stomach was found crammed full of corn-meal unmixed with other food.

It is notorious that insect-eaters turn putrid in a few hours, and yet this Shrew lay on my desk in warm weather from 6 P. M. July 21, to noon July 22, without showing any signs of corruption. From this I argue that she had lived on corn for many days previously.

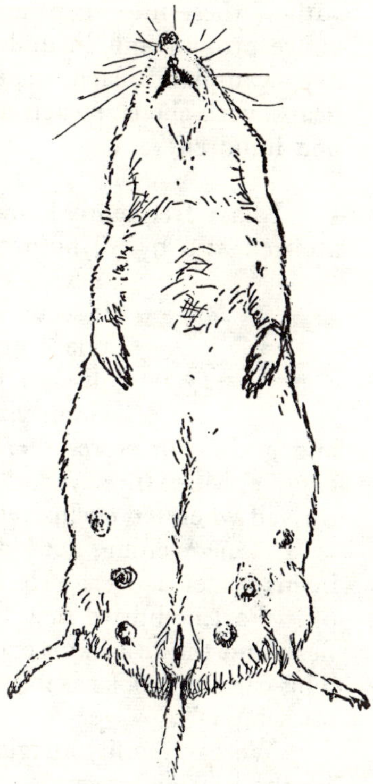

FIG. 259—Blarina, Cos Cob, July 22, 1904. Stomach crammed with grain—corn or wheat.

Numerous experiments and observations on captive animals prove that the Blarina, like its smaller kin, has an enormous appetite which must be satisfied or in a very few hours the creature succumbs. It makes no pretence at hibernation—is as active, indeed, all winter under the snow, as in summer under the grass. How, then, does it support life when living food is so scarce? The answer is not simple.

Dormant insects undoubtedly form a large part of its sustenance. As Dr. Merriam says,[7] and I have often proved, "the rigours of our northern winters seem to have no effect in diminishing its activities, for it scampers about on the snow during the severest weather, and I have known it to be out when the thermometer indicated a temperature of —20 Fahr. (—29 C.). It makes long journeys over the snow, burrowing down whenever it comes to an elevation that denotes the presence of a log or stump, and I am inclined to believe that at this season it

[7] Mam. Adir., 1884, p. 164.

must feed largely upon the chrysalides and larvæ of insects that are always to be found in such places."

The interesting experiments made at Ann Arbor, Mich.,[8] by A. F. Shull, prove that, in winter, *Blarina brevicauda* habitually feeds on snails of the genus *Polygyra*—that it gathers these into heaps as a sort of store, keeping them alive above ground and in underground dens. In two patches of dry swampland, each less than 20 acres, he found 40 of these heaps, with shells in each numbering from two or three to over one hundred.

STORAGE

I have frequently known it to kill and eat Field-mice, as is attested also by all naturalists who have studied the species. Yet, in view of the Mouse's superior agility, it seems unlikely that a Blarina could catch one every day, and the only stop-gap I can discern is that already suggested by Merriam, viz., this Shrew not only eats nuts and grain, but stores them up for winter use. Writing of his captive, Merriam says:[9] "He is very fond of beechnuts, and thrived when fed exclusively on them for more than a week.

FIG. 260—Excrement of the *Blarina brevicauda*. (Life size, after Shull.)

"One evening, not long ago, I put a handful of beechnuts in his water saucer. He soon found them and carried them off. Part he buried in a hole under the saucer, part under his nest, and the rest in an excavation near one corner of the box. This certainly looks as if the animal was in the habit of hoarding for winter."

We can readily imagine, then, that it is a perfect windfall for the Blarina when he discovers a fat colony of fat Mice with a fat store of food already laid up; for then, between murder and robbery, he can live happily for a month on a single find.

The most curious case of storage among small mammals is ascribed by Bachman to this creature.[10] After mentioning the long branching tunnels evidently made by the form *caro-*

[8] *Loc. cit.*, p. 514.

[9] *Loc. cit.*, p. 169.

[10] Quad. N. A., 1849, Vol. II, p. 177.

linensis, he adds: "We observed on the sides of one of these galleries a small cavity containing a hoard of coleopterous insects, principally composed of a rare species (*Scarabæus tityus*), fully the size of the animal itself; some of them were nearly consumed, and the rest mutilated, although still living."

All the Shrews are heavy drinkers, which is one reason for their choice of habitat. The Blarina is no exception to this rule. DRINK

To our ears the cries of Bats and Shrews are much alike—a sharp, high-pitched twittering, or a screech as sharp and fine as a needle. Doubtless each emotion of each species has an individual and characteristic sound to express it, but such things are as yet beyond our poor powers of discrimination. When our inventors give us a practical field microphone we shall be able, doubtless, to enter a hitherto unexplored wonderland of sound, experiencing many new delights and, doubtless, also some new sorrows. VOICE

Kennicott says[11] of the Mole-shrew which he had captive: "When hurt or irritated, it uttered a short, sharp, tremulous note, like *zee-e*, and when it was much enraged this note became longer, harsher, and twittering, like that of some buntings or sparrows. Sometimes a short, clear cry was uttered, the voice calling to mind that of the common Mink (*Putorius vison*), but softer and lower."

Correlating with its vocal powers, the Blarina is possessed of exquisite hearing, associated with which is a fine discrimination that stands between the creature and a world of false-alarms. A. F. Shull found that a captive specimen soon became oblivious to sounds that were often repeated. No matter how heavy, harsh, or piercing the noise, the Shrew soon learned to hear it with indifference. There was, however, one notable exception. "The flutter of wings of a pigeon kept in the same vivarium, on the other hand, always sent the Shrew skurrying HEARING

[11] Quad. Ill., 1858, p. 96.

into its burrows. I observed this more than twenty times, at intervals throughout the five weeks of the Shrew's captivity, and the last flutter produced as much disturbance as the first. This particular sound must have been heard hundreds of times during that period, yet even at the last could not be heard with equanimity by the Shrew."[12]

In this we read a story of perpetual danger from birds of prey, through past ages in the history of Blarina's race.

TOUCH AND EYESIGHT

Its eyes are of use apparently only to distinguish light from shade, but its exquisite hearing and sense of touch seem to compensate for the lack of vision, and to render it equally at home in the blazing sun, on the snow, in the midnight woods, or pursuing the Field-mouse to its lair far underground. It has, indeed, exchanged its sight for better touch, nor has it lost by the bargain; that I learned at Rat Portage, when collecting long ago. One line of traps consisted of pitfalls made by sinking pickle-jars in the mouse-runs. The run was fenced and roofed and smoothed to lend all baleful aid, and I caught many Mice and some Cooper Shrews, but never once a Blarina did I get; its senses, though limited in number, totalled up far better than those of the bead-eyed Mice it preyed on. The only Short-tailed Shrews I got were taken in cage-traps baited with fresh meat.

Kennicott, after capturing a pair of these for observation, wrote:[13] "While alive, the minute black eye is distinctly seen and always open; but, though the sense of sight may be possessed in the dark, it certainly is not used in the full light. Upon waving different objects before one, or thrusting my finger or a stick close to its face, no notice was taken of it whatever; but if I made any noise near by, it always started. If the floor was struck, or even the air disturbed, it would start back from that direction. I observed no indication that an acute sense of smell enabled it to recognize objects at any considerable distance; but its hearing was remarkable. An exceedingly delicate sense of touch was exhibited by the

[12] *Loc. cit.*, p. 513.

[13] Quad. Ill., 1858, pp. 95–6.

whiskers, and if, after irritating a Shrew, I placed a stick against it, in even the most gentle manner, the animal would instantly spring at it. I could see that, in running along the floor, it stopped the moment its whiskers touched anything; and often, when at full speed, it would turn aside just before reaching an object against which it seemed about to strike and which it certainly had not seen. Unless enraged by being teased, it endeavoured to smell every new object with which its whiskers came in contact, turning its long flexible snout with great facility for this purpose.

"My caged specimens, both male and female, exhibited great pugnacity. When I touched one several times with a stick, it would become much enraged, snapping and crying out angrily. When attacked by a Meadow-mouse (*Arvicola scalopsoides*), confined in a cage with it, one fought fiercely; and though it did not pursue its adversary when the latter moved off, neither did it ever retreat; but the instant the Mouse came close, it sprang at him, apparently not guided in the least by sight. It kept its nose and whiskers constantly moving from side to side, and often sprang forward with an angry cry when the Mouse was not near, as if deceived in thinking it had heard or felt a movement in that direction. In fighting, it did not spring up high, or attempt to leap upon its adversary, as the Mouse, but jerked itself along, stopping firmly, with the fore-feet well forwards, and the head high. On coming in contact with the Mouse, it snapped at him, and, though it sometimes rose on its hind-feet in the struggle, I did not observe that it used its fore-feet as weapons of offence, like the *arvicolæ*. Its posture, when on guard, was always with the feet spread and firmly braced, and the head held with the snout pointing upwards, and the mouth and chin forwards, in which position its eyes would have been of no use, could it have seen. The motions of this animal, when angry, are characterized by a peculiar firmness; the muscles appear to be held very rigid, while the movements are made by quick, energetic jerks. Short springs, either backwards, forwards, or sidewise; appear to be made with equal readiness.

PUGNACITY

"This Shrew is quite active as well as strong; the snout and head are powerful, and seem to be much used in burrowing; the tough cartilaginous snout received no injury from the rough edge of a pane of glass, under which that of a caged specimen was forcibly thrust in endeavouring to raise it. When liberated, upon a smooth floor, it runs rapidly, without ever leaping, placing only the toes on the surface; though in moving slowly the whole tarsi of the hind-feet are brought down. By placing an ear of corn over two inches in diameter at the edge of the room, and chasing a Shrew towards it by striking the floor behind the animal, I have seen one several times spring over it, apparently without great effort; but if not much frightened it would always go around objects an inch high, running close along them, as it did beside the wall, invariably feeling its way. One would never leave the side of the wall to run across the room, and would always run around the side of its cage, rather than go across the middle."

HABITS The Short-tailed Shrew is incapable of climbing or running fast. It is, as we have seen, practically blind, does not smell well, is vulnerable at all points of its body, and yet it is an admitted success in life. It offsets all its shortcomings by a superlative development of hearing and touch, and a restless energy combined with indomitable courage, great muscular powers, and tireless activity, an equipment that makes it a fearsome beast of prey; a terror to all wild creatures of its small world, that are less than double its weight.

The earliest account I can find of its exploits as a hunter of big game is by John Morden, of Hyde Park, Ont.[14] "In a trap set for Mice he found, at one time, a [Mole-]shrew and two White-footed Mice, one of the latter being dead and about half eaten. 'The evening of that day the Mole was placed in an old laundry boiler and the entire dead Mouse given to it, which by morning was entirely eaten, bones and all, except the hair. We then gave the Mole a large rat just killed, when it at

[14] Can. Sport and Nat., December, 1883, p. 283, quoted by Merriam, Mam. Adir., 1884, p. 165.

once proceeded to eat out its eyes, and by four o'clock next afternoon one side of the rat's head, bone together with the brains, were eaten, and, strange to say, the Mole looked no larger. * * * Our curiosity was aroused to know by what means a Mole or Shrew could kill Mice which were larger than itself; so four large Meadow-mice being procured, they were placed in the boiler with the Mole, which as soon as it met a Mouse showed fight, but the Mouse knocked it away with its front feet and leaped as far away as it could. The Mole from the first seemed not to see very plainly and started around the boiler at a lively rate, reaching and scenting in all directions with its long nose, like a pig that has broken into a backyard and smells the swill-barrel. The Mice seemed terror-stricken, momentarily rising on their hind-legs, looking for some place to escape, leaping about, squeaking in their efforts to keep out of the way of the Mole, which pursued them constantly. The Mole's mode of attack was to seize the Mouse in the region of the throat. This it did by turning its head as it sprang at the Mouse, at the same time uttering a chattering sound. The Mice would strike at and usually knock the Mole away with their front feet, but if the latter got a hold of the Mouse it would then try to bite, and they would both tumble about like dogs in a fight. The little chap at last attacked one Mouse and kept with it, and in about ten minutes had it killed; but even before it was dead the Mole commenced eating its eyes and face. About ten minutes later the Mole had devoured all the head of the Mouse and continued to eat. I have captured and caged several Moles this winter, and they all display the same untiring greedy nature. According to my observation, the little mammal under consideration eats about twice or three times its own weight of food every twenty-four hours, and when we consider that their principal food consists of insects, it is quite bewildering to imagine the myriads one must destroy in a year.'"

Dr. Merriam repeated these experiments and found that a small Blarina weighing 11.20 grammes could tire out and overcome a vigorous male Deer-mouse weighing 17 grammes.

The Blarina ate the brain, one side of the head, and part of the shoulder in fifteen minutes after the death of the Mouse, and immediately after the meal he weighed 12 grammes—an increase of .80 gramme.[15]

"The Shrew was half an hour in tiring the Mouse, and another half-hour in killing him. But it must be remembered that he was not fully grown, and was doubtless, on this account, longer in capturing and killing his victim than would have been the case had he been an adult. Still, it is clear that a Shrew could never catch Mice on open ground. His small size, however, enables him readily to enter their holes and to follow them to their nests and the remotest ramifications of their burrows, where, having no escape, he can slay them with fearful certainty."

At Toronto on February 4, 1888, I collected an old Blarina whose tail was gone, probably in battle, so that they do not always go scot-free.

The most desperate exploit accredited to one of this species is recorded by Professor E. D. Cope in the *American Naturalist* for August, 1873:[16]

"I recently [says he] placed a water snake (*Tropidonotus sipedon*) of two feet in length in a fernery which was inhabited by a Shrew, either a large *Blarina carolinensis* or a small *Blarina talpoides*. The snake was vigorous when placed in the case in the afternoon and bit at everything within reach. The next morning the glass sides of his prison were streaked with dirt and other marks, to the height of the reach of the snake, bearing witness to his energetic efforts to escape. He was then lying on the earthen floor, in an exhausted state, making a few ineffectual efforts to twist his body, while the Blarina was busy tearing out his masseter and temporal muscles. A large part of the flesh was eaten from his tail, and the temporal and masseter muscles, and eye of one side, were removed, so that the under jaw hung loose. The temporal was torn loose from the cranium on the other side, and as I watched him the Blarina cut the other side of the mandible loose and began to

[15] *Loc. cit.*, pp. 166–8.

[16] Vol. VII, No. 8, pp. 490–1.

tear the longicolli and réctus muscles. His motions were quite frantic, and he jerked and tore out considerable fragments with his long anterior teeth. He seemed especially anxious to get down the snake's throat (where some of his kin had probably gone before), and revolved on his long axis, now with his belly up, now with his sides, in his energetic efforts. He had apparently not been bitten by the snake and was uninjured. Whether the Shrew killed the snake is of course uncertain, but the animus with which he devoured the reptile gives some colour to the suspicion that he in some way frightened him to exhaustion."

Had it been a Mouse of the same size instead of the Shrew, the incident would undoubtedly have terminated the other way; but the strength, ferocity, activity, and courage of the Blarina are such that if it were increased to half the size of a tiger it might quite logically make tigers its habitual prey.

ENEMIES

But even this valiant one has foes to fear. Hawks and owls of all kinds are ready to kill the Blarina and swallow it whole; while Lynxes and Weasels, dogs and cats rarely lose a chance of giving it a fatal nip or a crushing death-blow, although they are deterred from eating it by the rank odour that it emits, doubtless as a protection.

Yet another class of foes it has, one that is too small for it to master, for the nursing female, already mentioned as taken July 21, was swarming with three different kinds of fur-lice.

It is to be hoped that the farmer will never enlist himself against the Blarina. It may kill good bugs and bad bugs indiscriminately; it may take a little grain when nothing better is at hand, but the balance of benefit is far in the farmer's favour. All the evidence goes to show that its favourite food is Mice. For mouse-meat it will hunt and struggle without wearying, eschewing all other diet, when this is at all a possibility; and just so surely as the Mouse is the farmer's foe, so surely is the Blarina his good friend and worthy of active protection.

LIII.

Star-nosed Mole.

Condylura cristata (Linnaeus).

(*Condylura*, from Gr. *kondylos*, a knob; *oura*, tail; L. *cristata*, crested.)

Sorex cristatus Linn., 1758, Syst. Nat., X Ed., I, p. 53.
Condylura cristata Desmarest, 1819, Journ. de Phys., LXXXIX, p. 230.

Type Locality.—Pennsylvania.

French Canadian, *le Condylure à longue queue; la Taupe du Canada; le Condylure à museau étoilè.*

The Family *Talpidæ* or Moles are like the Shrews in their soft, velvety fur, their apparent lack of eyes, ears, and neck, their scanty-haired, scaly tails, and also in the general style of their teeth; but they differ in being much larger, and in having the front feet enormously developed for digging.

The genus *Condylura*, founded by Illiger (1811), for the present, its only known species, has, in addition to the Family characters, a remarkable fringe of 22 fleshy points or feelers around the nose; its tail is longer than in most Moles, being over half as long as the head and body. The teeth are:

$$\text{Inc. } \frac{3-3}{3-3}; \text{ can. } \frac{1-1}{1-1}; \text{ prem. } \frac{4-4}{4-4}; \text{ mol. } \frac{3-3}{3-3} = 44$$

SIZE Total length, snout to tail-bone tip, about 7 inches (178 mm.); tail, 3 inches (76 mm.); hind-foot, $1\frac{1}{16}$ inches (27 mm.).

COLOUR Above, dull brownish-black, becoming paler and browner on chin, throat, and lower parts.

An adult male, taken at Cos Cob, Conn., July 7, 1909, weighed 24 grammes.

LIFE-HISTORY.

The range of the species is fairly well worked out in the south and east, but the northern and western boundaries, as shown on the map, No. 62, are sure to be greatly modified by fuller investigation. RANGE

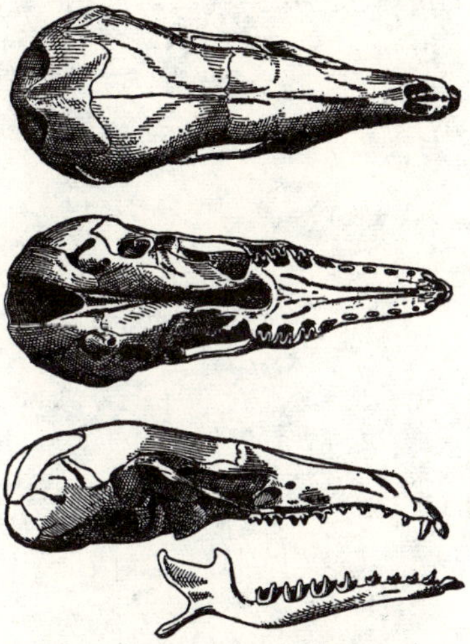

FIG. 261—Skull of Star-nosed Mole (*Condylura cristata*). (1½ times natural size.)
From True's Rev. Am. Moles, P. U. S. N. M., 1897. Plate IV.

Dr. R. Bell gives it as common at Moose Factory;[1] F. W. True records one from Moose Factory and one from Rupert House, James Bay;[2] C. L. Herrick secured a single specimen in Minnesota;[3] R. Kennicott mentions[4] its occurrence at Fort Ripley, Minn.; Bailey gives it as scarce at Elk River, Minn.[5] At Duluth it is not uncommon; there is a local specimen in the High School Museum.

It is entered as Manitoban, on the authority of W. R. Hine, who assures me that specimens have been brought to his taxidermist shop in Winnipeg; unfortunately, they were not kept.

At Nipigon it was taken by G. S. Miller, who found it in seemingly abundance at Peninsula Harbor also.[6]

Outram Bangs records[7] the capture of a specimen at Lake Edward, south of Lake St. John, Quebec. But elsewhere[8]

[1] Mam. H. Bay (1884), App. II, p. 48, D. D. Geol. Surv. Can., 1885.
[2] Proc. U. S. Nat. Mus., Vol. XIX (1896), 1897, p. 84.
[3] Mam. Minn., 1892, p. 56.
[4] Quad. Ill., 1858, p. 101.
[5] Rep. Orn. and Mam., U. S. Dept. Agr. (for 1887), 1888, p. 435.
[6] Mam. Ont., Proc. Bost. Soc. Nat. Hist., April, 1897, p. 39.
[7] Proc. Bi. Soc., Wash., March 9, 1896, p. 51.
[8] Mam. Labrador, Am. Nat., July, 1898, p. 497.

MAP 62—RANGE OF THE STAR-NOSED MOLE.

Condylura cristata (Linn.).

The outlines are the theoretical range, the spots are actual records given by Audubon and Bachman, S. F. Baird, R. Kennicott, J. E. DeKay, C. Hart Merriam, S. N. Rhoads, O. Bangs, G. S. Miller, and E. T. Seton.

he gives a yet farther afield occurrence—a Star-nose from Rigoulette, on the coast of Labrador. His statement is as follows:

"Goldthwaite saw and fully identified a Star-nosed Mole that the dogs caught at Rigoulette. As he assures me there is not the slightest chance of a mistake in his identification, the species must be included."

Its environment is fairly well understood. It is the least molish of its family, preferring swimming to digging, and a low meadow, a mossy bog, or even a wet marsh, to the most alluring of upland pastures. ENVIRONMENT

Commenting on the Lake Edward specimen, Bangs says:

"No work of this Mole was seen anywhere. The one taken was caught in a cyclone trap set under an old log. Probably the animal lives below the deep layer of moss, with which everything is covered, and therefore gives no sign of its presence."

Nevertheless, it does not entirely avoid the drier fields, the ploughed lands, or even the garden, when making a side trip or seeking a new range.

The home-range of the individual is probably about the same size as that of other Moles—an acre of swamp is sufficient. Here it will dwell in comfort for weeks or months, till famine, flood, or a strong invader compels it to depart to some other swamp a hundred long yards off. HOME-RANGE

The Star-nose is remarkable among Moles in being sociable, or at least gregarious, as these animals are known to live in colonies, which, judging from the results of their united labours, should contain at least a dozen individuals. Merriam speaks[9] of capturing 8 in one colony, and evidently did not exhaust their number. Elsewhere[10] he refers to their being "in large colonies," and says he considers it one of the commonest Moles in the Adirondack region. SOCIABILITY

[9] Mam. Adir., 1884, p. 148.

[10] *Ibid.*, p. 150.

MATING The species is supposed to pair. The mating takes place in November. It is signalized in a unique manner. As the neck of a Deer or the throat feathering of a ruff become greatly enlarged in the rut, so the tail of the Star-nose swells to double its usual size during the time of ardent passion.

Dr. Harlan, not knowing of this periodic change, supposed the thick-tailed one to be a new species, and named it *macroura*.[11]

NEST A nest of this animal dug out by Audubon and Bachman[12] was approached by a long winding burrow, and situated in a large excavation under a stump; it was quite "spacious and composed of withered grasses."

YOUNG It contained "3 young, apparently a week old. The radiations in the nose were so slightly developed that, until we carefully examined them, we supposed they were the young of the common Shrew-mole."

Other authorities set the number of young from 4 to 6. The history of their development is not further known. "Two or more litters are produced each season."[13] (*Merriam.*)

HABITS "If we may judge by its remarkable resemblance to that of the Muskrat [says Rhoads],[14] his tail is often brought to play in swimming. I have no doubt that the anatomy of this species, as well as its chosen habitat, infallibly indicates a much more aquatic life than we have yet been able to prove by actual observation."

This accords very well with my own experience. Two specimens which I got from Toronto marsh were taken while swimming in the water *under the ice*.

On July 7, 1909, at Cos Cob, Conn., I received an adult male Star-nosed Mole captured alive by a stream, not in the water, but running along a mossy bank.

[11] Fauna Americana, 1825, p. 39.
[12] Quad. N. A., 1849, Vol. II, p. 142.
[13] Mam. Adir., 1884, p. 152.
[14] Mam. Penna, 1903, p. 207.

With the help of a commodious cage I made a number of observations. When put in deep water he swam swiftly and strongly. He progressed largely by the sculling action of his tail, but also swam with all four feet, striking alternately, never by striking with both hind-feet at once, as most truly aquatic quadrupeds do occasionally. He did not dive, and

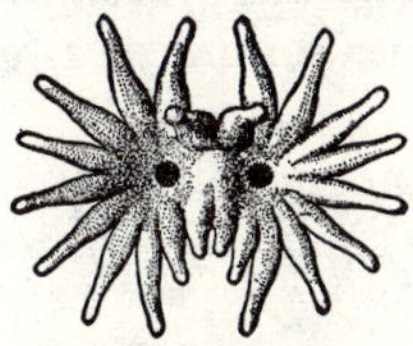

Nasal Disk of Condylura.
(Twice natural size.)

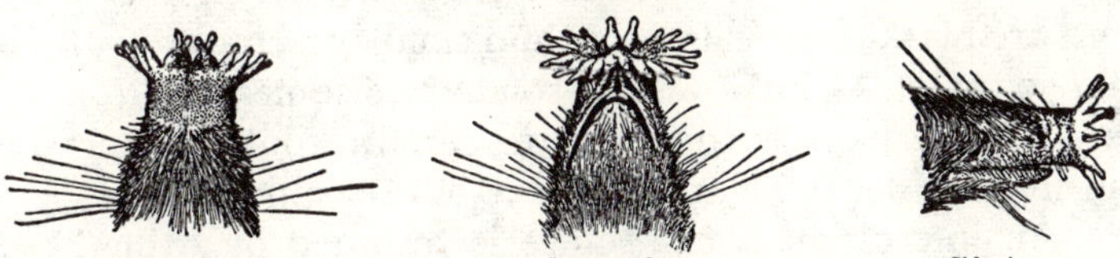

Upper surface. Lower surface. Side view.

FIG. 262—Snout of *Condylura cristata.* (Natural size.)
From True, Proc. U. S. N. M., XIX, 1897.

always endeavoured to get out of the water as quickly as possible. His fur got very wet and there was little to suggest adaptation to a truly aquatic life.

He climbed fairly well in the corner of the cage where helped by the wire netting, but could not get up where the corner was of rough boards. While hanging from the wires his hind-feet were his chief support, but he was greatly helped by the tail, which was held woodpecker-fashion tight against the wall. He was much less of a climber than, for example, the *Microtus*.

In the middle of each day he curled up and slept for two or three hours. At night he was very active.

When given a pile of loose earth in which were many worms he showed great delight, and dived again and again through the pile, sometimes coming out with a worm, and suggested an Otter diving in a salmon river.

His fur, previously rough and ill-kempt, was left beautifully velvety by this operation.

NON-HIBER-NANT

Of course the species is non-hibernant and is often captured during its very active winter life.

Merriam mentions[15] having seen it travelling over the deep snow in March; when so discovered, it tries to dive down out of sight, and readily does so, unless a hard crust bars the way.

TUNNELS

Its manner of tunnelling and heaving up hills is precisely like that of the Pocket-gopher, in whose biography the subject is fully treated.

When the Star-nose dwells in or crosses dry ground, it seems not easy to distinguish its work from that of other Moles; both are marked by hills, large and small, on crooked galleries, that sometimes go below, and sometimes so near the top as to be ridges of loose earth. But in certain kinds of country it can live and leave behind no trace of excavation.

An interesting circumstance is recorded by Miller.[16] At Peninsula Harbor, Ont., in early October, he found the remains of a Star-nose in the stomach of a rough-legged hawk.

A similar record appears in "Fisher's Report on the Food of Hawks and Owls."[17]

A screech owl, taken at Washington, D. C., on June 2, 1889, had in its stomach a Mole of the present species.

FOOD

Commenting on its food, Rhoads says:[18] "As the boggy nature of its haunts is distasteful to earthworms and other animals on which the upland Moles subsist, we must conclude that these form but a small part of its diet, but the numerous aquatic and subaquatic insects and crustaceans which harbour in wet meadows and stream banks would form bountiful supply."

As soon as the above-named captive was caged I gave him 12 grammes of common worms. He paid no heed for half

[15] Mam. Adir., 1884, p. 152.
[16] Mam. Ont., 1897, p. 39.
[17] Bull. 3, 1893, U. S. Dep. Agr., p. 171.
[18] Mam. Penna., 1903, p. 207.

an hour, but then aroused himself and fell on the worms with great demonstration, continually twiddling them with his 22 nose-fingers. Though avid, he ate them slowly, holding them with his fore-claws and tearing them up before devouring. In half an hour all were gone. This was at noon; at 1:45 he seemed ravenous again. I gave him a similar amount of worms, also 3 cutworm grubs; these latter he ignored while the former lasted. Towards night I gave the Mole about 2 ounces of raw beef, of which ⅓ only was lean, the rest fat. In the morning all the lean was eaten and all the fat rejected.

FIG. 263—Scatology of Star-nosed Mole

Now a newly killed Deer-mouse was offered to him. He sprang on this with much demonstration and little effect. After twiddling it all over, he began on the eyes and then ate the brains where the head had been crushed by the trap, turning back the skin. By next morning the Deer-mouse (it weighed more than the Mole) was devoured, except the skin, which was neatly turned inside out, and the bones—even the smallest ribs were left intact and quite clean. During the previous evening he ate also 8 grammes of worms. I found, however, that he preferred the large fat white grubs that are found under manure piles (*Lachnosterna fusca*); for these he neglected both worms and Mouse. A large blue wasp he would not touch; also a stag-beetle and he lived amicably together till the end. He refused several kinds of farinaceous food.

VOICE, ETC.

During the second night he escaped. I was awakened in my room (one flight down) by hearing the patter of small feet on the floor; as I stood near the window the sound came towards me and I felt a furry creature pushing under my naked instep. I stooped and seized it in the dark; a strong musky smell and a faint husky squeak informed me that I had recaptured my Star-nose. This was the only time I heard him utter a sound. Possibly he was injured at this time, as he died next day.

USE

As the Star-nose feeds only on insects and worms, utterly eschewing seeds, roots, and all things vegetable, there can be no very serious charge against it. The disfigurement of lawns is the only one that has ever given it local outlawry; but most naturalists at least consider this fully offset by the good it does as a tiller of the soil on lands where there is no other husbandman.

BATS
ORDER CHIROPTERA

LIV.

Little Brown-bat or Blunt-nosed Bat.

Myotis lucifugus (Le Conte).

(Gr. *mys*, mouse; *otis*, ear; L. *lucis*, of the light; *fugus*, one who flies from, a fugitive.)

Vespertilio lucifugus LE CONTE, 1831, McMurtrie's Cuvier, An. King. I, p. 431.
Myotis lucifugus MILLER, 1897, N. Am. Fauna, No. 13, p. 59.
TYPE LOCALITY.—Georgia; probably near Riceboro, Liberty County.

FRENCH CANADIAN, *la Chauve-souris brunette.*
CREE, *Pee-kwa-nah-djee'* (applied to all Bats).
OJIB., *Ah-pe-kwa-nah-djee'* (applied to all Bats).

The whole Family, *Vespertilionidæ*, have simple noses, that is, without a leaf membrane; turbinal bones folded; palate deeply notched in front; molars with W-shaped cusps; tail, long and enclosed to the tip in a membrane or tail-web.

The sub-family, *Vespertilioninæ*, have 6 lower incisors, and ears separate at base.

All of the Manitoban Bats are in this sub-family.

The genus *Myotis* (Kaup, 1829) comprises small, slender Bats, with hairy faces, tapering tragus, long tails, naked wings and tail membranes, and the following tooth formula:

$$\text{Inc. } \frac{2-2}{3-3}; \text{ can. } \frac{1-1}{1-1}; \text{ prem. } \frac{3-3}{3-3}; \text{ mol. } \frac{3-3}{3-3} = 38$$

The combination of 4 upper incisors with 6 upper premolars is important.

SIZE Length, from snout to tail-tip, 3 to $3\frac{1}{2}$ inches (76 to 89 mm.); forearm, from elbow to wrist, $1\frac{3}{8}$ to $1\frac{5}{8}$ (35 to 41 mm.); tibia, $\frac{9}{16}$ to $\frac{11}{16}$ (14 to 17 mm.); spread across wings, 9 to 10 inches (229 to 254 mm.); ears, "short and pointed, reaching, when laid forward, barely to tip of the nose."[1] (*Miller*.)

COLOUR In colour it is everywhere of a dull brown, paler below.

In discussing a highly coloured phase of the British Barbastelle, Sir Harry Johnston says:[2] "In some other Vespertilionid Bats there is a tendency in the breeding season for the males to develop a rich yellow tinge in the lower half of the hair of the under parts. It may be the same tendency which tinged the fur of this example with a purplish tone."

This species closely resembles *M. subulatus*, but may be distinguished by its shorter ears, which do not reach to the end of the nose (instead of considerably over), and by its shorter, more rounded tragus (see Fig. 264.).

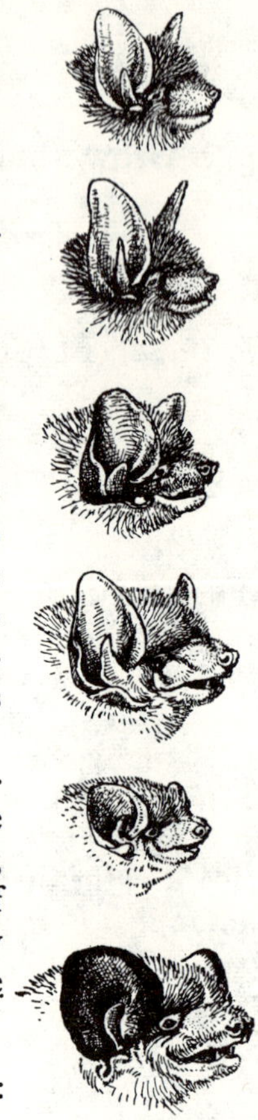

Spread 9 to 10 inches. Colour, dull brown. *M. lucifugus.*

Same size and colour, but ear and tragus longer and slenderer. *M. subulatus.*

Spread about $11\frac{1}{2}$ inches. Colour, black, with silvery tips. *L. noctivagans.*

Spread 12 to 13 inches. Colour, dull brown. *E. fuscus.*

Spread about 12 inches. Colour, bright clear orange, with some silver tipping. *L. borealis.*

Spread 15 to 17 inches. Colour, yellowish, with silver tipping; the ear-rims and snout black. *L. cinereus.*

FIG. 264—The Bats found in Manitoba. (All life size.)
To serve as a key in diagnosing the species.

Three races are recognized:

lucifugus (LeConte), the typical form.
alascensis Miller, darker in colour, and with longer ears.
longicrus (True), like *lucifugus*, but larger, and with proportionately longer legs, and shorter ears and arms.

[1] N. A. Fauna, No. 13, 1897, p. 60.   [2] Brit. Mam., 1903, p. 104.

MAP 63—RANGE OF THE LITTLE BROWN-BAT AND ITS THREE RACES.

Myotis lucifugus (Le Conte).

The outline shows the theoretical range. The spots are actual records; they are chiefly from H. Allen's Monograph and G. S. Miller's Revision, with others by Le Conte, O. Bangs, W. H. Osgood, S. N. Rhoads, C. Hart Merriam, J. A. Allen, E. A. Preble, and E. T. Seton.

Life-history.

RANGE The range of this well-known, yet mysterious, little creature is nearly all of North America, excepting the tropics and the polar regions. The map (No. 63) sets forth these facts with black spots for the actual records. The only Manitoban specimen I have is from Poplar Point.

ENVIRONMENT This is a cave Bat—the common species indeed of the Mammoth Cave of Kentucky—and when no caves are available, it finds acceptable substitutes in hollow trees and farm buildings, rarely frequenting towns.

In old-established, happily placed hiding-places it is usual for many to congregate. Thus the species is gregarious and to some extent sociable, since the individuals profit by each other's company in the matter of warmth.

SOCIABILITY In flight, the Little Brown-bat may be distinguished from its remote kin by its small size, its early evening appearance, and its erratic course. I do not know how to tell it from *subulatus*.

FLIGHT Oftentimes the chimney-swifts are out so late as to fly with this species. There are points of resemblance in their flight, though that of the swift is without the erratic dodging. Furthermore, it is usual to see 2 or 3 swifts careering along side by side—a sociable style that I have never seen in any Bat.

VOICE The voice of this, and indeed of all our Bats, is an exceedingly fine squeak, finer than that of a Mouse, and often heard as they fly close overhead at night. When captured they utter a volley of these squeaks, varied with a hissing and fizzing sound. If a 'battery' be disturbed, they combine in a deafening and unpleasant chorus of squeaking and chirring, that reminds one of a nest of young swifts at food-time.

It is a well-known fact that some persons, with otherwise perfect hearing, cannot hear a Bat's squeak. Millais writes:[3]

[3] Mam. G. B. & I., Vol. I, 1904, p. 54.

"There is an odd superstition in Sussex that persons over 40 years of age are unable to hear the cry of a Bat."

According to Tyndall,[4] "the human ear is limited in its range of hearing musical sounds. If the vibrations number less than 16 a second, we are conscious only of the separate shocks. If they exceed 38,000 a second, the consciousness of sound ceases altogether.

"The range of the best ear covers about 11 octaves, but an auditory range limited to 6 or 7 octaves is not uncommon.

"The sounds available in music are produced by vibrations comprised between the limits of 40 and 4,000 a second. They embrace 7 octaves.[5]

"While endeavouring to estimate the pitch of certain sharp sounds, Dr. Wollaston remarked in a friend a total insensibility to the sound of a small organ-pipe, which, in respect to acuteness, was far within the ordinary limits of hearing. The sense of hearing of this person terminated at a note 4 octaves above the middle E of the pianoforte. The squeak of the Bat, the sound of a cricket, even the chirrup of the common house-sparrow, are unheard by some people who for lower sounds possess a sensitive ear. A difference of a single note is sometimes sufficient to produce the change from sound to silence.

"'Nothing can be more surprising,' writes Sir John Herschel, 'than to see two persons, neither of them deaf, the one complaining of the penetrating shrillness of a sound, while the other maintains there is no sound at all.' Thus, while one person mentioned by Dr. Wollaston could but just hear a note 4 octaves above the middle E of the pianoforte, others have a distinct perception of sounds full 2 octaves higher. The chirrup of the sparrow is about the former limit; the cry of the Bat, about an octave above it; and that of some insects, probably, another octave. In 'The Glaciers of the Alps' I have referred to a case of short auditory range noticed by myself, in crossing the Wengern Alps in company with a friend. The grass at each side of the path swarmed with insects, which to

[4] Sound, p. 81.

[5] Each note has double the vibrations of its octave below.

me rent the air with their shrill chirruping. My friend heard nothing of this, the insect music lying beyond his limit of audition."

Birds, we know, are gifted with a great range of sounds to express their varying emotions. Shrews and Bats, being in general much more highly organized than birds, have probably a much greater range of sounds.

There is every reason to believe that they have at least as many varied calls as a crow or a magpie, *including even a song* for the season of love; and yet, because of our dull ears, these things are in a sealed book, and the older we grow the more curtailed is our power of peeping even at the covers.

MATING There is little evidence on the mating of Bats. Each year in late August or early September they are seen in unusual numbers, rushing about and chasing each other in great excitement, for one or two evenings. Specimens collected at such a time show, by the enlarged sexual organs, that now is the rutting season. I saw this in a marked degree at Owen Sound, Ont., on August 17, 1889. About nine in the evening the Bats appeared in hundreds, and circled around the electric lights like swarming bees. They were chasing each other in the greatest excitement. I think that without doubt this was the rut. Next evening it seemed to be over, and the number of Bats went down to 3 or 4 in sight at one time. I am not sure what kinds predominated, but have little doubt that the present species was represented. I saw a similar demonstration at Cos Cob, Conn., September 7, 1906.

GESTATION Gestation is unusually long in all of this group, and apparently 10 months in the present species, as the young are not born till June. C. L. Herrick says:[6] "Although northern species mate ordinarily in autumn, eggs are not fertilized until spring, when impregnation takes place." The observations on which this is founded are not given; it implies a number of unusual modifications.

[6] Mam. Minn., 1892, p. 22.

The sum of many observations shows that the young are usually 2 in number, but occasionally 1, and, on rare occasions, 3. YOUNG

We have no direct light on the parturition of any of our Bats, but the observations of M. Rolinat and Dr. E. Trouessart on *Vespertilio murinus*, a common Bat in France,[7] will afford valuable side-light. PARTURITION

On June 9 they saw a captive female clinging to the wire of the cage, head up. The hind-legs were much spread; the tail was curved up in front. Her flanks were heaving and she seemed in pain. At 10 the left knee of the little one appeared. The mother made violent efforts to lick it and uttered a feeble cry like rapidly opening and shutting the lips. At 10:20 the body of the little one appeared. It was at once licked by the mother. At 10:30 the body, and immediately afterwards the head and front limbs came forth, and the new-born dropped into the sack made by the interfemoral membrane. The mother had been much agitated, but now grew calmer, and she licked the young one vigorously. Thus stimulated, it climbed out of the pouch and hung on to the old one's fur. At 10:55 the mother turned nearly head down, and the young one seized the left teat, to which it continued attached for several days. It was blind at birth, but its eyes opened on the fifth day. On the thirteenth day it quit its mother's protecting wing, and thenceforth roosted much alongside.

These observers conclude that it is the habit of the mother to carry her young one with her as she flies, until it is about 2 weeks old, after which she leaves it at home in the den.

At 35 days the one described above was still nursing.

At 50 days it was eating cockroaches.

At 2 months it no longer nursed, and would eat 34 to 37 cockroaches each night.

Rhoads says:[8] "These [the young of *lucifugus*] cling by their mouths to the teats of the mother until large enough to

[7] Sur la reproduction des Chauve-Souris, Mem. Soc. Zoöl. de France, 1896, IX, pp. 230–1 and 234.

[8] Mam. Penn., 1903, p. 209.

grasp her body. Thus laden, she pursues her nightly avocations until they can be left 'hung up' in some secret place till her return."

None of our Bats is known to make anything in the nature of a nest. During the second period of their infancy, that is, till they are able to fly, the young remain at home, and are fed by the mother. When about three months old they fly and forage for themselves.

The father is not known to help in rearing the young.

HABITS

Some curious observations on this Bat appear in Stone and Cram's "American Animals."[9] A number of the species used to spend the day behind the blind of an open window in their New Hampshire house. During August, 1898, an exact count was kept of their number, resulting in proof that they regularly congregated there Sunday nights, and grew fewer towards the middle of the week. The four weeks stood thus (Sundays are in heavy type):

?	?	4	**16**	9	4	2
5	10	10	**18**	10	2	0
0	1	1	**?**	3	2	0
0	0	1	**2**	3	0	0

This periodicity may have been chance; it may have been that their alternate roost was in a church which was made intolerable to them by Sunday service, or it may point to some rotary food-habit, for which there are many precedents. It is indeed a rule for some predaceous creatures to hunt in different places at regular times, taking them in something like serial order, on a calendar dictated by the local attractions.

The above record would have still greater value if it stated what proportion of each sex was represented each time.

POWERS

Its envied conquest of the realms of air seems the Bat's most wonderful gift. And yet more marvellous, though less spectacular, is its astonishing sense of touch, which, perhaps

[9] Am. Animals, 1902, p. 199.

more often than its eyes, averts the wreck and ruin that impetuous flight in gloomy woods might bring.

It is needful, indeed, that these two fairy gifts should dwell in complement of each other.

Writing of the wing-membrane, Harrison Allen says:[10]

"It is in this latter structure that the sense of touch chiefly resides. The bones of the extremities being covered on either side with an enduplication of skin, form a framework upon both sides of which the papillæ of touch are extensively distributed. This function, in many places, is probably aided by the delicate hairs which are sparsely distributed linearly upon the under surfaces of the membranes. These may perform a function analogous to that observed in the labial whiskers which are so prominent in the *Felidæ*. Spallanzani was the first to notice the high development to which this sense had been brought in these animals. His experiment is well known, but will bear repetition here:

SENSE-POWER

" 'In 1793, Spallanzani put out the eyes of a Bat, and observed that it appeared to fly with as much ease as before, and without striking against objects in its way, following the curve of a ceiling, and avoiding with accuracy everything against which it was expected to strike. Not only were blinded Bats capable of avoiding such objects as parts of a building, but they shunned with equal address the most delicate obstacles, even silken threads stretched in such a manner as to leave just space enough for them to pass with their wings expanded. When these threads were placed closer together, the Bats contracted their wings in order to pass between them without touching. They also passed with the same security between branches of trees placed to intercept them, and suspended themselves by the wall, etc., with as much ease as if they could see distinctly.'[11] "

Cuvier's explanation (1796) was that during the flight of the blinded Bat, the waves of air set in motion by its wings reacted against their acutely sensitive surface in proportion,

[10] Monog. Bats. N. A., 1864, pp. xv–xvi.

[11] Godman's Am. Nat. Hist., I, 1826, p. 57.

as it approached the solid object from which these had recoiled, and thus the creature had timely warning.

A comprehensive abstract of an article by Dr. Joseph Schöbl, of Prague, regarding the sense-power of Bats, appeared in the *American Naturalist* for May, 1871:[12]

Dr. Schöbl repeated the early experiments, but for the putting out of the eyes he substituted the more humane method of covering them with adhesive plaster. He kept Bats thus treated for a year alive in his room, and entirely confirmed Spallanzani's observations.

The experiment was repeated in detail recently by A. Whitaker, of Barnsley, Eng. His observations are thus detailed by Millais:[13] "He obtained a Natterer's Bat and covered its closed eyes with wax, fastening it with a little patch of rubber and solution, and then released the Bat in a room in which it had not been before. When, on previous occasions, other Bats had been at liberty in this room, they had usually circled round close to the ceiling; but this blinded Bat at first flew in a hesitating manner, and then, gaining confidence, went straight towards the closed door. It stopped when about six inches away, and hovered slowly along the line of the top and right down the side, without doubt keeping its position through its sensitiveness to the slight draught which came through the tiny chink. Finding no opening large enough to get through, it flew quickly down the room towards the fireplace, no doubt again feeling the draught, but turned away when it felt the heat of the fire. It flitted then close to the wainscot, hesitating repeatedly at a spot where the woodwork was a little sprung, and there was again a sensible draught. It flew quickly, passing under chairs, of which there were twelve in the room besides other furniture, and never even touched anything with the tips of its wings. An attempt to catch it showed that, although incapable of sight, it was well able to dodge; but it constantly stopped in its flight, hovered, and scratched at the covering over its eyes. When a stick was held in its direct path, it avoided it

[12] Am. Nat., Vol. V., No. 3, May, 1871, pp. 174-5.
[13] Mam. G. B. & I., Millais, 1904-5-6, Vol. I, pp. 96-7.

when three or four inches away. When it wanted to rest it settled on one of the weights of a gas chandelier in quite an orthodox manner, and when a hand was stretched out to capture it, it flew off again before it was touched. This experiment is interesting, especially when we hear that the animal's sight was in no way injured when the wax was removed; it shows that the power of ascertaining the presence of an object does not depend entirely upon sight, and that the difference of air pressure was perceptible by the animal."

"To account for these phenomena [says Merriam][14] the wings of Bats have been examined for peculiar nerve endings, by Cuvier, Leydig, and Krause, but without any success. The author's discoveries are therefore quite new to science. The following is a short abstract of his [Dr. Schöbl's] results. The Bat's wing-membrane consist of 2 sheets of skin, the upper derived from that of the back, the lower from that of the belly. The epidermic and Malpighian layers in each sheet remain separate, while the true skin is inseparably fused. In this fused medium layer are imbedded the muscles, nerves, vessels, etc., of the wing. * * * The whole wing is covered, both on the upper and under surface, with extremely fine, sparsely scattered hairs. * * * Each hair-sac has from 2 to 7 sebaceous glands, according to the species, and one sweat gland opening into its sac. The 2 outer fibrous layers of the hair-sac have no sharp line of demarcation to separate them from the surrounding connective tissue, but the inner, or hyaline coat, is highly developed, and, after being constricted beneath the hair bulb, widens out and encloses the sense-bodies (Tastkörperchen), one of which organs is connected with each hair. SENSE-BODIES

"The nerves of the wings may be considered to consist of 5 layers, *i. e.*, there is one occupying the centre of a transverse section of the wing, which gives off on each side of it 4 others, and these are successively finer and finer as they approach the opposite surfaces. The inner layer and the one immediately on each side of it consist of nerve fibres with dark borders, the other layers of pale fibres only. The tastkörperchen are con-

[14] Mam. Adir., 1884, pp. 185–7.

nected with the second layer. The fifth layer, of finest fibres, ends as a network between the innermost layer of cells of the Malpighian layer of the epidermis. The tastkörperchen are shaped like a fir-cone with a rounded apex turned inwards. They lie immediately below the root of the hair; and their core, or central substance, is formed of a prolongation of the cells forming the 2 root-sheaths, of the hair. Their length is 0.0259, and their breadth 0.0175 mm. A nerve containing about 6 dark-edged fibres is distributed to each körperchen. Just before the nerve reaches this organ it splits into 2, and 3 fibres pass to one side of it, 3 to the other. The fibres are then wound round the body so as to sheath its cellular core. Dr. Schöbl thinks it probable that the fibres on one side are continuous with those on the opposite side, and that there is thus a bipolar arrangement here. He attributes to the fine network of pale nerve fibres belonging to the fifth layer the appreciation of temperature, pain, etc.; to the tastkörperchen the highly exalted sense of touch. It is curious that both kinds of nerve endings are connected with the Malpighian layer of skin."

Some have sought to prove that the greatly developed ears were the organs of this far-feeling, but the sum of evidence goes to show that in nearly all parts of its skin the Bat has these highly developed nerve endings which endow it with such exquisite sensibility.

The purpose of the antitragus has been supposed by some to be closing the tragus, to protect its ear, while the animal slept. But Millais points out that during sleep it hangs outwards, leaving the ear still wider open.

FLIGHT The flight of all Bats is ideal; their mastery of the air is perfect; far better, indeed, than that of most birds. I am almost tempted to say, than that of *any* bird. Nothing but recollections of swallow and falcon restrain the phrase. In one way at least the Bat excels even these—his flight is absolutely silent. He skims and darts and turns within a foot of one's head, but never a swish of his wings is heard. The only bird that can approach him in silence is the slow, lumbering owl. I have heard it passed as a compliment to a certain owl—his flight recalled that of a big Bat.

A curious circumstance that at first sight looks like an exception to this rule is thus commented on by Merriam:[15]

"In localities where we had hunted Bats for some time Dr. Fisher and I have on several occasions heard a Bat, when swooping overhead, produce a sound which was distinctly audible at a distance of several paces. But in each instance, if the Bat rose against the clear western horizon, we saw the light shine through numerous perforations in its wings, and the noise was unquestionably produced by the whistling of the air through these shot holes."

During a recent motor trip, near Naples, I had an opportunity of gauging the speed of certain Bats. About sundown a considerable number of them appeared; they were of two sizes. The larger easily kept pace with the motor-car which was going 20 miles an hour; the smaller dropped behind. **SPEED**

So far as known, all our Bats are exclusively insectivorous. The Shrew may eat fruit, the Blarina may vary his diet of worms with nuts and grain, but the Bat at all times and places is found destroying only the little foemen that men hate. **FOOD**

The evidence relating to the British Long-eared Bat may help us to a knowledge of Bat foods in general. This species is known to feed on moths and the largest insects, and J. G. Millais says[16] that "when capturing its prey, the Long-eared Bat, in common with several other species, uses the interfemoral pouch as a trap or bag in which to hold its captive until it is firmly gripped."

He cites several instances to show that Bats do this either when taking insects on the wing or picking them off flowers and twigs.

All our Bats are great drinkers, going twice a day at least to the water and indulging in copious draughts. It is probable

[15] Mam. Adir., 1884, p. 188.

[16] P. 47.

that this and all our species have 2 feed times each day—one at sundown, one before sunrise. Never by any chance do they work between meals, but return to their dens to repose, thus living a life that, according to some of our prophets, is ideal.

TOILET

The behaviour of this Bat when performing its toilet is thus described by Stone and Cram:[17]

"One near the middle of the row was wide-awake; washing himself after the manner of a cat, he would lick his foot or a portion of his wing and rub his head with it the wrong way of the fur, and scratch himself rapidly behind the ear with one of his little thumb nails at the bend of his wing, the long bone of his forearm beating a tattoo on the glass beside him as he did so. The elasticity of the wing-membrane is truly astonishing; he would seize an edge of it in his mouth and stretch it into all kinds of grotesque shapes in his endeavour to get it clean enough to suit his fancy, and sometimes, when at work on the inside, he would wrap his head up in it entirely, the thin rubbery stuff conforming to the general outline of his skull in the most startling manner."

SLEEPING

"The method of alighting is first by the wing or arm-hooks, head upward, assisted by the hind-feet. As soon as the latter are firmly implanted, the Bat turns head downwards and hangs by the sharply recurved nails of the hind-feet."[18] (*Rhoads.*)

Millais says[19] of the Greater Horse-shoe Bat in England:

"When this Bat is preparing to sleep, it begins to doze gradually, nodding its head a little and ceasing to look about; finally, its head falls and hangs straight, and the whole animal commences to shiver. This muscular movement soon ceases, and the animal is asleep."

STRANGE IMMUNITY

The singular specialization of Bats is further evidenced in the following remarks by Sir Harry Johnston on the Noctule, a British kinsman of the present species:[20] "It would seem to possess a relative insensibility to the effects of poison.

[17] American Animals, 1902, p. 199.
[18] Mam. Penn. and N. J., 1903, p. 209.
[19] Mam. G. Br., 1904–6, Vol. I, p. 30.
[20] Br. Mam., 1903, p. 89.

One living specimen had a drop of prussic acid placed on its tongue, and was some time dying. Meantime, its parasites (all Bats are much afflicted with fleas and lice) dropped off dead from the poisoned blood."

MIGRATION AND HIBERNATION

With flight equal to that of birds, why should not these insect-eaters migrate when cold weather cuts off the food supply?

Such light as we have indicates that all of our 6 species of Bats are migratory, and yet hibernate. If we divide the range of the present species in 3 equal parts along the lines of equal temperature, we shall find that in summer the bulk of it is in the middle and north parts; in winter, is found in the south and middle parts; and, wherever it is, the individuals hibernate during frosty weather, but are always ready to come forth from cave and hollow tree and resume active life as soon and so long as the temperature permits.

Instances of this I noted at Wyndygoul Park in the fall of 1908. One or two small Bats were about each night after sundown through September and early October. In the middle of the latter month were two or three frosty nights, during which no Bats were seen. On the 16th the weather was warm again and I saw a small Bat on the wing; another on October 18; none afterwards. This, it seems to me, almost proves that these Bats went into a temporary cold-sleep during the frosty spell, but took advantage of the later warm days to migrate southwards, since they are not known to torpify for all winter.

ENEMIES

From the nature of their haunts, the Bats have little to fear from larger enemies by day, and through the power of their flight they are safe from most flyers by night. The fact that bat-remains have been found in owl-pellets is of sinister signification, but must have been exceptional, for I should as soon expect a Bear to catch a Jack-rabbit as an owl to capture a Bat in open space.

The worst known enemies of the group are undoubtedly fur-lice. Dr. Harrison Allen remarks:[21]

[21] Monog. Bats. N. A., 1864, p. xxi.

"The fact mentioned above of the numerous parasites infesting Bats is perhaps the most revolting feature in these creatures. The enormous population of *Acari* found upon their bodies is due to the great generation of animal heat in their close haunts, a condition conducive to a rapid increase of all kinds of vermin. In this country the common bed-bug (*Cimex lectularis*) is frequently found upon their fur. The entrance of a Bat with its precious burden into the open window of a farm-house is the solution of that frequently propounded question of the despairing housewife—'Where *can* the bugs come from?'"

It is only fair to these much misunderstood and maligned creatures of the night to add that no one else has remarked it, and many observers, including so good an authority as S. N. Rhoads, discredit the bed-bug theory altogether.

LV.

Say Bat.

Myotis subulatus (Say).

(L. *subulatus*, from *subula*, an awl, noting the awl-shape of the tragus.)

Vespertilio subulatus SAY, 1823, Long's Exp. Rocky Mts., II, p. 65.
Myotis subulatus MILLER, 1897, N. Am. Fauna, No. 13, p. 75.
TYPE LOCALITY.—Arkansas River, near La Junta, Colo.

FRENCH CANADIAN, *la Chauve-souris de Say.*

This species so closely resembles *M. lucifugus* that naturalists did not recognize their distinctness until recently. The most obvious difference is in the relative length of the ears, which, when drawn forward, do not reach the end of the muzzle in *lucifugus*, while in *subulatus* they overlap it by $\frac{1}{16}$ to $\frac{3}{16}$ of an inch (2 to 5 mm.). But the long thin tragus is equally distinctive (see Fig. 264, p. 1148).

Two races are recognized:

subulatus Say, the typical form.
keeni Merriam, with longer ears and tail, and darker colour.

LIFE-HISTORY.

It has not been taken in Manitoba, but it will be seen by the map (No. 64) that the Province falls within the probable range of the species.

MAP 64—RANGE OF THE SAY BAT.

Myotis subulatus (Say).

The outline in Eastern America is the theoretical range of the typical form. The outline on the West Coast is the theoretical range of the dark race, *keeni* (Merriam); it is supposed to meet and mingle somewhere with *subulatus*.

The spots are all the actual records that seem to be reliably distinguished from those of *lucifugus*. They are chiefly from G. S. Miller's Revision, and H. Allen's Monograph, with others by S. N. Rhoads, C. Hart Merriam, J. Richardson, J. Rowley, and O. Bangs.

It is much rarer than *M. lucifugus*, and we have but little light on its habits.

At Lake Nipissing, Ont., it seems to be common, which is somewhat of a surprise. Gerrit S. Miller, Jr., who found it there, says:[1]

"At North Bay I shot 5 of these Bats as they flew along the roadways through the woods at dusk. One evening I saw several feeding among the tops of some tall birches, to the twigs of which they would cling for an instant while picking off their prey."

[1] Mam. Ont., Proc. Bost. Soc. Nat. Hist., April, 1897, p. 39.

LVI.

Silvery-bat, Silver-haired Bat or Black-bat.

Lasionycteris noctivagans (Le Conte).

(*Lasionycteris*, from Gr. *lasios*, hairy; *nycteros*, a bat, and L. *noctivagans*, from *noctis*, of the night; *vagans*, wandering.)

Vespertilio noctivagans LE CONTE, 1831, McMurtrie's Cuvier, An. King., I, p. 431.

Lasionycteris noctivagans PETERS, 1865, Monatsb. Akad. Berlin, p. 648.

TYPE LOCALITY.—Eastern United States.

FRENCH CANADIAN, *la Chauve-souris argentée.*

In addition to the Family and sub-family characters, the genus *Lasionycteris* (Peters, 1865) has short broad ears, broad tragus, and partially furred tail-flap, 2 mammæ, and teeth as follows:

$$\text{Inc. } \frac{2-2}{3-3};\ \text{can. } \frac{1-1}{1-1};\ \text{prem. } \frac{2-2}{3-3};\ \text{mol. } \frac{3-3}{3-3} = 36$$

The present species is the only member of the genus.

SIZE In length it is $3\frac{3}{4}$ to $4\frac{1}{8}$ inches (95 to 105 mm.); forearm, about $1\frac{5}{8}$ inches (42 mm.); tibia, about $\frac{5}{8}$ of an inch (16 mm.); tail, about equal to forearm; spread, about $11\frac{1}{2}$ inches (292 mm.). For head, see Fig. 264, p. 1148.

COLOUR Fur, very dark brown, silver-tipped with white; this frosting confined chiefly to the back, and more conspicuous in the young than in the adult. I have in my collection an old female (New York State) without any silver tipping—brownish-black everywhere.

LIFE-HISTORY.

This and certain other Bats are the only mammals in America that range from the Atlantic to Pacific in a belt 1,000 miles wide, without splitting up into several races. Obviously, their wonderful powers of flight combine with wandering habits to keep the population thoroughly mixed. RANGE

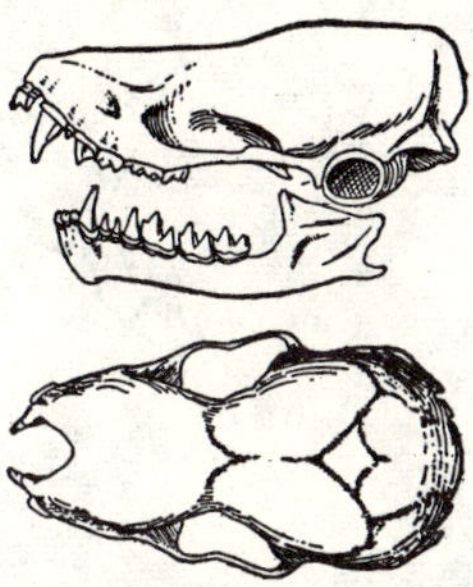

FIG. 265—Skull of Silvery-bat. (Twice life size.)

From Miller's Revision, N. A. Vesp. Bats, p. 85. Fauna 13, Biol. Surv. U. S. Dept. Agr.

All of Manitoba falls within its breeding range; the spots on the map show the actual records. IN MANITOBA

"Like many other Bats, it has a decided liking for water-ways, coursing up and down streams and rivers, and circling around lakes and ponds. In some places its habit of keeping directly over the water is very marked. At Lyon's Falls [on Black River, N. Y.] it is exceedingly abundant, particularly just below the Falls. I have stood, gun in hand, on a point on the east bank of the river, and have seen hundreds passing and repassing, flying over the water, while during the entire evening not more than two or three strayed so far that if shot they would fall on the land."[1] ENVIRONMENT

Over the Red River where it runs through Winnipeg City is a favourite flying place for this Bat. All the summer long, in the evenings, they may be observed hawking for their prey in this inviting open place.

At Calgary, I was told by G. F. Dippie, it is abundant.

In the vicinity of Toronto, the Don Valley and the Credit River Valley were evidently much to its taste. There I found the species in swarms during the August twilight. But any

[1] Merriam, Mam. Adir., 1884, p. 188.

MAP 65—RANGE OF THE SILVER-HAIRED BAT.

Lasionycteris noctivagans (Le Conte).

The unbroken outlines show the theoretical range. It is probable that it breeds in this area north of the dotted line, but winters in the belt that is south of it. The spots show the actual records; they are chiefly from H. Allen's Monograph, G. S. Miller's Revision, and O. Bangs's papers on Labrador and Florida, with others by R. MacFarlane, C. H. Townsend, W. H. Osgood, V. Bailey, S. N. Rhoads, and E. T. Seton.

opening near the woods seemed in some degree an acceptable hunting ground.

INDIVIDUAL RANGE

Nothing is known of the individual range of any of the Bats, but, gifted with such powers, they doubtless emulate those birds that range for miles to seek their favourite food. All the circumstances of their environment, habits, food, and powers, emphasize their analogy to the swallows. The Bats are the swallows of the night, and light on the home-range of the swallows will be side-light on the home-range of the Bats.

Like most of our mammals, they are subject to fluctuations of their numbers in periods covering several years. Thus the Silvery-bat was common at Winnipeg in 1905 and 1906, but only one was reported all through the season of 1907. These variations of the population are wholly uncomprehended; indeed, the first step towards comprehension—exact observations on their time, etc.—has not yet been taken.

VOICE

The remarks on the voice of the Little Brown-bat apply equally to this one.

GREGARIOUS

All of our Bats, except the Hoary, are known to gather together in numbers when lured by some especially desirable and commodious dormitory. In the Seneca Point 'battery,' described later, 9,640 Bats, by actual count, were killed, besides nearly 2000 not included, before the colony was extirpated. It is probable that most were of the present species.

Thus, while highly gregarious, they are also slightly sociable, since they profit by each other's company.

RUT

The rut is believed to be in late August or early September, as with others of the family, but there are several unwritten chapters in this department of the vespertilion history, and attempts to write them have resulted in the discovery of new mysteries to be solved.

The male takes no part in the rearing of the young; in fact, he is conspicuously absent at all times when the female does not actively interest him. Dr. Merriam, commenting on this remarkable circumstance, says:[2]

"Out of 85 adult specimens killed in Lewis County during the past summer (1883) there was but a single male. Two other males were killed in the early autumn. Of 32 young killed during the same period, there were 19 males and 13 females, showing that the disproportion does not exist at birth. I am at a complete loss to explain this enormous preponderance of females among the adults. At first I was inclined to think that the sexes separated during the period of bringing forth and caring for the young, but, although we visited a number of different localities, we were never able to find the males. Thinking that they might not fly until early morning, I several times went out before daylight, but females only were killed."

Since it is desirable to offer a theory, I suggest first, that the species is eminently polygamous, and second, that, like Wapiti, Mountain Sheep, pheasants, and sundry other polygamous species, the male associates with the female only during the procreative season, passing the interval in distant regions, usually higher and further north or otherwise less crowded with their own kind, as well as freer from insect plagues.

The great apparent increase of the Bat population in late August, therefore, is a real increase caused by the arrival of the flocks of males.

If this theory is sound, we should discover far north, or up in the mountains, great numbers of males that are there all summer, unassociated with females.

GESTATION

Gestation lasts apparently *ten months*. Young are usually 2 in number, but sometimes only 1. In the Adirondack region, according to Merriam, they are born during the first week of July.

"Females," he says,[3] "killed during the latter part of June, were heavy with young, but up to July 1 not one had given birth to its offspring. All that were killed after July 4 had already been in labour and were then suckling their young.

[2] Mam. Adir., 1884, p. 190.

[3] Mam. Adir., 1884, p. 190.

Of 3 females shot June 30, 1883, one contained but a single embryo, and the others, 2 each. All were nearly ready for extrusion, and would doubtless have been born within 48 hours."

This is a Bat of hollow trees, but I have known it to harbour in a thicket of spruce boughs high up from the ground, as well as in the lumber piles at Winnipeg.

The mother Bat does not prepare any nest for the young, but avails herself of such ready-made accommodations as she can find.

The most unexpected lodging of the species and nursery for the young is that detailed by Dr. Merriam, as follows:[4]

"Frank Hough tells me that when looking for young crows some years ago in the deep ravine that runs through the village of Lowville, in Lewis County, he espied a crow's nest in a large and densely foliaged hemlock. On climbing the tree, he found the nest to be an old one, and commenced tearing it to pieces, when, to his astonishment, he discovered 13 young Bats embedded in the sticks and litter of which it was composed. These Bats were taken home and shown to several members of the family. Their eyes were not yet open. They were, of course, the progeny of a number of females, and presumably were of the species now under consideration, because it is by far the most common in the region. IN CROW'S NEST

"The young * * * commence to fly when three weeks old. Those killed on the first evening of their appearance averaged 90 mm. [$3\frac{3}{5}$ inches] in length by 261 mm. [$10\frac{1}{4}$ inches] in stretch, but weighed only half as much as their parents. The adults average about 104 mm. [$4\frac{1}{8}$ inches] in length by 302 mm. [$11\frac{7}{8}$ inches] in stretch. When on the wing, the young may be distinguished from the old by the weakness and hesitancy of their flight, rather than by the difference in size. The young are much more beautiful than the adults, and they alone possess the perfect silvery tips to the hairs YOUNG

[4] Mam. Adir., 1884, p. 190.

from which the species derives its name. Even before going into winter quarters their soft silvery backs have given place to the grizzly coats that characterize the adults."

They are full grown by September; thus their very long gestation is offset somewhat by a rapid growth to adolescence.

HABITS

From sunset till black dark is the evening twilight, approximately equalled by the corresponding time before sunrise. The duration of this half-light varies of course with each day and at each latitude. In illustration Professor F. L. Blake, of Toronto Observatory, has supplied me with the following table of evening twilight at his station. Sunset is given in standard time, 75th meridian. The table is for latitude 43°, 40′ N.

March 1st,	Sun sets	6.05	P.M.	Twilight lasts for	1 h. 39 m.
April 1st	" "	6.44	"	" " "	1 h. 44 m.
May 1st,	" "	7.19	"	" " "	1 h. 56 m.
June 1st,	" "	7.52	"	" " "	2 h. 24 m.
July 1st,	" "	8.03	"	" " "	2 h. 30 m.
August 1st,	" "	7.41	"	" " "	2 h. 06 m.
September 1st,	" "	6.55	"	" " "	1 h. 48 m.
October 1st,	" "	6.00	"	" " "	1 h. 41 m.

The morning period is of about the same duration. These two twilights are the time of the Bats, and all the evidence we have goes to prove that it is the exception when they are abroad in black darkness.

It is probable that each species is adapted to a certain measure of light, and those that come out early do not stay late.

All must vary their habits with the ever-varying twilight, but, broadly speaking, we can arrange the species in a scale of crepuscularity, giving each an hour for its food-hunt and leaving out the Say Bat, as not reliably observed. It is worthy of note that the Bats come out nearly in order of size, and that a similar scale and remark will probably be found to fit our night-flying insects.

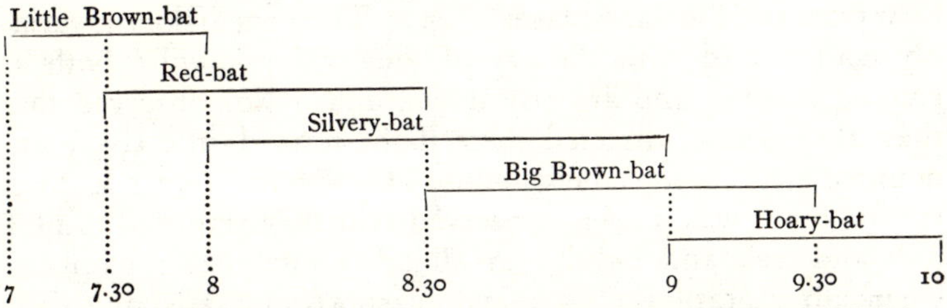

Bat scale for evening early in August, near Toronto, Can., sun setting just before 8. In the morning it is probably reversed. (This is, of course, diagrammatic rather than literal.)

MEALS

There seems little doubt that each of our Bats gathers all its food in the two twilights, retiring between times to its lurking place all day and all night. Moonlight probably has a modifying effect.

HABITS

The following interesting record, by M. Figaniere, appeared in Allen's "Monograph."[5] It is not by any means certain which kind of Bat was meant; the probabilities are that there were several species, with the present one predominant, since he twice calls them very small and very black:

"In the winter of 1859, having purchased the property known as Seneca Point, on the margin of the North-east River, near Charlestown, in Cecil County, Md., we took possession of it in May of the next year. * * * Having been uninhabited for several years, it exhibited the appearance, with the exception of one or two rooms, of desolation and neglect. * * * The weather which was beautiful, balmy and warm, invited us towards evening to out-door enjoyment and rest after a fatiguing day of travel and active labour; but chairs, settees, and benches were scarcely occupied by us on the piazza and lawn, when, to our amazement, and the horror of the female portion of our party, small black Bats made their appearance in immense numbers, flickering around the premises, rushing in and out of doors and through open windows.

"* * * Evening after evening did we patiently, though not complacently, watch this periodical exodus of dusky wings into

[5] H. Allen, Monog. Bats N. A., 1864, pp. xvii–xviii.

light from their lurking places. * * * Their excursions invariably commenced with the cry of the 'whippoorwill,' both at coming evening and at early dawn, and it was observed that they always first directed their flight towards the river, undoubtedly to damp their mouse-like snouts, but not their spirits, for it was likewise observed that they returned to play hide-and-seek and indulge in all other imaginable gambols, when, after gratifying their love of sport and satisfying their

EXTERMINATE MOSQUITOES

voracious appetites (*as the absence of mosquitoes and gnats testified*),[6] they would re-enter their habitations, again to emerge at the first signal of their feathered trumpeter. I thus ascertained one very important fact, namely, that the Bat, or the species which annoyed us, ate and drank twice in twenty-four hours."

So far as known, all our Bats live on insect food captured and eaten while the Bat is on the wing, but the insect itself may be either flying or perched on a flower or leaf.

Some general remarks on methods of capture are given in the chapter on *M. lucifugus*, but are equally applicable here.

There is, however, one little ceremony with which it preludes every meal; and this observation applies certainly to *Myotis* and *Lasionycteris*, and probably to all the others. On leaving its den it flies first and straight to the nearest river or pond and there drinks copiously as it skims over the surface. Merriam says of this species, as he observed it in the Adirondacks:[7]

"In the early dusk the Silver-haired Bat emerges from its hiding place. After a few turns about the immediate neighbourhood it generally takes a pretty direct course for water. I have seen it start from the summit of a high densely wooded hill, circle around for a few minutes, and then, keeping far above the tree-tops, sail leisurely towards a distant river till lost from sight in the valley below. And, standing on the banks of the large stream that winds along the foot of this hill, I have seen the Bats flying over at a height of several hundred feet, all moving in the same direction—towards a more distant river."

[6] Italics mine.—E. T. S.

[7] Mam. Adir., 1884, p. 189.

SWIMMING

A past master in the air and fairly active on tree trunk and ground, the Silvery-bat rounds out its accomplishments by swimming fairly well. Dr. Merriam speaks of shooting some of them over the river at Lyon's Falls, N. Y.:

"Several that were wounded and fell into the water[8] [says he], at a distance of 15 or 20 feet from the bank, swam ashore. They swam powerfully and swiftly, for the current is here quite strong, and would otherwise have carried them some distance down stream."

This remark will doubtless be found applicable to all our Bats, as their European relatives, without exception, are known to be strong swimmers, though they never voluntarily take to the water.

ENEMIES

There is no doubt a host of creatures that would destroy the Bat if they could catch it, but safe in its cave by day and safe on its wings by night, it stands in awe of very few. A record in Fisher's "Food of Hawks and Owls" (p. 178) shows that the winged tiger of the woods does indeed prey on this Bat, but there is no telling how it secured the squeaker of the lightning wing. It may have been an accident or it may be there is a weird unwritten chapter of owl audacity awaiting the careful student of birds.

HORNED OWL

MOONLIGHT

It is well known that an exceptionally dull day or afternoon will temper the light down to the requisites of the Bats, and bring many forth long before their appointed hour. There is every reason to suppose that moonlight may similarly change their habits; but hitherto we have no observations to prove it.

DEPARTURE

Beginning at the vernal equinox and continuing all summer long, the Silvery-bats go skimming over the broad Red River where it mirrors the tall buildings of Winnipeg; darting and wheeling like swallows of the gloom, enlivening bank and sky, and retiring between times to the shelter of the lumber

[8] *Ibid.*, p. 188.

piles, that rise in yellow foursquare towers by hundreds on the banks.

But again a change comes with the equinox. Each year, about September 21, the flittering host of eventide is gone—has disappeared as mysteriously as it came. The workmen no longer discover them crouching during the day beneath the sheltering boards. They have totally vanished, and all the evidence I have goes to show that this disappearance is due to migration.

How much farther south do they go?

ROUTE

Miller says:[9] "The Silver-haired Bat occurred in spring and fall about the light-house on Mount Desert Rock, 30 miles off the coast of Maine, a treeless islet where Bats were at other times unknown."

He states also that though the species was unknown in early summer at Highland Light, Cape Cod, Mass., "the animals * * * suddenly became numerous shortly after the middle of August, and remained abundant for about a month, when they as suddenly disappeared."

Thus they are being traced on their route and marked down in their seasonal homes. Before long the fragmentary observations of many naturalists put together will spell the truth and show us that the Bats are as migratory as the birds, and, though long despised, may be also as interesting and beautiful.

[9] N. A. Fauna, No. 13, 1897, p. 11.

LVII.

Big Brown-bat, House-bat, Serotine Bat or Carolina Bat.

Eptesicus fuscus (Beauvois).
(G. *Epten*, to fly; *oikos*, house, *i. e.*, house-flyer; L. *fuscus*, brown.)

Vespertila fuscus BEAUVOIS, 1796, Cat. Peale's Mus., Philadelphia, p. 14.
Eptesicus fuscus MÉHELY, 1900, Monogr. Chir. Hung., p. 208.

TYPE LOCALITY.—Philadelphia, Pa.

FRENCH CANADIAN, *la Chauve-souris brunne*.

In addition to the Family and sub-family characters, the genus *Eptesicus* (Rafinesque, 1820) has 2 mammæ, the basal third of tail-web hairy, ears medium and somewhat pointed, and teeth thus:

$$\text{Inc. } \frac{2-2}{3-3}; \text{ can. } \frac{1-1}{1-1}; \text{ prem. } \frac{1-1}{2-2}; \text{ mol. } \frac{3-3}{3-3} = 32$$

The size of *E. fuscus* is distinctive among American species of the genus. Length, $4\frac{3}{16}$ to 5 inches (106 to 127 mm.); tail, $1\frac{1}{2}$ to $2\frac{1}{16}$ inches (38 to 52 mm.); forearm, $1\frac{3}{4}$ to $1\frac{7}{8}$ inches (44 to 48 mm.); tibia, about $\frac{3}{4}$ of an inch (about 19 mm.); spread, 12 to 13 inches (305 to 330 mm.). SIZE

Weight of an adult male, $\frac{1}{2}$ ounce. WEIGHT

In colour it is wood-brown throughout, paler below; fur never silver tipped. (For head, see Fig. 264, p. 1148). COLOUR

The following races are recognized:

fuscus Beauvois, the typical form.
miradorensis H. Allen, larger and darker.
propinquus Peters, very small, colours dark.
bahamensis Miller, small and with narrow muzzle.
cubensis Gray, larger, resembling *miradorensis*, but ears smaller and more pointed.
peninsulæ Thomas, small and pale.
bernardinus Rhoads, like *fuscus*, but paler.
osceola Bangs, more cinnamon; otherwise like *fuscus*.
melanopterus Rehn, from Mt. Tallac; has blackish feet and wings.

Life-history.

DISTRIBUTION

This Bat is one of the widest of rangers, and although it scarcely enters the Canadian zone, it more than makes up by spreading far into Central America (Map 66). It is included in the Manitoban list solely on the strength of a specimen secured by Kennicott on Lake Winnipeg (No. 6192, U. S. N. M., alcoholic), identified by H. Allen in his 1893 Monograph, p. 121. My own acquaintance with it was made at Toronto and New York, where I found it at its old trick of entering the house, like a winged burglar, under cover of night.

HOME-RANGE

Nothing is known of the home or individual range of this or any other of our Bats.

ENVIRONMENT

The chosen environment of the species differs little from that of its congeners of the smaller kinds. Open ways between trees seem especially alluring. In primitive times this confined them largely to the rivers and ponds, but now the clearances have enlarged their opportunities, and every meadow and field near sheltering groves is an eligible hunting-ground for the Big Brown-bat.

It is a lower flyer than the Red-bat or the Hoary, and has a marked preference for town life.

MAP 66—RANGE OF THE BIG BROWN-BAT.

Eptesicus fuscus **Beauvois.**

The outlines show the theoretical range of the species; the spots are actual records. They are chiefly from G. S. Miller's Revision and H. Allen's Monograph, with others by J. A. Allen, C. Hart Merriam, G. S. Miller, C. H. Townsend, V. Bailey, and E. T. Seton.

Time is an important element in all Bat fly-ways, and Dr. A. K. Fisher's observations help us to place this species in the nightly time-table of the out-doors. Excepting that the Hoary-bat is still later and is probably not reckoned with in his observations, mine agree with those of Dr. Fisher.

Writing of the Big Brown-bats, he says:[1] "They are the last to make their appearance in the evening. In fact, when it gets so dark that objects are blended in one uncertain mass, and the bat-hunter finds that he is unable to shoot with any precision, the Carolina Bats make their appearance as mere dark shadows, flitting here and there, while busily engaged in catching insects. We have to make a snapshot as they dodge in and out from behind the dark tree-tops, and are left in doubt as to the result until in the gloom we may perchance see our little 'black-and-tan,' seemingly as interested in the result as we are, pointing the dead animal."

The breeding habits of this Bat are unknown.

BREEDING

Probably they rut in late summer and produce their 2 young in late spring, but I can find no positive evidence of it, which is the more remarkable since this is a conspicuous and abundant species, and was one of the earliest to be described.

FOOD HABITS

Their food no doubt differs chiefly in quantity from that of the smaller species. Dr. R. W. Shufeldt says:[2]

"They drink a good deal and have simply enormous appetites. One specimen, in the course of a single night, consumed 21 full-grown June-bugs (*Lachnosterna fusca*), leaving only a few legs and the hard outside wing-sheaths."

ENEMIES

By nature of their retreats and their wonderful powers of flight, the Bats have few enemies to fear, and it was a surprise to me when I read in Bailey's "Biological Survey of Texas"[3] that: "At Mr. C. O. Finley's ranch, at the west base of the

[1] Forest and Stream, July 21, 1881, p. 490.
[2] Chapters, Nat. Hist. United States, 1897, p. 440.
[3] N. A. Fauna, No. 25, 1905, p. 211.

Davis Mountains [south-western Texas], I found two lower jaws of this Bat among numerous other bones in pellets under the nest of a great horned-owl."

At Chilliwack Lake, in British Columbia, the rainbow trout are of great size, 8 pounds to 12 pounds, and these giants were often seen by Professor John Macoun, leaping after the Bats that skim the surface of the lake at evening. In one case he thinks he saw a Bat captured by the trout, and is satisfied that the fish would not jump so persistently if they did not frequently succeed.

The following interesting remarks on its habits in general are contributed by Rhoads:[4] HABITS

"Among American Bats this species may be said to correspond in its fondness for the homes of man to the Mouse and Rat, or to the robin and the wren among birds. During summer, they are as likely to hang up for day-dreams behind an unused shutter or door, or a crack in the wall, or a shady porch or out-house, as anywhere else. At night, they incessantly circle about the house and lawn and street lamps until some fleeing insect suddenly leads one into the kitchen or the bedchamber, and, 10 to 1, a panic ensues, resulting in no small noise, destruction of furniture, and the miserable death of the innocent, harmless, and useful Bat. Such an occurrence as this, related by Audubon, happened in his Kentucky home in 1818, in the bedroom occupied by the traveller Rafinesque. It resulted in the destruction of a favourite violin, etc. [which the guest used as a bat-club], and, so far as we are able to follow the sequel, in the immortality of the Bat as '*Eptesicus melanops* Raf.', which, being interpreted, is no less than a synonym of *Eptesicus fuscus* (Beauvois), the subject of this article. It is interesting to note that Rafinesque, in describing the genus *Eptesicus*, says: 'The name means house-flyer'; and of the species *melanops* he says: 'It comes often in the house at night'; recording in this way the indelible impressions of his midnight battle two years before. This Bat is accused of bringing bed-bugs and other insect

[4] Mam. Penn., 1903, p. 212.

vermin into houses. I have never found any vermin on them except lice of a species not parasitic on man."

In the Museum of the Geological Survey at Ottawa is a specimen found dormant behind a window of the building on December 3, 1894. This is the more interesting because it is nearly the north-most record and the species is supposed to be migratory.

LVIII.

Northern Red-bat or Tree-bat.

Lasiurus borealis (Müller).
(Gr. *lasios*, hairy; *oura*, tail; L. *borealis*, of the North.)

Vespertilio borealis MÜLLER, 1776, Natur. Syst., Suppl., p. 21.
Lasiurus borealis MILLER, 1897, N. Am. Fauna, No. 13, p. 105.
TYPE LOCALITY.—New York.

FRENCH CANADIAN, *la Chauve-souris rouge.*

In addition to the Family and sub-family characters, the genus *Lasiurus* (Gray, 1831) has the *tail densely hairy* above, continuous with the back furring; the ear broad, low, and round-topped; mammæ, 4; teeth:

$$\text{Inc. } \frac{1-1}{3-3};\ \text{can. } \frac{1-1}{1-1};\ \text{prem. } \frac{2-2}{2-2};\ \text{mol. } \frac{3-3}{3-3} = 32$$

Total length, 4 to $4\frac{3}{8}$ inches (102 to 111 mm.); tail, about 2 inches (about 51 mm.); forearm, $1\frac{1}{2}$ to $1\frac{5}{8}$ inches (38 to 41 mm.); tibia, about $\frac{3}{4}$ inch (about 19 mm.); spread, about 12 inches (305 mm.). SIZE

In size it is much like the Silver-haired Bat (see Fig. 264).

The colour of the Red-bat is usually described as a bright rufous or dull orange. This, however, conveys no idea of the exquisite tints that go to make up the general tone of red. The Manitoba specimen before me is, first, all over of a delicate, pale COLOUR sienna, but on the upper part and across the breast each hair has a dark orange-brown or sienna outer part that gives the general colour, with the pale under-fur showing through at places; finally, each hair on the fore-back and across the chest has a silvery-white tip; on the shoulder in front of their wings these are so large as to form a white spot on each side. In certain lights the back fur has the effect of rich golden-brown glossed with purple. Individuals vary from yellow to deep red, but the style is unmistakable; by its colour alone the Red-bat may be known.

In the air, its long, thin wings distinguish it from all but the Hoary-bat; the probabilities of time and place, as well as their sizes, will help in discriminating these two.

The following races are recognized:

borealis Müller, the typical form.
seminolus Rhoads, darker (mahogany brown).
pfeifferi Gundlach, larger and deeper coloured than *borealis*.
teliotis H. Allen, smaller and deeper coloured than *borealis*, with tail-web furred only on basal third.
mexicanus Saussure, like *teliotis*, but membranes and feet less hairy.

Life-history.

RANGE Its range covers the continent from the southern edge of the Hudsonian to the northern edge of the Tropical fauna, east of the Great Plains, as set forth on Map No. 67, and yet farther into Central or South America. Its claim to be in the Manitoban list rests on 3 specimens, 1 I received from Morden, where it was captured by D. Nicholson, and 2 taken at Winnipeg and shown to me by Edward Wilson. Another was taken by Edwin Hollis at Touchwood Hills, September 16, 1901; it was a female. G. F. Dippie assures me that he has taken it several times at Calgary, Alta.

MAP 67—RANGE OF THE RED-BAT AND ITS FIVE RACES.
Lasiurus borealis (Müller).

The outlines show the theoretical range, the spots are the actual records, chiefly from G. S. Miller's Revision, and H. Allen's Monograph with others by V. Bailey, S. N. Rhoads, O. Bangs, and E. T. Seton.

ENVIRONMENT The Red-bat is, above all its kin in our country, a tree Bat, never frequenting caves. Every specimen that I have seen, and all I have knowledge of, were found hanging from a branch in the woods, generally a very low one, and looking like a rumpled leaf that is prematurely dead but not yet fallen. Merriam speaks[1] of finding them "asleep in the daytime, hanging by their thumb-nails to small twigs or leaf stems within easy reach." If disturbed on such occasions, they flit away to some other lowly place, apparently not at all incommoded by the brightness of the daylight.

C. W. Nash tells me that in Manitoba he has seen them coming out of burrows in the moss.

SOCIABILITY This Bat, like its relative, the Great Hoary-bat, seems to be far from gregarious, in fact almost a solitary species, or seen only in pairs.

MATING It is not known whether our Bats are polygamous, polyandrous, or promiscuous. Rhoads speaks of finding this kind in pairs,[2] which is a mite of proof that the species has progressed. The only recorded observation on their mating is the following by Dr. J. A. Allen:[3]

"Very little seems to be known respecting the time of copulation or the period of gestation of the Bats. From Mr. J. G. Shute, of Woburn (Mass.), I learn a fact in reference to this point observed by him some few years since. Soon after sunset, one evening in October, he observed a strange object pass him in the air, which seemed to fall to the ground not far from where he was standing. Repairing immediately to the spot, he soon found it, which proved to be a pair of these Bats *in coitu*. They were captured and thrown into alcohol, and thus forwarded to the Museum of Comparative Zoölogy."

Miller mentions[4] that this Bat *breeds* at Brownsville, Tex., so that it may breed in all of its range north of that point.

[1] Mam. Adir., 1884, p. 182. [2] Mam. Penn., 1903, p. 213.
[3] Bull. Mus. Comp. Zoöl., Vol. I, No. 8, 1869, p. 208.
[4] N. A. Fauna, No. 13, 1897, p. 108.

As the young are found of tender age early in June and late in May, it is probable that the gestation lasts for 8 months. GESTATION

Although but 1 young one has been found with the mother in most cases observed, M. W. Lyon, Jr., calls attention[5] to the fact that this Bat has 4 mammæ, and that 4 unborn young have been taken from an adult female of *L. borealis salinæ*. YOUNG

He says: "While the rule for most Bats is 1, or sometimes 2, offspring at parturition, yet a careful examination of material and the literature shows the number of young produced at a time by members of the genus *Lasiurus*, and probably *Dasypterus*, is usually double that number. This might safely be inferred from the fact that 4 mammæ are found in Bats of this group, as has been noted by several writers. In all other Bats, so far as the writer is aware, there are 2 mammæ, each of which is placed near the middle of the outer border of the pectoral muscle. In the *Lasiurine* Bats, in addition to these 2, there is a second pair, located more posteriorly, each mamma of which is nearer the back and pretty well up under the wing.

"As to the number of young in *Lasiurus*, Professor Wilder found 3 embryos in each of 2 specimens of *L. borealis* from Massachusetts.

* * * * * * *

"A specimen of *L. borealis* from Illinois (No. 14,273), preserved in alcohol, contains 2 fetuses.

"The most interesting specimens in this connection are Nos. 114,044 to 114,048, an adult female nursing 4 young, brought into the National Museum alive by Mr. J. C. Lawson, of Washington, D. C., on June 18, 1902. * * * A young one was at each of the adult's nipples, where it held on with great tenacity, having in its mouth a good deal of its mother's hair, into which its hooked milk teeth firmly caught. As Dobson has suggested, it is probably for the purpose of holding securely

[5] Proc. United States Nat. Mus., Vol. XXVI, pp. 425–6, 1903.

to their mothers that the milk teeth of Bats differ in form from those of other mammals.

* * * * * * *

"The young ones were less than a third grown as to weight, and about half grown as to linear measurements. The combined weights of the four young amounted to 12.7 grammes, while the mother weighed but 11 grammes."

Three of the young were females, 1 a male.

Titian Peale, of Philadelphia, is responsible for the following interesting story of maternal instinct in a Red-bat:[6]

"In June, 1823, the son of Mr. Gillespie, keeper of the city square, caught a young Red-bat (*Vespertilio noveboracensis L.*), which he took home with him. Three hours afterwards, in the evening, as he was conveying it to the Museum in his hand, while passing near the place where it was caught, the mother made her appearance, followed the boy for two squares, flying around him, and finally alighting on his breast, such was her anxiety to save her offspring. Both were brought to the Museum, the young one firmly adhering to its mother's teat. This faithful creature lived two days in the Museum, and then died of injuries received from her captor. The young one, being but half grown, was still too young to take care of itself, and died shortly after."

"The young of this species continue to nurse till at least a month old. I shot a female on the 31st of July (1883) whose udders still contained milk, and whose long nipples were much drawn out. A week later (August 7) I killed a full-grown young flying over the same meadow."[7]

From the above I should rather infer that the young nurse till about two months old, or even more.

HABITS "The Red-bat [says Merriam][8] generally makes its appearance earlier in the evening than the other species, evidently fancying the dusk of twilight more than the increased darkness of advancing night, and I have killed it even on a cloudy after-

[6] Godman's Am. Nat. Hist., 1826, Vol. I., pp. 56–7.

[7] Merriam, Mam. Adir., 1884, p. 184.

[8] *Ibid.*, pp. 181–2.

noon, while flying to and fro in pursuit of insects near the border of a hardwood grove."

Whether it has a morning meal-hour as well is not ascertained.

Many observers comment on the readiness of this Bat to enter the house at night and pursue the insects that are attracted by the candles. Yarrow says[9] that in most portions of the United States it would be safe to say that, in any given instance of a Bat entering our rooms in the evening, the chances are a hundred to one of its being either the Red-bat or the Little Brown-bat.

Dr. Hornaday, after crediting this species with unusual cleverness, adds[10] that the only mistake it makes "is in flying into houses through open windows, and instantly forgetting the location of the means of escape. Once in a room, the Bat flies slowly, and frequently is so bewildered by the sudden change from semi-darkness to light, that it strikes a wall and falls to the floor."

In summer, as we have seen, the Bat invariably roosts in a tree. In Manitoba, at least, it certainly does not do so in winter. In summer it is solitary and not known to frequent caves. In winter it is known to gather in vast numbers in the caves of its more southerly range. As far as these facts go, they point to a migration from the northern part of its range and a hibernation in the southern part, a complete change of behaviour in each case accompanying the change of life. MIGRANT

The segregation of the sexes seems an important rule in Bat life. Dr. E. A. Mearns has supplied some remarkable observations on this head during the migration. Writing from the Hudson Highlands of New York, where this Bat is very abundant in summer, he says:[11]

"During the latter part of October and the first week of November I have seen great flights of them during the whole day. In 1876, I noted that all of the individuals shot from any

[9] Zoöl., Surv. West of 100th Merid., 1875, p. 89. [10] Am. Nat. Hist., 1904, p. 65.
[11] Vert. Faun. Hudson Highlands, Bull. Am. Mus. Nat. Hist., 1898, Vol. X, p. 345.

single flock were of the same sex, though another flock might yield all of the opposite sex. One year specimens are recorded on four days, on two days only males, and on two only females. So far as they go, these observations suggest that the sexes perhaps separate during autumnal flights, as birds commonly do."

Following along their route further south, we come into the observational region of another good naturalist, S. N. Rhoads, who writes:[12]

"I have observed this species returning from apparently extensive flights over the ocean on the New Jersey coast in the early morning before sunrise. On one or two occasions in September single individuals have been observed flying directly towards the shore, so exhausted as to make little progress against a land breeze, and alighting on the nearest object as soon as land was reached. It is possible that these had been blown to sea during their migrations along the coast."

A gale so trying to the Red-bat must have been disastrous to the birds. For this is one of the swiftest and strongest fliers of its tribe. Oftentimes in the evening one has the chance to compare the flight of the Little Brown-bat with that of the chimney-swift, and never does one incline to give inferior rank to the Bat. But the Red-bat is superior to its twilight brother as a flier; not only is it swift as the swift itself, but it can turn and twist and dash within a hair-breadth of destruction, or through a hole that is not half its wing-extent, and perform a hundred feats of power that are far beyond any but birds of the longest and strongest wings.

It has, indeed, achieved a consummate mastery of the realms of air, a conquest at least as complete as that attained by swallow, swift, or hawk, a fact that should have its meed of comfort for those hopeful human æronauts who have long been told in scorn that feathers are the only means to perfect flight.

[12] Mam. Penn., 1903, p. 213.

LIX.

The Hoary-bat or Great Northern Bat.

Lasiurus cinereus (Beauvois).
(L. *cinereus*, cindery or ash-coloured.)

Vespertilio cinereus BEAUVOIS, 1796, Cat. Peale's Mus., Phila., p. 15.
Lasiurus cinereus H. ALLEN, 1864, Monog. Bats N. A., p. 21.
TYPE LOCALITY.—Philadelphia, Pa.

FRENCH CANADIAN, *la Chauve-souris grisonnée ou grise.*

In addition to the generic characters given in the preceding article, this Bat has very obvious and specific features that make it easy to identify.

In size it exceeds all others in our list. Total length, 5 to 5½ inches (127 to 140 mm.); tail, 2 to 2⅜ inches (51 to 60 mm.); forearm, 1¾ to 2¼ inches (45 to 57 mm.); tibia, ⅞ to 1 inch (22 to 25 mm.); spread, 15 to 17 inches (381 to 432 mm.). SIZE

In style of colour it approaches the Red-bat. The fur next the body is dull blackish, but this does not ordinarily show; the general under-colour visible is a soft sienna or orange buff, much grayer on breast and belly; but on the chest and upper parts each hair has a dark-brown zone, and, finally, a silvery-white tip; the general effect is rich, deep brown, exquisitely frosted over with white; the chin is clear pale brown or yellow; *the ears have black rims*, and the muzzle is more or less black. COLOUR

When in air, this species may be distinguished from all by its long, pointed wings, great size, and swift zigzag flight.

Life-history.

DISTRIBUTION

Map No. 68 shows that this fine Bat ranges over nearly all the continent, breeding in the northern and wintering in the southern half. I have seen about a dozen specimens in Manitoba; these were taken at Carberry, Sourismouth, Morden, and Winnipeg. Edwin Hollis captured a female at

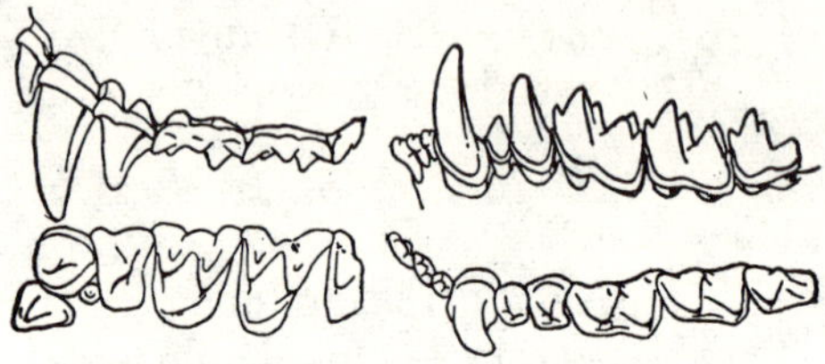

Fig. 266—Left side teeth of Hoary-bat; two views of each row. (Five times life size.) Those to the left are the upper; to the right the lower.
From Miller's Revision, N. A. Fauna No. 13, p. 114. Biological Surv., U. S. Dept. Agr.

Touchwood Hills, September 11, 1901.[1] G. F. Dippie reports it at Calgary, and on July 13, 1907, E. A. Preble saw one at Fort Resolution.

INDIVIDUAL RANGE

Nothing is known of the extent to which the individual will roam, but there is no reason why it should not cover as much ground as a small falcon, if its necessities should demand it.

ENVIRONMENT

All the Hoary-bats I have met were in half-open country, where both woods and water abounded. I see nothing peculiar in this, and yet its chosen surroundings differ somewhat from those of its congeners. I am not aware that any of our Bats habitually hunt in the thick woods; they can find a better, easier prey in the near-by openings. Some of the small species are content with the small openings and small prey near the ground; but the Great Hoary-bat soars high. Its proper place, as I understand it, is far above the tree-tops of the forest,

[1] Zoöl., Aug. 15, 1902, p. 297, and subsequent personal letter.

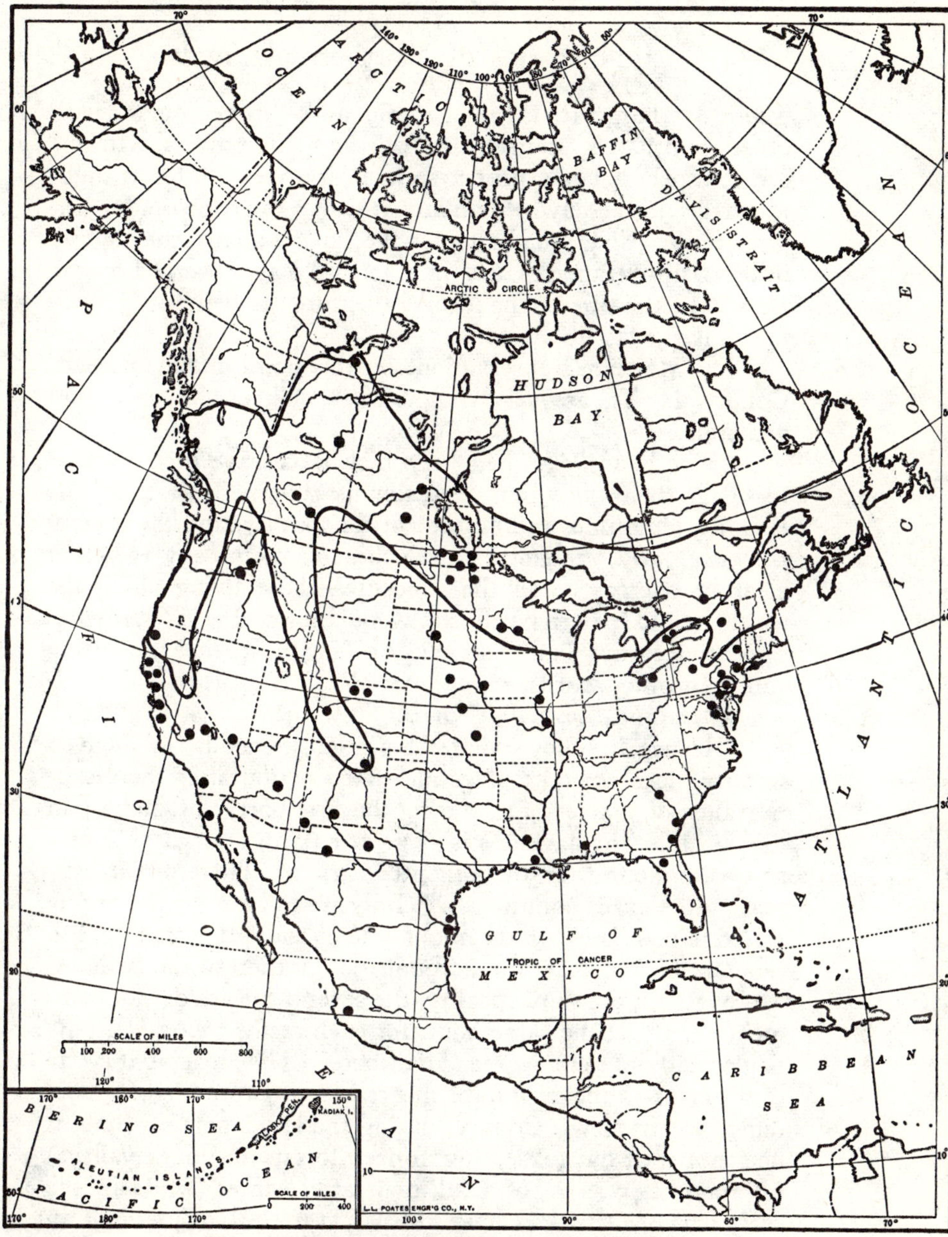

MAP 68—RANGE OF THE HOARY-BAT.

Lasiurus cinereus (**Beauvois**).

The outlined portion is the theoretical breeding-ground; over the rest of the continent, to the south of this, it is a migrant or winter resident only. The spots are actual records taken chiefly from H. Allen's Monograph and G. S. Miller's Revision, with others by J. A. Allen, J. Richardson, V. Bailey, O. Bangs, and E. T. Seton.

where fly the great, fat luna moths and the cecropias, with others half as big as itself and worthy of its powers. In one other respect it has a proper sphere—it appears later at night (or earlier in the dawn) than any others of the tribe on our list.

In the Silvery-bat chapter we have noticed the nightly order of appearance that is, in a general way, observed by the Bats, and the present handsome creature, as befits its importance, is the last to appear.

Dr. Merriam's observations on this point, made in Lewis County, N. Y., are very complete, as well as the only ones available.

"The hour," says he,[2] "at which Bats leave their retreats to begin their nocturnal excursion is governed, first, by the latitude, longitude, and altitude of the locality, and the time of the year; and, second, by the character of the sky (whether clear or overcast) and the exposure—those living along the southern and eastern borders of woodlands and in dark ravines appearing earlier than those whose hiding places face the setting sun. In other words, *the time at which Bats appear depends solely upon the degree of darkness.*[3]

"Hence it follows that their nightly exodus, in a given locality, does not take place in a fixed period after the disappearance of the sun, for, during the first part of October, in this latitude, the darkness is as great half an hour after sunset as it is an hour after, three months earlier. Therefore, in estimating the exact hour at which Bats are to be expected at any stated date, it is necessary not only to consider the time the sun sets, but also to take into account the duration of the twilight. Moreover, in the same locality, the several species do not commence to fly at the same hour, for each seems to await a particular and different degree of darkness. The Hoary-bat is one of the last to appear, and for this reason its capture is the most difficult. In Lewis County, during the latter part of June, it does not start out (excepting in deep forests and dark valleys) till about 8.45 P. M., or a full hour after sunset, while in the early part of October I have killed it at 6 P. M., or just half an

[2] Mam. Adir., 1884, pp. 180–1.

[3] Italics mine.—E. T. S.

hour after sundown. The following table is calculated to illustrate the above remarks:

TIMES OF EVENING APPEARANCES OF 'ATALAPHA CINEREA' [HOARY-BAT] AT LOCUST GROVE, N. Y., AT DIFFERENT DATES IN 1883.

DATE	SUNSET	FIRST BAT SEEN	TIME AFTER SUNSET
June 30	7.42 o'clock	8.45 o'clock	63 minutes
July 9	7.38 "	8.30 "	52 "
July 31	7.21 "	8.10 "	49 "
Aug. 3	7.17 "	8.00 "	43 "
Aug. 21	6.52 "	7.30 "	38 "
Oct. 8	5.30 "	6.00 "	30 " "

In Manitoba the den or lurking place of this species is among the thickest boughs of a spruce top. Whether it ever frequents caves or hollow trees when available I cannot say. LURKING PLACE

A specimen which I took near Carberry, August 21, 1883, bit viciously and screamed aloud, in the style of the Little Brown-bats, but in deeper, stronger tones. VOICE

I cannot learn that any one ever saw a great number of these Bats in one lurking place. The smaller species may gather in crowds at a convenient shelter nook, yet this, the king of the northern kinds, is, so far as known, solitary at all times save the breeding season. UNSOCABLE

Concerning its mating in the Adirondack region, Dr. Merriam says:[4] "That the species ruts about the 1st of August there can be no reasonable doubt, for I saw more of them from the 30th of July till the 6th of August [1883] than I have seen in all before and since, and 12 adult specimens killed during that brief period were all males. They were not feeding, but were rushing wildly about, evidently in search of the females. Many flew so high as to be entirely out of range, though directly overhead." RUT

[4] Mam. Adir., 1884, p. 179.

GESTATION The gestation is, as usual, prolonged, being 9½ or 10 months.

YOUNG As in the case of the Red-bat, 4 appear to be the normal number of young at birth, though experience would lead us to look for a lower average in the nursery. In the collection of the Manitoba Natural History Society is a female with 2 young at her breast.

J. S. Charleson, of MacDonald, Man., tells me that at Sourismouth he found a female hanging in a tree with 4 young clinging to her breast; they were so tightly fast to her teats that they could not be removed.

They are born, apparently, in the last week of May. Merriam records[5] that: "On the evening of the 30th of June last (1883) Dr. A. K. Fisher shot a large female, measuring 422 mm. (16⅝ inches), at my home in Lewis County. It had already given birth to its young, and each of its 4 mammæ bore evidence of having recently been nursed."

These might have been born in June, but the analogy of the closely related Red-bat weighs in favour of a date at least a week earlier than June 1. This would fit in nicely with another record by Merriam:

"The only young I have ever seen [he says][6] was shot here August 6, 1883, by Walter H. Merriam. It was nearly full grown, measuring 400 mm. (15¾ inches) in extent, and differed from the adults chiefly in being a little lighter coloured."

MIGRATION Like all northern species of the Family, this Bat is migratory. Not a surprising fact when one remembers that it hates the cold as much as swallows do, and is at least as well equipped to seek more genial climes when frosty nights come on to nip its tender wings, rob the forests of its food, and turn its favourite forage ground to bleak and barren wastes.

Observations on its movements are hard to make, and so are few to-day. An important record by G. S. Miller runs thus:[7]

[5] *Ibid.*, p. 179.
[6] *Ibid.*, p. 179.
[7] N. A. Fauna, No. 13, 1897, p. 11.

"In August and September, 1890 and 1891, I had an opportunity to watch the appearance and disappearance of three species of Bats, *Lasionycteris noctivagans*, *Lasiurus borealis*, and *Lasiurus cinereus*, at Highland Light, Cape Cod, Mass. The animals, which were not to be found during the summer, suddenly became numerous shortly after the middle of August and remained abundant for about a month, when they as suddenly disappeared. The regularity with which this phenomenon occurred on the two successive years over which my observations extended shows that the migration of Bats is probably as definite as to dates and paths as that of birds."

I find in one of my journals a Bat note that refers to this present question:

One day, early in September, 1885, I went at dawn to the marsh near Toronto, Ont., to see the autumnal departure of the swallows whose head-quarters for a week before had been the great reed-beds on the Don side of the Bay. As the sky was brightening in the east, they began to arise in a body, towering like a swirling column of smoke, to seek the elevation at which they make their daytime flight. Sunrise was near, and most of them were gone when a large Bat arose with the last swarm from the reed-bed, circling up in plain view. My companion called out, "Look at that big Bat," and made an attempt to collect it, but the Bat sailed away with the swallows towards the south. It was almost certainly a Hoary-bat.

This incident, if it proves anything, would seem to show that one of our large Bats migrates by day, and possibly also, that, like many migrants, it seeks the company of other travellers more likely than itself to know the way.

The uninitiated may ask whether the Bat could fly in company with such dashing coursers of the air as swallows without being a hopeless laggard. But none who know the Bat will think of such a thing. Reference to the incident of the old Red-bat carrying her young that weighed more than she did, and yet catching flies in the air for food, will show that these creatures have solved the problem of flight better than birds. POWERS

For it is very doubtful if any bird could fly and catch insects while bearing a load greater than its own weight. And the Hoary-bat we know to be at least the Red-bat's peer in flight. The latest date in my Journal for the Hoary-bat, at Toronto, is mid-September.

One of the most interesting facts connected with the far-flight of this Bat is its occasional visits to the Bermudas in autumn and early winter.[8]

As these islands are 580 miles in a straight line from the nearest land, which is Cape Hatteras, this is evidence of very high-class wing-power indeed.

HABITS Concerning its habits in general, I have little original information, nor do I know of any detailed account except that by Dr. Merriam:

"Imagine for the moment, sympathetic reader [he says],[9] that you are an enthusiastic Bat hunter, and have chanced to visit some northern forest where this handsome species occurs. The early evening finds you, gun in hand, near the border of a lonely wood. The small Bats soon begin to fly, and in the course of fifteen or twenty minutes you may have killed several, all of which prove to be the Silver-haired species (*Lasionycteris noctivagans*). The twilight is fast fading into night, and you are making a constant effort of searching its obscurity, when suddenly a large Bat is seen approaching, perhaps high above the tree-tops, and has scarcely entered the limited field of vision when, in swooping for a passing insect, he cuts the line of the distant horizon, and disappears in the darkness below. In breathless suspense you wait for him to rise, crouching low that his form may be sooner outlined against the dim light that still lingers in the north-west, when he suddenly shoots by, seemingly as big as an owl, within a few feet of your very eyes. Turning quickly, you fire, but too late! He has vanished in the darkness. For more than a week each evening is thus spent,

[8] J. M. Jones, Mammals of Bermuda, Bull. 25, U. S. N. Mus., 1884, p. 145.
[9] *Op. cit.*, pp. 176–7.

and you almost despair of seeing another Hoary-bat, when, perhaps, on a clear, cold night, just as the darkness is becoming too intense to permit you to shoot with accuracy, and you are on the point of turning away, something appears above the horizon that sends a thrill of excitement through your whole frame. There is no mistaking the species—the size, the sharp, narrow wings, and the swift flight serve instantly to distinguish it from its nocturnal comrades. On he comes, but just before arriving within gunshot he makes one of his characteristic zig-zag side shoots, and you tremble as he momentarily vanishes from view. Suddenly he reappears, his flight becomes more steady, and now he sweeps swiftly towards you. No time is to be lost, and it is already too dark to aim, so you bring the gun quickly to your shoulder and fire. With a piercing, stridulous cry, he falls to the earth. In an instant you are stooping to pick him up, but the sharp, grating screams, uttered with a tone of intense anger, admonish you to observe discretion. With delight you cautiously take him in your hand and hurry to the light to feast your eyes upon his rich and handsome markings."

The Bat is one of the masterpieces of Creation. It exemplifies, in high degree, the perfect beast with perfect senses, equipped with perfect flight, so there be few indeed that in the scale outrank it. And the Prince among these winged ones is the magnificent Hoary-bat, whose imperfect history is before us. To the general and generous gifts of its tribe it adds great size, with corresponding higher power, a furry robe of exquisite beauty—a combination indeed of Sable, seal, and Silver-fox—and last, a blameless life. Many of its kin have equal and different claims to admiration and respect, but all these beauties go for naught with the world; even as Cyrano's nose outweighed all other worth, so one external blemish damns the Bat, for on its face this noble creature bears the mark of Cain, and every man's hand is against it. Its face is its fortune, indeed, but alack! an ill-fortune proved, for none can discern angel grace in a creature cursed with a visage of such hobgoblin ugliness. To this, no doubt, the harmless Bat owes thanks for centuries of human cruelties and oppression, and for all its ill repute. Could we but feel and remember that this is the original of the winged Brownie, the sprite that dwells in caves, coming forth at the magic call of the dew-time, to sport and circle in the airy woods, the elf that sometimes truly sleeps the winter-long sleep ascribed by legend to the elfin folk, and that is, withal, the highest and most gifted of the earth-born beings in our northern woods—could we but keep these things in mind, how differently we should feel towards the Bat.

A LIST OF THE
CHIEF WORKS CITED

A LIST OF THE CHIEF WORKS CITED.

1795. **Samuel Hearne.** A Journey from Prince of Wales's Fort in Hudson Bay, to the Northern Ocean. Undertaken by order of the Hudson's Bay Company, for the discovery of copper mines, a North-west Passage, &c., in the years 1769, 1770, 1771, and 1772. 4to, xliv, 458 pp., 9 plates, 5 maps. London.

[1799–1814. **Alexander Henry II.** MS. Journal, a valuable work of travel and Natural History observations in the North-west, edited by Dr. Elliott Coues, and published 1897, *q. v.*]

1823. **John Franklin.** Narrative of a Journey to the shores of the Polar Sea, in the years 1819, 20, 21, and 22. Narrative by Franklin; Natural History appendix by J. Richardson, J. Sabine and others. 4to, 30 plates, 4 maps, 768 pp. London.

1823. **Stephen H. Long.** Account of an Expedition from Pittsburg to the Rocky Mountains, performed in the years 1819 and 1820. Notes by T. Say and Edwin James. 2 vols., 8vo, 5 and 503 pp.; 442 and xcviii pp., and atlas, 2 maps and 9 plates. Philadelphia.

1825. **William H. Keating.** Narrative of an Expedition to the source of St. Peters River, Lake Winnipeek, Lake of the Woods, etc., performed in the year 1823. (Under Stephen H. Long.) 2 vols., 8vo, 458, 248 pp. Nat. Hist. app., pp. 156. Pll. 8, incl. 1 map. London.

1825. **Richard Harlan.** Fauna Americana; being a description of the mammiferous animals inhabiting North America. 8vo, 320 pp. Philadelphia.

1826–8. **John D. Godman.** American Natural History. 3 vols., 8vo, 362, 331 and 264 pp.; 21, 19 and 9 pll. Philadelphia.

1828. **John Franklin.** Narrative of a second expedition to the shores of the Polar Sea, in the years 1825, 1826, and 1827. Including an account of the progress of a detachment to the eastward, by John Richardson, Surgeon and Naturalist. 4to, xxiv–320, clviii pp., 31 plates, 6 maps. London.

1829. **John Richardson.** Fauna Boreali-Americana, Part I, Quadrupeds. 4to, xlvi–300 pp., 23 plates. London.

1842. **James E. Dekay.** Zoölogy of New York, or the New York Fauna, Part 1, Mammalia. 4to, 146 pp. Albany, N. Y.

1846–54. **John James Audubon and John Bachman.** The Quadrupeds of North America. 3 vols., roy. 8vo, 383–334–348 pp., pll. clv. New York.

[All three volumes are dated 1849. As no other date appears, I give that only in references.]

1851. **John Richardson.** Arctic Searching Expedition: A Journal of a boat voyage through Rupert's Land and the Arctic Sea, in search of the Discovery ships under command of Sir John Franklin. 2 vols., 8vo, 413–426 pp. London.

1853. **Zadock Thompson.** Natural History of Vermont. 8vo, 224 pp., app. 64 pp. Burlington, Vt.

1857. **Spencer F. Baird.** General Report upon the Zoölogy of the several Pacific Railroad Routes. Part I, Mammals. About 40 plates. Washington. 4to, 757 pp. Reprinted in 1859 as Mammals of North America. 764 pp., 87 plates. Philadelphia.

1857–8–9. Robert Kennicott. The Quadrupeds of Illinois, injurious and beneficial to the farmer. Rep. Comm. Pat., Agriculture, for 1856–7–8. Washington.

1860. Henry Youle Hind. Narrative Canadian Red River Exploring Expedition of 1857, and of the Assiniboine and Saskatchewan Exploring Expedition of 1858. 8vo, 2 vols., 494 and 472 pp. London.

1861. Bernard Rogan Ross. A popular treatise on the Fur-bearing Animals of the Mackenzie River District. Can. Nat. and Geol., VI, pp. 5–36. Montreal.

[He defines Mackenzie River District as lying between Salt River and the Arctic Ocean; and he treats Lynx, Weasels, Dogs, Wolves, Foxes, and Bears.]

1861. Bernard R. Ross. List of Mammals, Birds, and Eggs observed in the Mackenzie River District, with notices. Can. Nat. and Geol., VIII, Art. XIII, pp. 137–155. Montreal.

1861. C. Birch Bagster. The progress and prospects of Prince Edward Island. Chaps. XV to XVIII on Natural History. 12mo, 139 pp., app. XXXVI pp. Charlottetown, P. E. I.

[A rare volume.]

1862. Archibald Hall. On the Mammals and Birds of the District of Montreal. Can. Nat. and Geol., VII, pp. 44–78; pp. 289–317; pp. 344–376; pp. 401–430. Montreal.

1862. A. E. Verrill. Notes on the Natural History of Anticosti. Proc. Boston Soc. Nat. Hist. Vol. IX, October, 1862, pp. 132–151. Notes on Mammals (pp. 135–6). Birds and plants.

[He gives Fox, Marten, and Blackbear as common, Otter not uncommon, White-hare unknown.]

1863. H. Y. Hind. Exploration in the Interior of the Labrador Peninsula. 2 vols., 8vo, 351 and 304 pp. London.

1868. **Lewis H. Morgan.** The American Beaver and his Works. 8vo, 330 pp., map, 23 ppl., and 26 cuts. Philadelphia.

1869. **Campbell Hardy.** Forest Life in Acadie. 8vo, 371 pp., 12 illustrations. London.
[A standard work on the natural history of Nova Scotia. Out of print.]

1873. **A. Leith Adams.** Field and Forest Rambles. 8vo, 333 pp. London.
[A standard work on natural history of New Brunswick.]

1874. **J. A. Allen.** Notes on the Natural History of portions of Dakota and Montana Territories. Proc. Boston Soc. Nat. Hist., XVII, pp. 33–91. Introduction and Mammals, pp. 33 to 43. Boston.

1874. **Thomas Bell.** A History of British Quadrupeds, including the Cetacea. 8vo, 474 pp. London.

1875. **Elliott Coues and H. C. Yarrow.** Rept. Expl. and Surv. West of 100th Meridian, Vol. V, Zoölogy. Chap. II. Rept. on Mammals by Coues and Yarrow, pp. 35–129. Washington.

1875. **George M. Dawson.** Report of the Boundary Commission. (Geology 49th Parallel.) 8vo, xi–387 pp. Many maps and illustrations. Montreal, London and New York.

1876. **Robert Bell.** Report on the country west of Lakes Manitoba and Winnipegosis, with notes on the Geology of Lake Winnipeg. Rept. Prog. Geol. Surv., Canada, for 1874–75. General description of region explored in summer of 1874. Contains notes on the Mammals, pp. 24–56. Montreal.

1877. **J. A. Allen.** History of the American Bison, *Bison americanus*. Ninth Ann. Rept. U. S. Geol. and Geog. Sur. Terr. for 1875. Pub. June, 1877, pp. 443–587. F. V. Hayden in charge. Washington.

[The original memoir had 1 map and 12 plates, also discussed extinct species. It was published in Geol. Mem., Kentucky, 1876, and Mem. Mus. Comp. Zool., Cambridge; 1876.]

1877. **J. D. Caton.** The Antelope and Deer of America. 8vo, 426 pp., many illustrations. New York.

1877. **Elliott Coues and Joel Asaph Allen.** Monographs of North American Rodentia. Vol. XI, U. S. Geol. Surv. Terr. F. V. Hayden in charge. 4to, 1,091 pp. Washington.

[Manitoba specimens of several species listed.]

1877. **Elliott Coues.** Fur-bearing Animals. Misc. Pub., No. 8, U. S. Geol. Surv. Terr. 8vo, 348 pp., 20 plates. Washington.

1878. **C. E. McChesney.** Notes on the Mammals of Fort Sisseton, Dakota. Bull. U. S. Geol. and Geog. Surv. Terr. Bull. I, Vol. IV, Art. VIII. Feb. 5, 1878. Washington.

1878. **Joel Asaph Allen.** The Geographical distribution of the Mammalia considered in relation to the principal ontological regions of the earth, and the laws that govern the distribution of animal life. Bull. U. S. Geol. Surv., Vol. IV, No. 2, pp. 313–377. Art. XV. May 3. Washington.

1882. **John Macoun.** Manitoba and The Great North-west. 8vo, 687 pp. Chap. XX, pp. 325–353, devoted to Mammals. Guelph, Ont.

1883. **Ernest E. T. Seton.** The Striped Gopher (*Spermophilus tridecemlineatus* Mitchill). Rep. Dep. Agr. Manitoba for 1882, pp. 169–172, 4 ills. A life-history of the species in Manitoba. Winnipeg.

1884. **Robert Bell.** Observations on the Geology, Mineralogy, Zoölogy and Botany of the Labrador Coast, Hudson Strait and Bay. Rept. Prog. Geol. and Nat. Hist. Surv., Canada. App. II. Mam., pp. 48 DD to 53 DD. Montreal, Canada.

1884. **Clinton Hart Merriam.** The Mammals of the Adirondack Region. 4to, 316 pp. Reprinted from Vols. I and II. Trans. Linn. Soc., N. Y., Sept. 1884. New York.

1885. **Miller Christy.** Notes on the Mammals of Manitoba. Nat. Hist. Journal and School Reporter. Vol. IX, May 15, 1885, No. 76, pp. 67–74. York, England.

1886. **Ernest E. T. Seton.** The Mammals of Manitoba. Hist. and Sci. Soc., Manitoba, Trans. No. 23, Season 1886. A paper of 15 pp., treating 49 species. Read before the Soc. on the evening of May 27, 1886. Pub. in *Manitoba Free Press*, May, 28, 1886. Winnipeg.

[A new edition appeared in the following year. It made a 26-pp. pamphlet, treated 52 species, and had 6 illustrations by the author.]

1886. **Lucien M. Turner.** Contributions to the Natural History of Alaska. No. II. Arctic series of Publications, Signal Service, U. S. Army. 4to, 226 pp., Mammals, pp. 197–208. Washington.

1887. **E. W. Nelson.** Report upon Natural History Collections made in Alaska between the years 1877 and 1881. No. III, Arctic series, Signal Service, U. S. Army, 4to, 337 pp., 21 pll. Mammals by Nelson and F. W. True, pp. 227–293. Washington.

1887. **Charles H. Townsend.** Field notes on the Mammals, Birds and Reptiles of Northern California. Proc. U. S. Nat. Mus., X, pp. 159–241. Mammals, pp. 164–190. Washington.

1888. Ernest Thompson Seton. Tracks in the Snow. *St. Nicholas*, March, 1888. Pp. 338–341, 6 diagrams. New York.

[Many tracks and track incidents are given.]

1888. Vernon Bailey. Report on some of the Results of a Trip through parts of Minnesota and Dakota. Ann. Rept. Dep. Agr., 1887. Pp. 426–454. Washington.

1888. Miles Spencer. Notes on the Breeding Habits of certain Mammals, from personal observations and enquiries from Indians. App. III, pp. 76–79. Rept. Expl. James Bay and country east of Hudson Bay, by A. P. Low. Ann. Rept. Geol. and Nat. Hist. Surv., Canada, Vol. III. Montreal.

1889. W. T. Hornaday. The Extermination of the American Bison. Ann. Rept. U. S. Nat. Mus. for 1887, pp. 367–548, with map and many illustrations. Washington.

1890. J. A. Allen. List of Mammals collected by Mr. Clark P. Streator in British Columbia, with descriptions of two new sub-species of *Sciurus*. Bull. Am. Mus. Nat. Hist., Vol. III, No. I, Art. IX, pp. 161–168, Nov. 14, 1890. New York.

1890. C. Hart Merriam. Results of a Biological Survey of the San Francisco Mountain Region, and the Desert of the Little Colorado, Arizona. N. Am. Fauna, No. 3, Div. Ornithology and Mammalogy, U. S. Dep. Agr., 136 pp., 13 plates, 5 maps. Washington.

1891. J. A. Allen. On a collection of Mammals from Southern Texas and North-eastern Mexico. Bull. Am. Mus. Nat. Hist., Vol. III, No. 2, Art. XV, pp. 219–228. April 29, 1891. New York.

1891. **Edgar A. Mearns.** Description of a new sub-species of the Eastern Chipmunk, from the upper Mississippi region, west of the Great Lakes. Bull. Am. Mus. Nat. Hist., Vol. III, No. 2, Art. XVI, pp. 229–233. New York.

1891. **J. A. Allen.** Notes on new or little known North American Mammals, based on recent additions to the Collection of Mammals in the American Museum of Natural History. Bull. Am. Mus. Nat. Hist., Vol. III, No. 2, Art. XX, pp. 236–310. June 30, 1891. New York.

1891. **C. Hart Merriam.** Results of a Biological Reconnaissance of South Central Idaho. N. A. Fauna, No. 5, Div. Ornith. and Mam., U. S. Dep. Agr., 127 pp., 4 plates. Washington.

1892. **J. A. Allen.** The Geographical Distribution of North American Mammals. Bull. Am. Mus. Nat. Hist., Vol. IV, No. 1, Art. XIV, pp. 199–243. (Read before N. Y. Acad. Sci. Jan. 26, 1891.) December 29, 1892. New York.

1892. **C. Hart Merriam.** The Geographical Distribution of Life in North America, with special reference to the Mammalia. Proc. Biol. Soc. Washington, Vol. VII, pp. 1–64 (with map). April 13, 1892. Washington.

1892. **J. B. Tyrrell.** Report on North-western Manitoba, with portions of the adjacent districts of Assiniboia and Saskatchewan. Ann. Rept. Geol. Surv., Canada, Vol. V, Part I, pp. 1E–235E. General report on explorations made during seasons of 1887, 1888, 1889, and 1890. A few notes on mammals. Montreal.

1892. **Horace T. Martin.** Castorologia, or the History and Traditions of the Canadian Beaver. 8vo, 238 pp., 54 ills. Montreal and London.

1892. **Henry Poland.** Fur-bearing Animals in Nature and Commerce. 8vo, 392 pp., 16 illustrations. Chiefly used for Statistics of Fur returns. Pp. xxii–xxxiii.

1893. **Harrison Allen.** A Monograph of the Bats of North America. U. S. Nat. Mus., Bull. No. 43, 198 pp. 38 pll. Washington.

1893. **J. A. Allen.** List of Mammals and Birds collected in North-eastern Sonora and North-western Chihuahua, Mexico, on the Lumholtz Archæological Expedition, 1890–92. Bull. Am. Mus. Nat. Hist., Vol. V, Art. III, pp. 27–42. March 16, 1893. New York.

1893. **J. A. Allen.** List of Mammals collected by Mr. Charles P. Rowley in the San Juan Region of Colorado, New Mexico and Utah, with description of New Species. Bull. Am. Mus. Nat. Hist., Vol. V, Art. VI, pp. 69–84. April 28, 1893. New York.

1893. **J. A. Allen.** On a Collection of Mammals from the San Pedro Martir Region of Lower California, with Notes on other Species, particularly of the Genus *Sitomys*. Bull. Am. Mus. Nat. Hist., Vol. V, Art. XII, pp. 181–202. August 18, 1893. New York.

1893. **C. C. Nutting.** Rept. Zoöl. Expl. Lower Saskatchewan. Bull. Laboratories Nat. Hist. University Iowa, Vol. II, No. 3, pp. 235–293. Iowa City, Ia.
[Contains a few notes on Manitoba Mammals.]

1893. **Vernon Bailey.** The Prairie Ground Squirrels or Spermophiles of the Mississippi Valley. Bull. No. 4, Div. Ornith. and Mam., U. S. Dept. Agr., 69 pp., 4 maps. Washington.

1894. **J. A. Allen.** Notes on Mammals from New Brunswick, with Descriptions of a New Species of *Evotomys*. Bull. Am. Mus. Nat. Hist., Vol. VI, Art. III, pp. 99–106. April 14, 1894. New York.

1894. J. A. Allen. On the Seasonal Change of colour in the Varying Hare (*Lepus americanus* Erxl.). Bull. Am. Mus. Nat. Hist., Vol. VI, Art. IV, pp. 107–128. May 7, 1894, New York.

1894. J. A. Allen. On the Mammals of Aransas County, Texas, with Descriptions of New Forms of *Lepus* and *Oryzomys*. Bull. Am. Mus. Nat. Hist., Vol. VI, Art. VI, pp. 165–198. May 31, 1894. New York.

1894. Frank M. Chapman. Remarks on Certain Land Mammals from Florida, with a list of the Species known to occur in the State. Bull. Am. Mus. Nat. Hist., Vol. VI, Art. XIV, pp. 333–346. November 30, 1894. New York.

1894. J. A. Allen. Remarks on a second Collection of Mammals from New Brunswick, and on the Rediscovery of the Genus *Neotoma* in New York State. Bull. Am. Mus. Nat. Hist., Vol. VI, Art. XVII, pp. 359–364. December 22, 1894. New York.

1895. J. A. Allen. List of Mammals collected in the Black Hills Region of South Dakota and in Western Kansas, by Mr. Walter W. Granger, with Field Notes by the Collector. Bull. Am. Mus. Nat. Hist., Vol. VII, Art. VII, pp. 259–274. August 21, 1895. New York.

1895. C. Hart Merriam. Revision of the Shrews of the American Genera *Blarina* and *Notiosorex*. N. Am. Fauna, No. 10, Div. Ornith. and Mam., U. S. Dep. Agr., pp. 5–34. Synopsis of the American Shrews of the Genus *Sorex* (same Fauna). Pp. 57–98. Washington.

1895. Gerrit S. Miller, Jr. The Long-tailed Shrews of the Eastern U. S. N. Am. Fauna, No. 10, Div. Ornith. and Mam., U. S. Dep. Agr., pp. 35–56. Washington.

1896. **C. Hart Merriam.** Synopsis of the Weasels of North America. N. Am. Fauna, No. 11, Div. Ornith. and Mam., U. S. Dept. Agr., 33 pp., 5 plates. Washington.

1896. **Outram Bangs.** A review of the Weasels of Eastern North America. Proc. Biol. Soc. Washington. Vol. X, pp. 1–23. February, 1896. Washington.

1896. **Outram Bangs.** On a small collection of Mammals from Lake Edward, Quebec. Proc. Biol. Soc. Washington, Vol. X, pp. 45–52. March 9, 1896. Washington.

1896. **C. Hart Merriam.** Preliminary Synopsis of the American Bears. Proc. Biol. Soc. Washington, Vol. X, pp. 65–83. April 13, 1896. Washington.

1896. **J. A. Allen.** List of Mammals collected by Mr. Walter W. Granger, in New Mexico, Utah, Wyoming, and Nebraska, 1895–96, with Field Notes by the Collector. Bull. Am. Mus. Nat. Hist., Vol. VIII, Art. XV, pp. 241–258. November 25, 1896. New York.

1896. **Outram Bangs.** A Review of the Squirrels of Eastern North America. Proc. Biol. Soc. Washington, Vol. X, pp. 145–167. December 28, 1896. Washington.

1896. **A. P. Low.** List of Mammalia of the Labrador Peninsula, with short notes on their distribution. Ann. Rept. Geol. Surv., Canada. Vol. VIII, App. I, pp. 313–321. Ottawa.

1896. **T. S. Palmer.** The Jack Rabbits of the United States. Bull. No. 8, Div. Ornith. and Mam., U. S. Dep. Agr., 84 pp., 6 plates. Washington.

1896. **Gerrit S. Miller, Jr.** Genera and Sub-genera of Voles and Lemmings. N. A. Fauna, No. 12, Div. Ornith. and Mam., U. S. Dep. Agr., 84 pp., 3 plates. Washington.

1897. **Vernon Bailey.** Revision of the American Voles of the Genus *Evotomys*. Proc. Biol. Soc. Washington, Vol. XI, pp. 113–138. May 13, 1897. Washington.

1897. **Gerrit S. Miller, Jr.** Notes on the Mammals of Ontario. Proc. Boston Soc. Nat. Hist., Vol. 28, No. 1, pp. 1-44. Boston.

1897. **Gerrit S. Miller, Jr.** Revision of the North American Bats of the Family *Vespertilionidæ*. N. Am. Fauna, No. 13, Div. Biol. Surv., U. S. Dep. Agr., 140 pp., 3 plates. Washington.

1897. **Alexander Henry, II.** The MS. Journals of Alex. Henry and David Thompson, 1799–1814. Edited by Elliott Coues. 3 vols., 8vo, 916 pp., with maps. New York.

[E. Coues, the editor, says (p. xx) "the main text consists solely of Henry's Journal, Thompson's contributions being, like my own, confined to the foot-notes." It is greatly to be regretted, therefore, that the title should have been complicated by the addition of Thompson's name.]

1898. **John Fannin.** A preliminary Catalogue of the collections of Nat. Hist. and Ethnology in the Provincial Museum, Victoria, British Columbia. 8vo, 196 pp. Victoria.

1898. **D. G. Elliott.** Lists of species of Mammals from Iowa, Wyoming, Montana, Idaho, Nevada and California. Field Col. Mus. Pub. 27, Zoöl. Ser., Vol. I, No. 10, pp. 193–221. March, 1898. Chicago.

1898. **J. A. Allen.** Revision of the Chickarees, or American Red Squirrels (sub-genus *Tamiasciurus*). Bull. Am. Mus. Nat. Hist., Vol. X, Art. XIV, pp. 249–298. July 22, 1898. New York.

1898. **Outram Bangs.** A List of the Mammals of Labrador. The Am. Nat., Vol. XXXII, pp. 489–507. July, 1898. Boston.

1898. **Outram Bangs.** The Land Mammals of Peninsular Florida, and the Coast Region of Georgia. Proc. Boston Soc. Nat. Hist., Vol. XXVIII, No. 7, pp. 157–235. Boston.

1899. **J. A. Allen.** The North American Arboreal Squirrels. American Naturalist, Vol. XXXIII, No. 392, pp. 635–642. August, 1899. Boston.

1899. **Gerrit S. Miller, Jr.** Preliminary List of the Mammals of New York. Bull. N. Y. State Museum. Vol. VI, No. 29, pp. 274–390. October, 1899. Albany.

1899. **Edward A. Preble.** Revision of the Jumping Mice of the Genus *Zapus*. N. Am. Fauna, No. 15, Biol. Surv., U. S. Dep. Agr., 41 pp., 1 plate. Washington.

1899. **C. Hart Merriam.** Result of a Biological Survey of Mount Shasta, Cal. N. Am. Fauna, No. 16, Div. Biol. Surv., U. S. Dep. Agr., 179 pp., 5 plates, 46 text figures. Washington.

1900. **Gerrit S. Miller, Jr.** Key to the Land Mammals of North-eastern North America. Bull. N. Y. State Mus., No. 38, Vol. VIII, 160 pp. October, 1900. Albany.

1900. **C. Hart Merriam.** Preliminary Revision of the North American Red Foxes. Proc. Wash. Acad. Sci., Vol. II, pp. 661–676. December 28, 1900. Washington.

1900. **Vernon Bailey.** Revision of American Voles of the Genus *Microtus*. N. Am. Fauna, No. 17, Div. Biol. Surv., U. S. Dep. Agr., 88 pp., 5 plates. Washington.

1900. **W. H. Osgood.** Results of a Biological Reconnaissance of the Yukon River Region. N. Am. Fauna, No. 19, Div. Biol. Surv., U. S. Dep. Agr., 100 pp., 7 plates. Washington.

1901. D. G. Elliott. A Synopsis of the Mammals of North America and the Adjacent Seas. Pub. No. 45, Zoöl. Ser., Vol. II, Field Columbian Museum. 8vo, 471 pp. Chicago.

1901. Gerrit S. Miller, Jr., and Jas. A. G. Rehn. Systematic results of the study of North American Land Mammals to the close of the year 1900. Proc. Boston Soc. Nat. Hist., Vol. XXX, No. 1, 352 pp. Boston. [Many of my references to ancient works are taken from this.]

1901. A. H. Howell. Revision of the Skunks of the Genus *Chincha*. N. A. Fauna, No. 20, Div. Biol. Surv., U. S. Dep. Agr., 62 pp., 8 plates. Washington. [Chincha=Mephitis.]

1901. W. H. Osgood. Natural History of the Queen Charlotte Islands, British Columbia. Natural History of the Cook Inlet Region, Alaska. N. A. Fauna, No. 21. Div. Biol. Surv., U. S. Dep. Agr., 87 pp., 7 plates. Washington.

1902. Ernest Thompson Seton. American Woodcraft. *Ladies' Home Journal*, May, 1902, pp. 15 and 41. Article on the tracks of animals. 14 ills. Philadelphia.

1902. Ernest Thompson Seton. American Woodcraft. *Ladies' Home Journal*, June, 1902, p. 15. Article on the tracks of animals. 11 ills. Philadelphia.

1902. Ernest Thompson Seton. American Woodcraft—'Freezing.' *Ladies' Home Journal*, November, 1902, p. 15. Article on 'freezing,' a protective device of animals. 4 ills. Philadelphia.

1902. Witmer Stone and W. E. Cram. American Animals. A popular guide to the Mammals of North America, North of Mexico, with intimate biographies of the more familiar species. 4to, 318 pp., with many illustrations. New York.

1902. **Edward A. Preble.** A Biological Investigation of the Hudson Bay Region. N. Am. Fauna, No. 22. Div. Biol. Surv., U. S. Dep. Agr., 140 pp., 14 plates. Washington.

1902. **Edwin Hollis.** Collection Small Mammals in N. W. T., Canada. The Zoölogist, August 15, 1902. Pp. 293–298. London.
[Notes on 22 species taken near Touchwood Hills, Sask.]

1903. **Samuel N. Rhoads.** The Mammals of Pennsylvania and New Jersey. 8vo, 266 pp., with plates and a faunal map. Privately published. Philadelphia.

1903. **T. Roosevelt, T. S. Vandyke, D. G. Elliott, and A. J. Stone.** The Deer Family. 334 pp., 7 maps by C. Hart Merriam, and 25 illustrations chiefly by C. Rungius. New York.

1904. **Ernest T. Seton.** The Master Plowman of The West. *Century Magazine*, June. A study of *Thomomys*. Pp. 299–307, 8 illustrations by the author. New York.

1904. **J. D. Figgins.** Field Notes on the Birds and Mammals of the Cook Inlet Region of Alaska. Proc. Linn. Soc. N. Y. December 19. New York.

1904. **William Temple Hornaday.** The American Natural History. 4to, 449 pp., with many illustrations. New York.

1904. **W. H. Osgood.** A Biological Reconnaissance of the Base of the Alaska Peninsula. N. Am. Fauna, No. 24, Div. Biol. Surv., U. S. Dep. Agr., 86 pp., 7 plates. Washington.

1904. **Woods Hutchinson.** Animal Marriage. *Cotemporary Review*, October, 1904. Pp. 485–496.

1904–5–6. **J. G. Millais.** The Mammals of Great Britain and Ireland. 3 vols., roy. 4to, many illustrations. London.

1905. **Ernest Thompson Seton.** The Secrets of the Trail. *Country Life in America*, June, 1905. Pp. 202–205. 15 ills. by the author. Gives the trails of many common animals. New York.

1905. **Vernon Bailey.** Biological Survey of Texas. N. Am. Fauna, No. 25, Bur. Biol. Surv., U. S. Dep. Agr., 222 pp., 16 plates, 23 text figures. Washington.

1905. **R. MacFarlane.** Notes on Mammals collected and observed in the northern Mackenzie River District, north-west Territories of Canada, with remarks on explorers and explorations of the Far North. Proc. U. S. Nat. Mus., Vol. XXVIII, pp. 673–764. No. 1405. Washington.

1906. **Ernest Thompson Seton.** The Wapiti and His Antlers. *Scribner's Magazine*, January, 1906. Pp. 15–33, 1 map, 16 illustrations, 6 photographs. New York.

1906. **Edward R. Warren.** The Mammals of Colorado. Colorado College Publ., Vol. XI, No. 46. Sci. Series, pp. 225–274. January, 1906. Colorado Springs.

1906. **Ernest Thompson Seton.** The Moose and His Antlers. *Scribner's Magazine*, February, 1906. Pp. 157–178, 1 map, 21 illustrations, 4 photographs. New York.

1906. **Ernest Thompson Seton.** The Caribou and His Kindred. *Scribner's Magazine*, April, 1906. Pp. 426–443, 1 map, 15 illustrations, 5 photographs. New York.

1906. **Ernest Thompson Seton.** The Prong-horned Antelope. *Scribner's Magazine*, July, 1906. Pp. 33–49, 1 map, 10 illustrations, 6 photographs. New York.

1906. J. A. Allen. Mammals from the States of Sinaloa and Jalisco, Mexico, collected by J. H. Batty, during 1904 and 1905. Bull. Am. Mus. Nat. Hist., Vol. XXII, Art. XII, pp. 191–262. July 25, 1906. New York.

1906. Ernest Thompson Seton. The White-tailed (Virginian) Deer and Its Kin. *Scribner's Magazine*, September, 1906. Pp. 321–341, 1 map, 20 illustrations, 1 photograph. New York.

1906. Ernest Thompson Seton. The American Bison or Buffalo. *Scribner's Magazine*, October, 1906. Pp. 385–405, 2 maps, 16 illustrations, 6 photographs. New York

1906. William T. Hornaday. Camp-fires in the Canadian Rockies, by William T. Hornaday and John M. Phillips. 353 pp., with 2 maps and many illustrations, chiefly photographs. New York.

1906. Ernest Thompson Seton. The Smallest of all Beasts of Prey. *Western Sportsman*. Pp. 316–318. December, 1906. Account of *Putorius rixosus*, with map and illustrations. Winnipeg.

1906. J. S. Talbot. Foxes at Home and Reminiscences. 8vo, 155 pp., 13 illustrations, mostly photographs. London.

1906. Charles C. Adams. An Ecological Survey in Northern Michigan. 8vo, 133 pp., 21 illustrations; many lists, including Mammals.

1907. Ernest Thompson Seton. The Merry Chipmunk. *Success Magazine*, May, 1907. Pp. 328–331, also pp. 368, 369 and 370; 4 illustrations by author. New York.

1907. Ernest Thompson Seton. The Snowshoe Rabbit. *Everybody's Magazine*, May, 1907. Pp. 599–608; 9 illustrations and map by author. New York.

1907. Ernest Thompson Seton. Dogs of Song. The Life-habits and Wonderful Vocal Abilities of the Coyote. *Success Magazine*, August, 1907. Pp. 537, 539 and 562–3, 1 map, 7 drawings, 1 photograph. New York.

1907. A. Franklin Shull. Habits of the Short-tailed Shrew *Blarina brevicauda* (Say). Am. Nat., August, 1907. Pp. 495–522, 10 illustrations. Boston.

1907. Ernest Thompson Seton. The Habits of Wolves. *American Magazine*, October, 1907. Pp. 636–645; 9 illustrations by the author. New York.

1907. Edgar Alexander Mearns. Mammals of the Mexican Boundary of the United States. Part I, *Didelphiidæ* to *Muridæ*. U. S. Nat. Mus. Bull. 56, 530 pp.; many maps and illustrations. Washington.

1908. Edward R. Warren. Further notes on the Mammals of Colorado. Colorado College Publication. General Series No. 33, Engin. Ser., Vol. I, No. 4, pp. 59–90. Colorado Springs, January, 1908.

1908. Edward A. Preble. A Biological Investigation of the Athabaska-Mackenzie Region. N. A. Fauna, No. 27, Bur. Biol. Survey, U. S. Dep. Agr., 574 pp., Plate XXV; text figures, 16. Washington, October 26, 1908.

1909. Wilfred H. Osgood. Revision of the Mice of the American Genus *Peromyscus*. N. A. Fauna, No. 28, Bur. Biol. Survey, U. S. Dep. Agr., 285 pp., 8 plates, 12 text figures (*i. e.*, maps). Washington.

1909. E. W. Nelson. The Rabbits of North America. N. A. Fauna, No. 29, Bur. Biol. Surv., U. S. Dep. Agr., 314 pp., Plates XIII; text figures, chiefly maps, 19. Washington, August, 1909.

SYNOPTIC INDEX

SYNOPTIC INDEX

THE main classification is *alphabetic*, but many of the sub-divisions are *synoptic*, as this plan seemed more likely to be of service. For example, the entry "Antelope" is strictly alphabetic, but the sub-headings of the 'Antelope chapter,' given next, are in their actual or synoptic order.

Under the name of each authority the alphabetic order is based on the capitalized name of the animal.

Every important fact is entered under at least two different headings.

NATURAL SCIENCES IN AMERICA

An Arno Press Collection

Allen, J[oel] A[saph]. **The American Bisons,** Living and Extinct. 1876

Allen, Joel Asaph. **History of the North American Pinnipeds:** A Monograph of the Walruses, Sea-Lions, Sea-Bears and Seals of North America. 1880

American Natural History Studies: The Bairdian Period. 1974

American Ornithological Bibliography. 1974

Anker, Jean. **Bird Books and Bird Art.** 1938

Audubon, John James and John Bachman. **The Quadrupeds of North America.** Three vols. 1854

Baird, Spencer F[ullerton]. **Mammals of North America.** 1859

Baird, S[pencer] F[ullerton], T[homas] M. Brewer and R[obert] Ridgway. **A History of North American Birds:** Land Birds. Three vols., 1874

Baird, Spencer F[ullerton], John Cassin and George N. Lawrence. **The Birds of North America.** 1860. Two vols. in one.

Baird, S[pencer] F[ullerton], T[homas] M. Brewer, and R[obert] Ridgway. **The Water Birds of North America.** 1884. Two vols. in one.

Barton, Benjamin Smith. **Notes on the Animals of North America.** Edited, with an Introduction by Keir B. Sterling. 1792

Bendire, Charles [Emil]. **Life Histories of North American Birds** With Special Reference to Their Breeding Habits and Eggs. 1892/1895. Two vols. in one.

Bonaparte, Charles Lucian [Jules Laurent]. **American Ornithology:** Or The Natural History of Birds Inhabiting the United States, Not Given by Wilson. 1825/1828/1833. Four vols. in one.

Cameron, Jenks. **The Bureau of Biological Survey:** Its History, Activities, and Organization. 1929

Caton, John Dean. **The Antelope and Deer of America:** A Comprehensive Scientific Treatise Upon the Natural History, Including the Characteristics, Habits, Affinities, and Capacity for Domestication of the Antilocapra and Cervidae of North America. 1877

Contributions to American Systematics. 1974

Contributions to the Bibliographical Literature of American Mammals. 1974

Contributions to the History of American Natural History. 1974

Contributions to the History of American Ornithology. 1974

Cooper, J[ames] G[raham]. **Ornithology.** Volume I, Land Birds. 1870

Cope, E[dward] D[rinker]. **The Origin of the Fittest:** Essays on Evolution and **The Primary Factors of Organic Evolution.** 1887/1896. Two vols. in one.

Coues, Elliott. **Birds of the Colorado Valley.** 1878

Coues, Elliott. **Birds of the Northwest.** 1874

Coues, Elliott. **Key To North American Birds.** Two vols. 1903

Early Nineteenth-Century Studies and Surveys. 1974

Emmons, Ebenezer. **American Geology:** Containing a Statement of the Principles of the Science. 1855. Two vols. in one.

Fauna Americana. 1825-1826

Fisher, A[lbert] K[enrick]. **The Hawks and Owls of the United States in Their Relation to Agriculture.** 1893

Godman, John D. **American Natural History:** Part I — Mastology and **Rambles of a Naturalist.** 1826-28/1833. Three vols. in one.

Gregory, William King. **Evolution Emerging:** A Survey of Changing Patterns from Primeval Life to Man. Two vols. 1951

Hay, Oliver Perry. **Bibliography and Catalogue of the Fossil Vertebrata of North America.** 1902

Heilprin, Angelo. **The Geographical and Geological Distribution of Animals.** 1887

Hitchcock, Edward. **A Report on the Sandstone of the Connecticut Valley,** Especially Its Fossil Footmarks. 1858

Hubbs, Carl L., editor. **Zoogeography.** 1958

[Kessel, Edward L., editor]. **A Century of Progress in the Natural Sciences: 1853-1953.** 1955

Leidy, Joseph. **The Extinct Mammalian Fauna of Dakota and Nebraska,** Including an Account of Some Allied Forms from Other Localities, Together with a Synopsis of the Mammalian Remains of North America. 1869

Lyon, Marcus Ward, Jr. **Mammals of Indiana.** 1936

Matthew, W[illiam] D[iller]. **Climate and Evolution.** 1915

Mayr, Ernst, editor. **The Species Problem.** 1957

Mearns, Edgar Alexander. **Mammals of the Mexican Boundary of the United States.** Part I: Families Didelphiidae to Muridae. 1907

Merriam, Clinton Hart. **The Mammals of the Adirondack Region,** Northeastern New York. 1884

Nuttall, Thomas. **A Manual of the Ornithology of the United States and of Canada.** Two vols. 1832-1834

Nuttall Ornithological Club. **Bulletin of the Nuttall Ornithological Club:** A Quarterly Journal of Ornithology. 1876-1883. Eight vols. in three.

[Pennant, Thomas]. **Arctic Zoology. 1784-1787.** Two vols. in one.

Richardson, John. **Fauna Boreali-Americana;** Or the Zoology of the Northern Parts of British America, Containing Descriptions of the Objects of Natural History Collected on the Late Northern Land Expeditions Under Command of Captain Sir John Franklin, R. N. Part I: Quadrupeds. 1829

Richardson, John and William Swainson. **Fauna Boreali-Americana:** Or the Zoology of the Northern Parts of British America, Containing Descriptions of the Objects of Natural History Collected by the Late Northern Land Expeditions Under Command of Captain Sir John Franklin, R. N. Part II: The Birds. 1831

Ridgway, Robert. **Ornithology.** 1877

Selected Works By Eighteenth-Century Naturalists and Travellers. 1974

Selected Works in Nineteenth-Century North American Paleontology. 1974

Selected Works of Clinton Hart Merriam. 1974

Selected Works of Joel Asaph Allen. 1974

Selections From the Literature of American Biogeography. 1974

Seton, Ernest Thompson. **Life-Histories of Northern Animals: An Account of the Mammals of Manitoba.** Two vols. 1909

Sterling, Keir Brooks. **Last of the Naturalists:** The Career of C. Hart Merriam. 1974

Vieillot, L. P. **Histoire Naturelle Des Oiseaux de L'Amerique Septentrionale,** Contenant Un Grand Nombre D'Especes Decrites ou Figurees Pour La Premiere Fois. 1807. Two vols. in one.

Wilson, Scott B., assisted by A. H. Evans. **Aves Hawaiienses:** The Birds of the Sandwich Islands. 1890-99

Wood, Casey A., editor. **An Introduction to the Literature of Vertebrate Zoology.** 1931

Zimmer, John Todd. **Catalogue of the Edward E. Ayer Ornithological Library.** 1926